Digital Image Analysis

All the papers included in
this book were presented at
the Second Conference on
Image Analysis and Processing,
held in Brindisi, Italy,
15–18 November 1982.

Digital Image Analysis

Edited by

S Levialdi
University of Rome

Pitman

PITMAN PUBLISHING LIMITED
128 Long Acre, London WC2E 9AN

PITMAN PUBLISHING INC
1020 Plain Street, Marshfield, Massachusetts 02050

Associated Companies
Pitman Publishing Pty Ltd, Melbourne
Pitman Publishing New Zealand Ltd, Wellington
Copp Clark Pitman, Toronto

© S Levialdi 1984

First published in Great Britain 1984

British Library Cataloguing in Publication Data

Digital image analysis.
 1. Image processing—Digital techniques—
congresses
 I. Levialdi, S.
 621.38'0414   TA1632
 ISBN 0-273-08616-2

Library of Congress Cataloging in Publication Data

Main entry under title:

Digital image analysis.
 Bibliography: p.
 Includes index.
 1. Image processing – digital techniques.
 I. Levialdi, S.
 TA1632.D492 1984   621.36'7   83-25147
 ISBN 0-273-08616-2

Printed and bound in Great Britain by
Biddles Ltd, Guildford and King's Lynn

621. 380414
DIG

Preface

Progress in any scientific field is achieved both by improving existing techniques, in order to reach better results, and by tackling sub-problems within the same field by means of a radically different approach. This applies just as well to the field of image processing and analysis, where theoretical questions and practical implications are constantly interacting — sometimes within the same research group — producing novel solutions and an overall enrichment of the field knowledge.

This field is vast and appealing: most of the information humans receive, learn, remember and use is in picture-like form. The largest projection area on the cortex of our brain corresponds to vision; this accounts for the extraordinary capacity that we have for processing pictorial information strategically, i.e., according to the specific contextual needs of the observer. As a consequence a great challenge is posed to the research community if computers must substitute (possibly with some advantage), partially or totally, the human operator in some of the tasks in which vision plays a significant role. Moreover, automation, whether in industry or commerce, requires fast and efficient artificial "eyes" if the tasks to be performed have a minimum degree of sophistication and take place in a noisy, dangerous or distant environment.

A better understanding of the perception mechanism of the visual system, of the significant information extracted at the retinal level and of the principles involved in the retrieving system used by man is also crucial for the development of new computer programs and systems that must outperform the existing ones. The background, experience and environment of each scientist generally determines his research philosophy, his goals, his approach and his results.

The nature of images is multifold: they can have a strong emotional impact and provoke fast responses, this accounts for their importance as a communication means, they may contain too much information to enable us to discover quickly and reliably what we require (for instance, in diagnosis with biomedical images) or, again, they might give us completely wrong cues, as in impossible figures, where a two-dimensional drawing provides conflicting evidence for the reconstruction of a three-dimensional object.

Active research in image processing and analysis is producing over 2000 papers yearly and a number of scientific journals specifically covering this area have already been published regularly for over 10 years; international and national meetings are run on a yearly basis and if we also include the field of computer graphics, where images are created artifically rather than analyzed, the number of active persons in research, industry and commerce would be of one order of magnitude greater than those acting mainly in image processing.

In Italy there are about 25 active groups in image processing and recognition. Since 1980 conferences, sponsored by the IAPR (International Association for Pattern Recognition) and promoted by the Italian Group of Cybernetics and Biophysics (GNCB) belonging to the Italian National Research Council (CNR), on image processing are organized every 2 years (Pavia 1980, Selva di Fasano 1982). The Second Conference on Image Analysis and Processing took place on 15–18 November 1982 and was attended by about 100 participants; there were eight invited papers and these form the first part of this book.

The invited speakers constitute a group of well-known leaders in the field of image processing and have been extensively active for many years; their work represents some of the most promising directions in this field today. The papers are organized to cover, first, the fundamental aspects of an image processing theory: coding and representation. Then the architecture for effective and efficient image processing is presented followed by some algorithms tailored to the most typical image processing operations and, at the end of this part, two areas of application are explored: an established one (remote sensing) and a new one (astronomy), which will surely benefit from the experience built up in other areas. Biomedical applications, optical character recognition and industrial inspection have been left out, although very important application areas, because a stronger emphasis is given on methodologies to enable the discussion of alternative approaches among the participants of the conference, rather than concentrating on the existing operative systems such as mail readers, blood cell analyzers, etc.

Let us now scan all the papers from the beginning. Shannon's information definition and measurement, based on a model of a communication channel where the symbol alphabet is well defined, has been used in image sampling, transmission and noise removal yet, for some other applications, the semantics of the information contained in an image still remains to be quantitatively evaluated: the first paper represents an interesting attempt toward this direction. The author introduces a new image information measurement, relating it to the grey level histogram of the picture and suggesting its use in picture encoding and retrieving.

The second paper considers a recent image representation method using variable resolution which also has a direct hardware counterpart in pyramid-like computational structures. The speed required for computing basic image properties on these structures is then shown to be much higher than the one required on conventional sequential machines; moreover segmentation — a basic process in image analysis — can be achieved using information on colour, texture, etc., with advantages of low noise sensitivity (at low resolution) yet preserving high resolution (at the base of the pyramid).

The third paper discusses the problems involved in building a three-dimensional object when only some two-dimensional projections are available; the extension of this problem to one of using a sequence (in time) of projections for the reconstruction is also included. The topic is closely related to the problems encountered in computer graphics but it is relevant to the notions which must be built in a program for scene analysis, sometimes in a changing environment, where objects must be detected and measured as well as displayed on the screen.

The fourth paper explores the different alternative image processing systems and, taking advantage of the fact that independent local operations are usually performed on digital images, shows that the class of SIMD machines well matches the image data structure and provides a convenient computational structure for performing such operations. The disadvantages for data-dependent operations are underlined and, as the title suggests, no absolute best architecture exists.

The field of mathematical morphology was cultivated by a restricted number of authors but has proven to be a valuable approach, especially if an adequate computing system is available. This fifth paper shows the two basic operations (dilation and erosion) which can be efficiently performed by a pipeline machine (the Cytocomputer) and a number of nice examples are shown, including one of national geographic relevance.

The sixth paper reviews all the elements which are required in order that the total amount of remotely sensed data may be properly used and emphasizes the need for integration not only of the collected data (from satellite, aeroplane, ground, etc.) but also considering specifically acquisition, storing, retrieving and processing as one global process of information management.

The seventh paper stresses the importance of the contributions of computer science toward achieving better recognition results; for instance, noting the problems coming from errors introduced by finite samples and discussing the performance of nearest neighbour classification. The need for improving both the algorithms (reducing complexity) and the programs (by means of a software science) is also mentioned as well as a number of key factors which affect performance: interactivity, pictorial database structure and computer architectures.

The last paper is a tutorial on the problems faced by astronomers today in their observations by means of special equipment and on the strategy that they are presently developing in order to guarantee efficient solutions to their problems. I believe that this paper is a good example in showing how a methodology used for correctly posing the problem will ensure against ill use of the rich and powerful technology of image processing. An appendix to this paper, by the same author, has been included to make available to the reader the present state of astronomical image processing in Italy.

The second part of the book contains the contributions presented at the conference. They follow the Chairman's opening address and are grouped in five sections according to content: image analysis and interpretation; image coding and filtering; a panel on the criteria for choosing a convenient image processing architecture; image processing systems; and, finally, image processing applications.

In order to speed up publication of the book and in view of the professional experience of the contributors, the papers have not been subject to refereeing.

I hope that after reading this book the impression of achievement which is conveyed in all these papers will stimulate research in this area in order to materialize the transition from subjective invention to objective knowledge.

S. Levialdi

Bari, 1983

ACKNOWLEDGMENTS

I wish to express my gratitude to a number of persons who helped me in organizing and running this conference so smoothly and pleasantly. Mr G. Piccirilli, with A. Fanelli, G. Perchiazzi and L. Rudd, and M. Chiumarulo and S. Sala as technical assistants, were the supporting group that solved all the problems that are inevitably generated by meetings of this sort.

The cover picture, a digital image of the 'trulli', typical houses of the area where the conference was held, was kindly provided by the IESI (Image and Signal Processing Institute of the Italian Research Council).

S.L.

Contents

Image Analysis and Interpretation

Image Coding and Filtering

Panel: "Which Computer Architecture for Image Processing?" 223

Image Processing Systems

List of contributors

PART ONE
INVITED PAPERS

1 Image information measures and encoding techniques

S K Chang

1. INTRODUCTION

A two-dimensional picture function f is a mapping, $f: N \times N \to \{0, 1, ..., L-1\}$, where N represents the set of natural numbers $\{1,...,N\}$ and $\{0, 1, ..., L-1\}$ is the *pixel value set* or *gray level set*. $f(x,y)$ then represents the pixel value at (x,y).

Given a picture function f and a point set S, f/S denotes the *restricted picture function* which is defined only for points in S, i.e., $f/S (x, y) = f(x, y)$ if x is in S. f/S is called the *support picture function* or *support picture* for S.

A measure for the amount of information contained in a picture f is based upon the minimum number of pixel gray level changes to convert a picture into one with constant gray level. Let $h: \{0, 1, ..., L-1\} \to N$ represents the *histogram* of f, where $h(i)$ is the number of pixels at gray level i. We define the *pictorial information measure* PIM(f) as follows:

$$(1) \quad PIM(f) = (\sum_{i=0}^{L-1} h(i)) - \max_{i} h(i)$$

We note that PIM(f) = 0, if and only if f is a constant picture (i.e., $f(x, y)$ = constant for all (x, y) in $N \times N$). On the other hand, PIM(f) is maximum if and only if f has a uniform histogram (i.e., $h(i)$ = constant, $0 \leq i \leq L-1$). Let the total number of pixels in f be N(f). f has a uniform histogram if and only if $PIM(f) = N(f)(L-1)/L$. In other words, PIM(f) is minimal when f is least informative, and maximal when f is most informative.

Suppose a picture point set S is divided into two disjoint subsets S_1 and S_2. We then have,

$$(2) \quad PIM(f/S) \geq PIM(f/S_1) + PIM(f/S_2)$$

Therefore, if we use disjoint sets S_i to cover a picture f, then the sum of $PIM(f/S_i)$ is always less than or equal to PIM(f). We can also define a *normalized picture information measure* as $NPIM(f) = PIM(f)/N(f)$. In picture encoding [1, 2], we can use PIM or NPIM to decide whether a picture f should be decomposed. For example, if NPIM(f) is less than a threshold, then f need not be further decomposed. On the other hand, if NPIM(f) is

This research was supported by the National Science Foundation under Grant ECS–8005953 and by the Naval Research Laboratory under Contract N00014–82–C–2156.

close to maximum, and for every subpicture f/S, NPIM(f/S) is close to maximum, then the picture f is almost random and also need not be further decomposed. If we define p_i as $h(i)/N(f)$, then we have

(3) $NPIM(f) = 1 - \max_i p_i$

Furthermore, if we define w_i as $N(f/S_i)/N(f)$, then we can prove

(4) $NPIM(f/S) \geqq w_1 \times NPIM(f/S_1) + w_2 \times NPIM(f/S_2)$

We can define a more general measure, PIM_k, as follows,

(5) $PIM_k(f) = (\sum_{i=0}^{L-1} h(i)) - (\sum_{\substack{i \text{ is one} \\ \text{of the k} \\ \text{largest } h(i)\text{'s}}} h(i))$

and $NPIM_k$ is accordingly defined as,

(6) $NPIM_k(f) = 1 - (\sum_{\substack{i \text{ is one} \\ \text{of the k} \\ \text{largest } p_i\text{'s}}} p_i)$

We can also prove that PIM_k satisfies inequality (2), and $NPIM_k$ satisfies inequality (4). The picture information measures introduced above can be used to select subpictures in picture encoding [2–4]. In what follows, we demonstrate that these information measures satisfy a number of axioms and useful inequalities (Section 2), and generalize the concept to a Lorenz information measure (Section 3). Based upon these information measures, a structural information measure can also be defined for logical pictures (Section 4). Applications to picture encoding (Section 5) and picture tree construction (Section 6) are discussed.

2. A FAMILY OF INFORMATION MEASURES

The picture information measures NPIM and $NPIM_k$ introduced above belong to a family of information measures. We will use p to denote a probability vector $(p_1, ..., p_n)$, where the p_i's are nonnegative real numbers.

Let p, r, and q be three probability vectors. We write $p = r + q$, if $p_i = r_i + q_i$, $1 \leqq i \leqq n$. Let w_1 be the sum of the r_i's, and w_2 be the sum of the q_i's. We call $IM(p)$ an *information measure*, if the following inequality holds.

(7) $IM(p) \geqq w_1 IM(q) + w_2 IM(r)$

Theorem: Let $IM(p) = F(1,0,...,0) - F(p)$. If the function F satisfies the following:

(B1) $F(p)$ is continuous in p_i for all i,
(B2) $F(p)$ is symmetric in p_i for all i,
(B3) $F(0,...,0) = 0$,
(B4) F is convex, i.e., $f(w_1 \times r + w_2 \times q) \leqq w_1 F(r) + w_2 F(q)$,

then $IM(p)$ is an information measure. Moreover, the minimum of $IM(p)$ is at $IM(1, ..., 0)$ $= 0$, and the maximum of $IM(p)$ is at $IM(1/n, ..., 1/n)$.

We give the following examples of information measures.

(8) $F(p) = \max_i p_i$

is a convex symmetric function, and $IM(p)$ in this case is the NPIM defined above.

(9) $F(p) = \sum_{\substack{i \text{ is one} \\ \text{of the } k \\ \text{larger } p_i\text{'s}}} p_i$

is a convex symmetric function, and $IM(p)$ in this case is the $NPIM_k$ defined above.

(10) $\quad F(p) = \sum_i p^a$

is convex symmetric if $a > 1$. Therefore, we can use

(11) $\quad IM(p) = 1 - \sum_i p^a$

as an information measure. Intuitively, this information measure behaves similarly to NPIM when a is large, and behaves similarly to $NPIM_k$ when a is closer to 1.

(12) $\quad F(p) = \sum_i p_i \log p_i$

is convex symmetric, and $F(1,0,...,0) = 0$. Therefore, $IM(p)$ in this case is the entropy function $H(p)$, the standard measure for information.

Comparing the pictorial information measure with the standard entropy function, we note that in axiomatic information theory, $H(p)$ is shown to satisfy the following three axioms:

(C1) $H(p)$ is continuous in p_i for all i,

(C2) $H(p)$ is symmetric in p_i for all i,

(C3) If $p_n = q_1 + q_2$, then

$H(p_1, p_2, ..., p_{n-1}, q_1, q_2) =$

$H(p_1, ..., p_{n-1}, p_n)$

$+ p_n H(q_1/p_n, q_2/p_n)$

The minimum of $H(p)$ is also at $H(1, 0, ..., 0) = 0$, and the maximum of $H(p)$ is at $H(1/n, ..., 1/n)$. (C3) is not satisfied by NPIM or $NPIM_k$, and this axiom is replaced by (B3) and (B4) given above. If the function F is as given in equation (10), we can show that

(C3') $IM(p_1, ..., p_{n-1}, q_1, q_2) =$

$IM(p_1, ..., p_n) + p_n{}^a IM(q_1/p_n, q_2/p_n)$

In general, if $F(p)$ is the summation of $f(p_i)$, where f is a continuous convex function in p_i, then F is a symmetric convex function, and $IM(p)$ thus defined is an information measure. More details and proofs of theorems can be found in [5].

3. LORENZ INFORMATION MEASURE

The reasons for using NPIM or $NPIM_k$ as information measures instead of using the usual entropy function, are (1) they have an intuitively meaningful interpretation with respect to pictures, (2) they are easy to compute, and (3) they represent a family of picture information measures, so that for a given application, a desirable one can be selected by adjusting the various thresholds and constraints. Furthermore, the picture information measure described above has a natural extension to a *Lorenz information measure*, as will be discussed in this section.

Let $NPIM_k$ denote the normalized picture information measure where $NPIM_k$ is the minimum number of pixel gray level changes to convert a picture to k gray levels. Suppose the p_i's are ordered such that

$$p_0 \leqq p_1 \leqq ... \leqq p_{L-1}$$

It can be seen that $NPIM_k(f)$ is 1 minus the sum of the last k terms in the sequence p_0, p_1, ..., p_{L-1}, which is equal to the sum of the first L—k terms in the sequence. In particular, $NPIM(f)$ or $NPIM_1(f)$ is the sum of the first L—1 terms in the sequence p_0, p_1 ..., p_{L-1}. The following inequalities hold:

$$0 = NPIM_L(f) \leqq NPIM_{L-1}(f) \leqq ... \leqq NPIM_1(f) \leqq NPIM_0(f) = 1$$

If we define s(k) to be $NPIM_{L-k}(f)$, we have

$$s_0 = 0$$

$$s_L = 1$$

$$s_k = \sum_{i=0}^{k-1} p_i$$

By plotting the points $(k/L, s_k)$, k=0, 1, ..., L—1, we obtain a piecewise linear curve, as illustrated in Figure 1, which is called the Lorenz curve [6]. We note note this curve represents the information content of a picture. To find out the value of $NPIM_k$, we simply check the point $((L—k)/L, s(L—k))$ on the Lorenz curve.

If the gray levels of the pixels are uniformly distributed, then the curve becomes a straight line from (0,0) to (1,1). Otherwise the curve will be convex piecewise linear curve under this straight line. In Figure 1, curve C_f for picture f is always above curve C_g for picture g. That is to say, $PIM_k(f) \geqq PIM_k(g)$ for every k, or picture f is more informative than picture g. Another way of describing this relationship is that picture g is less complex than picture f. In other words, "as the bow is bent, concentration increases", and

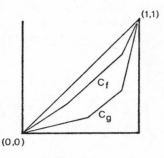

Figure 1.

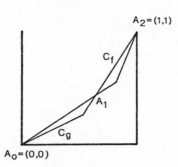

Figure 2.

the corresponding picture is less complex and consequently less informative. Using the concept of majorants, if $C_f > C_g$, then f is more informative than g.

The Lorenz curve derived from the picture information measures $NPIM_k(f)$ is called the *Lorenz information curve*. It can be seen that once the histogram h is given, the Lorenz information curve is completely specified. Conversely, if the Lorenz information curve is given, we know the histogram in its permutation equivalent class.

The *Lorenz information measure* $LIM(p_1, ..., p_n)$ is defined to be the area under the Lorenz information curve. Clearly, $0 \leq LIM(p_1, ..., p_n) \leq 0.5$. For any probability vector $(p_1, ..., p_n)$, $LIM(p_1, ..., p_n)$ can be computed by first ordering the p_i's, then calculating the area under the piecewise linear curve. Since $LIM(p_1, ..., p_n)$ can be expressed as the sum of $f(p_i)$, and $f(p_i)$ is a continuous convex function in p_i, $LIM(p_1, ..., p_n)$ is also an information measure. Intuitively, the Lorenz information measure is the weighted sum of the PIM_k's, so that LIM can be regarded as a global measure of information content.

4. STRUCTURED INFORMATION MEASURE

Since the Lorenz information curve is always normalized, we can plot such curves for pictures having different sizes and gray level sets and compare them. As illustrated in Figure 2, two Lorenz information curves may intersect at points $A_0 = (0,0)$, A_1, A_2, ..., $A_m = (1,1)$. We can define a similarity measure between f and g, d(f, g), as the summation of the polygonal areas enclosed by the two curves C_f and C_g. Clearly, $0 \leq d(f, g) \leq 0.5$. If d(f, g) is below a preset threshold t, the two pictures can be considered as *information-ally similar* in the sense of having similar Lorenz information curves.

It should also be noted that this approach can be generalized to handle not only physical pictures defined by a picture function f, but also logical pictures consisting of logical objects and relational objects [1]. Suppose there are N objects in the logical picture, and these objects are classified into L different types: T_1, T_2, ..., T_L. We can define h: $\{1, 2, ..., L\} \rightarrow N$ to be the *logical histogram*, where h(i) is the number of objects having type T_i. The Lorenz information curve for logical pictures can then be computed.

As an example, suppose the picture object set is $V = \{v_1(A), v_2(A), v_3(B), v_4(B), v_5(B)\}$. The relational object set is $R = \{r_1(X, v_1, v_2), r_2(Y, v_1, v_2), r_3(X, v_1, v_3), r_4(Y, v_1, v_5), r_5(X, v_2, v_4)\}$. The picture objects and relational objects are illustrated in Figure

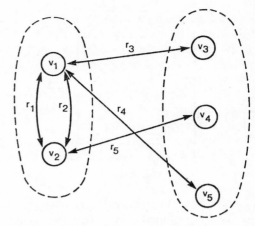

Figure 3. Logical picture with objects and relations.

Type A objects Type B objects

3. Objects v_1 and v_2 are of type A, and v_3, v_4, v_5 of type B. Relations r_1, r_3, r_5 are of type X, and r_2, r_4 of type Y. We can define the *structural information measure* SIM as the weighted sum of three parts:

(S1) *Object information measure* OIM = IM$(a_1, ..., a_n)$ where a_i is the probability of occurrence for objects of type T_i.

(S2) *Intraset information measure* IIM(i) = IM$(b_{i1}, ..., b_{ik})$ where b_{ij} is the probability of occurrences for relations of type R_j in the object set of type T_i.

(S3) *Interset information measure* TIM(i,j) = IM$(c_{ij1}, ..., c_{ijk})$ where c_{ijk} is the probability of occurrence for relations of type R_k between object sets T_i and T_j.

The structural information measure SIM is defined to be:

$$(13) \quad \text{SIM} = w_0\,\text{OIM} + \sum_{i=1}^{L} w_i\,\text{IIM}(i) + \sum_{i=1}^{L}\sum_{j=1}^{L} u_{ij}\,\text{TIM}(i, j)$$

where w_0, w_i, u_{ij} are nonnegative weights. For example, if there are L types of objects, we can set w_0 to 1, w_i to $1/L$, and u_{ij} to $2(L \times (L-1))$.

In (S1), (S2), and (S3), we can use any information measure IM. If we use LIM as the information measure, we have OIM = LIM(0.4, 0.6) = 0.45, IIM(A) = LIM(0.5, 0.5) = 0.5, IIM(B) = LIM(0) = 0, and TIM(A,B) = LIM(1/3, 2/3) = 0.4167. Therefore, SIM = $0.45 + 0.5 \times 0.5 + 0.5 \times 0 + 1 \times 0.4167 = 1.1167$. If the summation of w_i is 1, and the summation of u_{ij} is 1, it is clear that $0 \leq \text{SIM} \leq 1.5$.

Since the structured information measure is defined with respect to a given object set V and relational object set R, we should write SIM(V,R) to indicate SIM is always calculated for given V and R. Therefore, SIM can be calculated for any subpicture of a logical picture, or any subpicture of a physical picture, by changing V and R.

To define similar pictures, we can use a combination of the following criteria: (1) their physical (and/or logical) histograms are similar, (2) their Lorenz information curves are similar, (3) their Lorenz information measures are similar, and (4) their structured information measures are similar.

The picture information measure introduced above is based upon the minimal number of gray level changes to convert a picture to one having a desirable histogram. In the case of PIM_1, the desirable histogram consists of a peak at a single (arbitrary) gray level. For PIM_k, the desirable histogram consists of k peaks at k (arbitrary) gray levels. For similarity measurement, the exact shape of the desirable histogram can also be specified, and the algorithms for optimal histogram matching for both the L_1 norm [7] and the L_n norm [8] have been developed. As suggested in [8], we can also use the computed minimal number of mismatches to measure the similarity between two pictures or subpictures.

5. APPLICATION TO PICTURE ENCODING

Figure 4(a) illustrates a Seasat image (image No. 1) of the Los Angeles area, and Figure 4(b) another Seasat image (image No. 2) of a certain coastal area. Both images were quantized into 128 × 128 pixels of 64 gray levels. Figures 5(a) and (b) illustrate the Lorenz information curves for Seasat image No. 1 and 2, respectively.

We can apply picture decomposition techniques (see [1, 2]) to decompose these two pictures into pages of size 10 × 10 and construct picture trees, using the Primitive algorithm, the PIM-guided algorithm, and the LIM-guided algorithm. The experimental results are summarized in Table 1.

In the above experiments, each picture is divided and subdivided by finding "subtractors", which are those areas with NPIM less than or equal to a threshold value. The subtractor areas are discarded, and the remaining areas are decomposed into rectangular blocks. The decomposition algorithms are based upon a heuristic to minimize the expected number of pages (Primitive algorithm), the minimization of the sum of PIM for the decomposed areas (PIM-guided algorithm), or the minimization of the sum of LIM

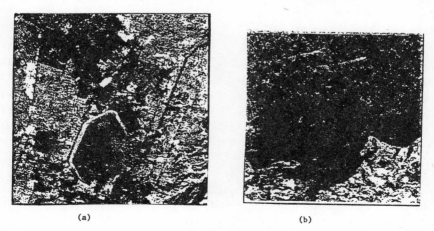

(a)   (b)

Figure 4. Seasat image No. 1 (a) and
image No. 2 (b).

Table 1 Application of the decomposition technique to two Seasat images

Threshold Value	Number of Nodes	Number of Blocks	Number of Pages	Primitive Type
Seasat image No. 1:				
0.0	52	31	135	Primitive
0.0	51	30	136	PIM
0.0	49	29	134	LIM
0.01	52	30	130	PIM
0.01	47	28	129	LIM
0.02	59	34	116	PIM
0.02	57	31	114	LIM
0.03	55	28	107	PIM
0.03	69	33	106	LIM
0.04	66	30	90	PIM
0.04	71	34	85	LIM
0.05	67	29	55	PIM
0.05	71	33	72	LIM
0.10	16	3	3	PIM
0.10	15	3	3	LIM
Seasat image No. 2:				
0.0	49	28	69	Primitive
0.0	50	28	69	PIM
0.0	50	30	70	LIM
0.01	50	28	69	PIM
0.01	54	32	71	LIM
0.02	44	23	55	PIM
0.02	46	24	54	LIM
0.03	43	21	47	PIM
0.03	43	21	45	LIM
0.04	41	17	28	PIM
0.04	41	17	24	LIM
0.05	35	10	12	PIM
0.05	42	14	15	LIM
0.10	15	2	3	PIM
0.10	5	0	0	LIM

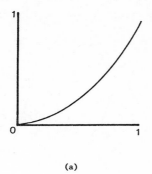

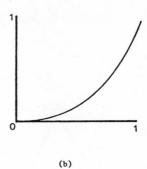

(a)   (b)

Figure 5.   Lorenz information curves for
image No. 1 (a) and image No. 2 (b).

for the decomposed areas (LIM-guided algorithm). In the seven experiments for Seasat image No. 1, the LIM-guided algorithm is strictly better than the PIM-guided algorithm in five cases. In the seven experiments for Seasat image No. 2, the LIM-guided algorithm performs better in four cases. Therefore, using the LIM-guided algorithm to construct picture trees usually gives good results.

6. PICTURE TREE OPERATIONS

In Section 5, we described the decomposition technique to encode a picture into a picture tree. In this section, we describe a set of *picture tree operations* for the retrieval and manipulation of picture trees.

Given a picture object set $V = \{v_1, ..., v_n\}$, we can construct a picture tree $PT = \{H_1, ..., H_m\}$, where the H_i's are subsets of V, satisfying the following:

(A1) $H_1 = V$.

(A2) For any two sets H_i and H_j, either they are disjoint, or H_i contains H_j, or H_j contains H_i.

The picture tree is constructed as follows: If $H_i \supset H_j$, and there is no H_k such that $H_i \supset H_k \supset H_j$, then there is a directed arc from H_i to H_j, written as $H_i \rightarrow H_j$.

Example 1: The H_i's are:

$H_1 = \{v_1, v_2, v_3, v_4, v_5\}$
$H_2 = \{v_1, v_2\}$
$H_3 = \{v_3, v_4, v_5\}$
$H_4 = \{v_1\}$
$H_5 = \{v_2\}$
$H_6 = \{v_3, v_4\}$
$H_7 = \{v_3\}$
$H_8 = \{v_4\}$

The picture tree is illustrated in Figure 6(a).

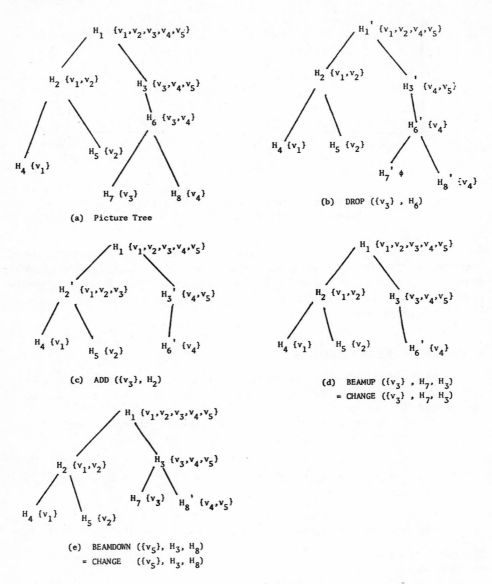

Figure 6.

Each node H_i in a picture tree is associated with a *depth* $D(H_i)$, which is the distance from the root node H_1. For example, in Figure 6(a), $D(H_2)$ is 1, and $D(H_8)$ is 3.

In the picture tree induced by PT, a node H_i is a *leaf node*, if there is no H_j such that $H_i > H_j$. We define the *ancestor set* of H_k as $A(H_k) = \{H_i: H_i \supseteq H_k\}$. We define the *follower set* of H_k as $F(H_k) = \{H_i: H_k \supset H_i\}$.

In Example 1, $A(H_6) = \{H_1, H_3, H_6\}$, and $F(H_6) = \{H_7, H_8\}$.

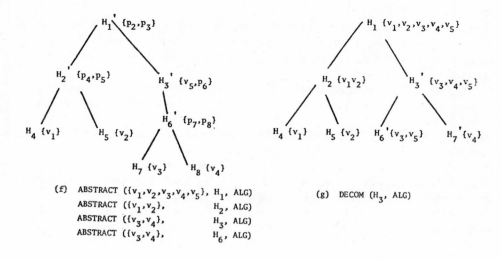

(f) ABSTRACT $(\{v_1,v_2,v_3,v_4,v_5\}, H_1, ALG)$
ABSTRACT $(\{v_1,v_2\}, H_2, ALG)$
ABSTRACT $(\{v_3,v_4\}, H_3, ALG)$
ABSTRACT $(\{v_3,v_4\}, H_6, ALG)$

(g) DECOM (H_3, ALG)

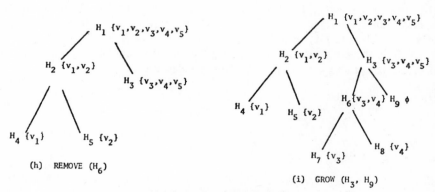

(h) REMOVE (H_6)

(i) GROW (H_3, H_9)

Figure 6. (Continued)

To restructure a picture tree, the following operations on picture tree H are introduced.

(a) DROP(Y, H_k)

The set Y is contained in H_k. We want to drop Y from H_k. To preserve (A1) and (A2), DROP(Y, H_k) has the following effects: every H_i in $A(H_k) \cup F(H_k)$ is changed to $H'_i = H_i - Y = \{v: v$ in H_i and not in $Y\}$. An example of performing DROP($\{v_3\}$, H_6) on the tree of Figure 6(a) is illustrated in Figure 6 (b).

(b) ADD(Y, H_k)

We want to add Y to H_k. To preserve (A1) and (A2), ADD(Y, H_k) has the following effects: every H_i in $A(H_k)$ is changed to $H'_i = H_i \cup Y$. An example of performing ADD ($\{v_3\}$, H_2) on the tree of Figure 6(b) is illustrated in Figure 6(c).

(c) CHANGE(Y, H_i, H_j)

This operation is equivalent to a DROP(Y, H_i), followed by an ADD(Y, H_j). In other words, it drops Y from one node H_i, and then adds Y to another node H_j. For example, the tree of Figure 6(c) can be obtained from the tree of Figure 6(a) by CHANGE({v_3}, H_6, H_2).

In performing DROP, ADD, and CHANGE operations, some node H_i may become the empty set, and some H_i may become identical to another node H_j. Such nodes should be removed.

Fact: Suppose H_j is in $F(H_i)$, and H'_j and H'_i become identical after a DROP operation, then H'_k becomes empty for all other H_k in $F(H_i)$.

Proof: Suppose $H_i = A \cup B \cup C$, $H_j = A \cup B$, and $Y = B \cup C$ is dropped from H_i and H_j. By (A2), all other H_k in $F(H_i)$ must be disjoint from $H_j = A \cup B$. Therefore, H_k is contained in C. When Y is dropped from H_k, it becomes empty. Q.E.D.

The CHANGE(Y, H_i, H_j) operation can be used to restructure a picture tree. It can be seen that CHANGE(Y, H_i, H_j) only affects nodes in $F(H_i) \cup (A(H_i) \cup A(H_j) - A(H_i) \cap A(H_j))$. In other words, CHANGE(Y, H_i, H_j) affects only nodes in a subtree having the common ancestor of H_i and H_j as the root node. The common ancestor node itself is not affected. Therefore, CHANGE is a *local operation* on the picture tree.

In a picture tree, information (picture objects) carried by a node is also always carried by its ancestor nodes. The converse is not true. As illustrated in Figure 6(a), an ancestor node (such as H_3) may carry some information (such as v_5) not carried by its follower nodes (such as H_6, H_7, and H_8).

Let $B(H_k)$ be the set of picture objects that can be found only at H_k, and $C(H_k)$ the set of picture objects that can be found at follower nodes. Formally,

$$C(H_k) = \bigcup_{i:H_k \supset H_i} H_i$$

and

$$B(H_k) = H_k - C(H_k)$$

H_k is called the *host node* of the picture object set $B(H_k)$ or any picture object v in $B(H_k)$.

In Figure 6(a), H_4 is host node of v_1, and H_3 is host node of v_5. It can be seen that all leaf nodes in a picture tree are host nodes. If only the leaf nodes are host nodes, then the picture tree is called a *skirting tree*. The classical Quad-tree can be regarded as a skirting tree, if we consider the pixels to be the set V, and the partitions to be the H_i's.

The host node of a picture object set Y is where information concerning Y can be retrieved. We can change the host node of Y by performing the following operations, both being special cases of the CHANGE operation.

(d) BEAMUP(Y, H_i, H_j)

If H_j contains H_i, CHANGE(Y, H_i, H_j) will move Y from a node H_i to its ancestor node H_j. H_j becomes the host node for Y. An example of performing BEAMUP({v_3}, H_7, H_3) on the tree of Figure 6(a) is illustrated in Figure 6(d). This example also illustrates the elimination of empty nodes (such as H_7) and identical nodes (such as H_8).

(e) BEAMDOWN(Y, H$_i$, H$_j$)

If H$_i$ contains H$_j$ and Y is contained in B(H$_i$), CHANGE(Y, H$_i$, H$_j$) will move Y from a node H$_i$ to its follower node H$_j$. H$_j$ becomes the host node for Y. An example of performing BEAMDOWN({v$_5$}, H$_3$, H$_8$) on the tree of Figure 6(a) is illustrated in Figure 6(e).

Intuitively, the BEAMUP and BEAMDOWN operations move information up and down a picture tree. In a picture tree, only host nodes actually store information. Information (picture objects) in nonhost nodes is *abstracted* and *condensed* into picture indexes.

(f) ABSTRACT(Y, H$_k$, ALG)

Y is in C(H$_k$), i.e., H$_k$ is not the host node of Y. The picture objects in Y are abstracted and condensed into picture indexes using algorithm ALG. An example of performing ABSTRACT operations on the tree of Figure 6(a) is illustrated in Figure 6(f). We note that since X$_k$ is not the host node of Y, condensed abstractions of Y can be allowed, so that picture indexes to other host nodes can be constructed.

Finally, to perform complete restructuring of a picture subtree, we have the decomposition operation.

(g) DECOM(H$_i$, ALG)

DECOM(H$_i$, ALG) replaces the subtree with root node H$_i$ by another subtree. The new subtree is constructed by decomposition of H$_i$, using decomposition algorithm ALG. An example of performing DECOM(H$_3$, ALG) on the tree of Figure 6(a) is illustrated in Figure 6(g).

In addition to the above picture-tree operations, there are two usual tree operations, REMOVE and GROW.

(h) REMOVE(H$_i$)

REMOVE(H$_i$) removes node H$_i$ and its follower nodes from the picture tree. An example of performing REMOVE(H$_6$) on the tree of Figure 6(a) is illustrated in Figure 6(h).

(i) GROW(H$_i$, H$_j$)

GROW(H$_i$, H$_j$) adds a new node H$_j$ as the follower node of H$_i$, where H$_j$ must be a new node name, and H$_j$ is initially empty. An example of performing GROW(H$_3$, H$_9$) on the tree of Figure 6(a) is illustrated in Figure 6(i).

Given a picture tree PT and a query Q, the pictorial retrieval problem is to find QS, the picture query set, by searching the picture tree PT. To facilitate pictorial retrieval, we describe the technique of *structured retrieval* as follows.

Given a picture query Q and a node H$_i$ in picture tree PT, it is assumed that we can compute the following:

We can derive (TS', Q', S') from (Q, H$_i$), where TS' is the *partial picture query set* obtainable from (Q, H$_i$), Q' is the *modified query* specifying the remaining query after the processing of (Q, H$_i$), and S' is the set of descendent nodes H$_j$ of H$_i$ to be visited next.

We can now describe informally the structured retrieval technique as follows. We start from node H$_1$ and compute (TS', Q', S'). If the query Q cannot be fully answered, we will have nonempty Q'. We then retrieve all nodes H_i in S'. We now use (Q', H') to compute (TS'', Q'', S''), where H' is a descendent node in S_k. If Q' cannot be fully answered, we will have nonempty Q'', and we should retrieve all nodes H_i in S''. We proceed recursively as above, and QS is $Q(TS)$, where TS is the union of all the TS's. We have the following RETRIEVE operation.

(j) RETRIEVE(Q, QS)

RETRIEVE(Q, QS) is the *structured retrieval* operation, which provides a query set QS for a given query Q, computed as follows:

  STEP 1: LIST $\leftarrow \{(Q, H_1)\}$, TS $\leftarrow$ empty set.
  STEP 2: Remove one tuple (Q, H_i) from LIST. Compute (TS', Q', S') from (Q, H_i).
    TS $\leftarrow$ TS $\cup$ TS$'$.
  STEP 3: For each successor node H_j of H_i, if $\{H_j\} \cap S'$ is nonempty, add (Q', H_j) to LIST.
  STEP 4: If LIST is nonempty, go to STEP 2, or else go to STEP 5.
  STEP 5: QS $\leftarrow$ Q(TS).

For a Quad-tree, the transformation $(Q, H_i) \rightarrow (TS', Q', S')$ is as follows:

(1) TS$'$ $\begin{cases} = PS_i, \text{ if } m_i = 0 \\ \\ = 0, \text{ otherwise} \end{cases}$

(2) Q' $\quad = Q$
(3) S' $\quad = \{H_j: H_j$ is a descendent of H_i and $(P_{kj} \wedge Q)$ is nonempty$\}$

  In summary, once the picture tree is constructed, we can retrieve a subpicture by searching the picture tree to find similar pictures according to the similarity criteria based upon a structured picture information measure. The structured picture information measure allows for the consideration of both the physical picture and the logical picture. The subpictures retrieved according to the similarity criteria can then be processed to determine whether they satisfy the detailed picture query [1]. Therefore, inexact similarity retrieval techniques can be combined with exact query processing techniques. Finally, picture trees can be changed dynamically, using the operations described above, thus providing a flexible approach for picture retrieval and manipulation.

REFERENCES

1. S. K. Chang, "A Methodology for Picture Indexing and Encoding", in *Picture Engineering* (T. Kunii and K. S. Fu, Eds.), Springer-Verlag, 1982.
2. S. K. Chang and C. C. Yang, "Picture Encoding Techniques for a Pictorial Database", Technical Report, Naval Research Laboratory, 1982.
3. J. Reuss, S. K. Chang and B. H. McCormick, "Picture Paging for Efficient Image Processing", in *Pictorial Information Systems* (S. K. Chang and K. S. Fu, Eds.), Springer-Verlag, 1980, 228–256.
4. S. H. Liu and S. K. Chang, "Picture Covering by 2-D AH Encoding", Proceedings of IEEE Workshop on Computer Architecture for Pattern Analysis and Image Database Management, Hot Springs, Virginia, November 11–13, 1981, 76–87.
5. H. Silver and S. K. Chang, "Picture Information Measures", Technical Report, University of Illinois at Chicago Circle, 1982.

6. A. W. Marshall and I. Olkin, *Inequalities: Theory of Majorization and Its Applications*, Academic Press, 1979.

7. S. K. Chang and Y. Wong, "Optimal Histogram Matching by Monotone Gray Level Transformation", Communications of the ACM, Vol. 22, No. 10, ACM, October 1978, 835–840.

8. S. K. Chang and Y. Wong, "Ln Norm Optimal Histogram Matching and Application to Similarity Retrieval", *Journal of Computer Graphics and Image Processing*, Vol. 13, 1980, 361–371.

9. R. C. Gonzalez et al., "A Measure of Scene Content", CH1318-5/78/0000-0385500.75, IEEE 1978, 385–389.

Shi-Kuo Chang
Information Systems Laboratory
Department of Electrical Engineering
Illinois Institute of Technology

2 Multiresolution image representation
A Rosenfeld

1. INTRODUCTION

The first part of this paper reviews methods of representing or approximating a digital image based on recursive subdivisions into quadrants; such methods give rise to trees of degree 4 ("quadtrees") whose leaves represent homogeneous blocks of the image. The second part deals with image "pyramids", which are exponentially tapering stacks of arrays, each half the size of the preceding, and shows how such pyramids can be useful in image segmentation. We assume in this paper that all images are of size $2^n \times 2^n$.

2. VARIABLE RESOLUTION IMAGES: "QUADTREES"

2.1. Image segmentation by recursive splitting

Suppose we are given a criterion for deciding that a digital image is uniform or homogeneous, e.g., that the standard deviation of its gray levels is below a given threshold t. Based on this criterion, we can recursively subdivide a given image into homogeneous pieces. For concreteness, let us assume that the subdivision at each step is into quadrants. If the image is homogeneous to begin with, we are done. If not, we split it into quadrants, and test each of them for homogeneity. If a given quadrant is homogeneous, we are done; if not, we split it into quadrants; and so on. Note that some quadrants may get split and others may not; we split only those that are not homogeneous.

The results of the subdivision process can be represented by a tree of degree 4 (a "quadtree"). The root node of this tree represents the entire image, and the children of a node represent its quadrants. Thus the leaf nodes represent blocks (sub. . .subquadrants) that are homogeneous. Suppose we associate with each leaf node the mean gray level of its block; then the resulting quadtree completely specifies a piecewise-constant approximation to the image, in which each homogeneous block is approximated by its mean; note that the rms error of this approximation is less than t.

This method of constructing piecewise approximations to an image was first proposed in the early 1970s. Such an approximation has the advantage of being very compactly specifiable by its quadtree. Note, however, that the homogeneous blocks are not necessarily maximal homogeneous regions in the image; it is likely that there will be unions of blocks that are still homogeneous. To obtain a segmentation of the image into maximal homogeneous blocks, we must allow merging of adjacent blocks (or unions of blocks) as long as the resulting region remains homogeneous. Note that the resulting segmentation can no longer be compactly represented by a quadtree. This "split-and-merge" approach to image segmentation is treated in detail by Pavlidis [1].

2.2. Exact image representation by quadtrees

The case where t=0 is of special interest; here a block is not regarded as homogeneous unless its value is perfectly constant, so that the image can be exactly reconstructed from its quadtree representation. This method of image representation was introduced by Klinger [2–3].

It should be pointed out that the quadtree representation of an image may be less compact than the array representation; its primary advantage is in cases where the image is composed of large regions of constant gray level. The size of the quadtree grows with the total length of the boundaries between regions of constant gray level in the image. An analysis of quadtree size for some simple types of images can be found in [4]. Note also that the quadtree size depends strongly on the positions and sizes of the constant-level regions. For example, a constant block of size $2^k \times 2^k$ whose coordinates are multiples of 2^k is represented by a simple quadtree node; but if its position is shifted by one pixel in the x and y directions, it requires on the order of 2^{k+2} nodes to represent it. On optimizing quadtrees with respect to translation see [5]; on region representation using forests of quadtrees see [6].

In the remainder of this section we will be concerned primarily with exact quadtree representations, and we will usually assume for simplicity that the given image is two-valued, i.e., it consists of 0's and 1's. In the following subsections we discuss conversion between quadtrees and other image representations; operations on images represented by quadtrees; and some related types of hierarchical representations.

2.3. Conversion between quadtrees and other representations

The quadtree construction process sketched in Section 2.1 operates "top-down"; starting from the entire image, we test each (sub . . .)-quadrant for homogeneity, and if it is not homogeneous, we split it. In general, this requires us to examine parts of the image repeatedly; we examine a block to check for homogeneity, and if it is not homogeneous, we must re-examine its quadrants. More economical algorithms exist [7] that allow us to build the quadtree representation of an image "bottom-up", by examining the image in an order such as

$$
\begin{array}{cccccccc}
1 & 2 & 5 & 6 & 17 & 18 & 21 & 22 \\
3 & 4 & 7 & 8 & 19 & 20 & 23 & 24 \\
9 & 10 & 13 & 14 & 25 & 26 & 29 & 30 \\
11 & 12 & 15 & 16 & 27 & 28 & 31 & 32 \\
33 & \ldots
\end{array}
$$

Here we create a leaf node corresponding to a 2x2 block whenever we find four consecutive pixels (with numbers 1,2,3,0 modulo 4) having only one value, and when this happens, we discard the pixels themselves. Similarly, we create a leaf node corresponding to a 4×4 block whenever we find four consecutive 2×2 blocks (with upper left corner pixels numbered 1,5,9,13 modulo 16), and we discard the 2×2 blocks themselves; and so on.

It should be mentioned that a bottom-up approach can also be used to construct piecewise approximations to an image. If we know the means and standard deviations of the gray levels in the four quadrants of a block, we can directly compute the mean and standard deviation of the entire block, without having to refer to the original gray levels. We can proceed in this way, starting from the individual pixels (mean=gray level, standard deviation=0), and continuing to merge quadrants as long as the standard deviation of the resulting block does not exceed the given threshold t. A generalization of this method can be used to construct piecewise least-squares approximations to the image by polynomials of any given degree (note that the mean is the least-squares approximation by a polynomial of degree 0, i.e., by a constant); if we know the coefficients of these approximations, and the rms errors, for the four quadrants of a block, we can directly compute the coefficients and error for the entire block [8].

For a different bottom-up algorithm that constructs a quadtree by a row-by-row scan of the image see [9]; this algorithm is appropriate if the image is given in the form of a row-by-row representation such as a run length code. Conversely, algorithms that construct a row-by-row image representation from a quadtree representation are described in [10]. Algorithms for directly converting from a quadtree representation to the codes representing the borders between the regions of constant gray level, and vice versa, are given in [11-12]. Algorithms could also be defined for converting between the quadtree and medial axis representations; on quadtree-related distance transforms, and a quadtree "medial axis transformation", see [13-15].

2.4. Operations on images represented by quadtrees

Algorithms exist for computing various properties of an image directly from its quadtree representation. For example, computation of moments [16] or of various types of discrete transforms is straightforward, since these are linear transformations and can be computed blockwise, and the block sizes and positions are determined by the tree structure. It is less trivial to compute the total perimeter (=sum of border lengths) directly from the quadtree [17]; here the task is related to that of constructing border codes from quadtrees. On labeling the connected components of an image directly from its quadtree representation see [18], and on computing the genus of an image from its quadtree see [19]. An extremely easy task is that of determining from the quadtree representation whether a pixel having given coordinates is 0 or 1; we simply use the coordinates to define a series of moves down the tree until a leaf is reached. On the detection of symmetries using quadtree representations see [20].

A key step in many of these tasks is to find the neighbors (in the image plane) of a given quadtree leaf [21]. This can be done by a straightforward tree traversal process. For example, if we want to find the east neighbors of a given leaf ℓ, we move upward from ℓ in the tree until we arrive at a node, say ℓ', from its northwest or southwest son. (If this does not happen, and we reach the root, ℓ must be on the east edge of the image and has no east neighbors.) We then move downward from ℓ', using mirror images of the moves that were made in going from ℓ to ℓ', for example, if the last upward move was from a northwest son, the first downward move is to the northeast son, and so on. If we reach a leaf before (or at the same time as) finishing this sequence of moves, that leaf is the sole east neighbor of ℓ, and represents a block at least as large as ℓ's block. Otherwise, when the sequence of mirror-image moves is finished, we continue to move downward through a sequence of northwest sons until we reach a leaf; that leaf is then the northern-most of ℓ's east neighbors. It turns out that the average number of moves required to find a neighbor of ℓ in this way is independent of the tree size. An alternative [22] is to represent the block adjacencies explicitly using pointers ("ropes"); but this requires additional storage.

It is quite straightforward to compute quadtree representations of Boolean combin-ations of images (AND, OR, etc.) from the representations of the images themselves [16, 22]. For example, to compute the OR (which has 1's wherever either of the given images has 1's), we synchronously traverse the two given trees, and construct a new tree as follows: whenever we reach a leaf labeled 1 in either tree, the new tree gets a corres-ponding leaf labeled 1; and whenever we reach a leaf labeled 0, the new tree gets a copy of the other tree's subtree whose root corresponds to that leaf. A more difficult task is that of computing the quadtree corresponding to a translation, magnification, or rotation (by an arbitrary angle) of the image defined by a given quadtree; see [23].

2.5. Some alternatives to quadtrees

For "digital images" represented by triangular or hexagonal arrays, rather than square arrays, we can also define analogs of the quadtree representation. In the case of a tri-angular grid [24], the blocks are equilateral triangles, and we split a block by subdividing it into four triangles; note that there are two types of blocks, one with base down, the other with apex down. In the case of a hexagonal grid [25] *, a regular hexagon cannot be exactly subdivided into regular hexagons, but "rosettes" of seven hexagons can be built up, and these in turn can be combined into larger rosettes, etc. (with the orientation changing by about 20° at each step). Algorithms analogous to those developed for quad-trees can also be designed for "tritrees" or "hextrees".

Analogs of the quadtree representation can also be defined for higher-dimensional arrays. For example, in three dimensions we can define an "octree" representation using recursive subdivision into octants having constant value (or below-threshold standard deviation) [26–29]. Algorithms analogous to those for quadtrees can be designed in three or more [30] dimensions.

In the quadtree representation, we split a block by subdividing both dimensions at once. An alternative is to subdivide one dimension at a time. This gives rise to a binary tree whose leaf nodes represent rectangular blocks of constant value (or below-threshold standard deviation). Here each nonleaf node must carry a label indicating whether its sons correspond to a vertical or horizontal subdivision. For a discussion of the quadtree and binary tree approaches see [31–32].

---

*Much work on the hexagonal-grid case has also been done by D. Lucas, but most of this work is unpublished.

3. MULTIPLE-RESOLUTION IMAGES: "PYRAMIDS"

3.1. Pyramid construction

The simplest type of image pyramid is based on recursive subdivision into quadrants, just as in quadtree construction, except that we always keep subdividing until we reach the individual pixels. Thus the leaves (i.e., the bottom layer or base of the pyramid) represent single pixels. The nodes just above the leaves represent nonoverlapping 2×2 blocks of pixels, constituting a $2^{n-1} \times 2^{n-1}$ array, where the value of a node is just the average of its block. The nodes at the next level form a $2^{n-2} \times 2^{n-2}$ array, and represent nonoverlapping 4×4 blocks of pixels (or 2×2 blocks of 2×2 blocks); and so on, until we reach the root node, whose value is the average gray level of the entire image.

For some purposes, it is desirable to define pyramids based on overlapping blocks of pixels. Here one of the simplest schemes is to use 4×4 blocks that overlap 50% horizontally and vertically. It is easily verified that in this scheme, each block at a given level is contained in four blocks at the level above; thus the containment relations between blocks no longer define a tree. To avoid border effects, it is convenient to regard each level as cyclically closed in both directions, i.e., as having its top row adjacent to its bottom row and its left column to its right column. This readily implies that the levels above the base contain exactly $2^{n-1} \times 2^{n-1}$, $2^{n-2} \times 2^{n-2}$, ..., nodes. Here again, we assume that the value of a node is the average of the values of the nodes in its block on the level below.

In constructing pyramids, we can use weighted rather than unweighted averaging. It turns out that for a certain simple class of weighting schemes, the resulting weights are very good approximations to Gaussian weights [33]. On the reconstruction of approximations to an image from such Gaussian-weighted pyramids see [34].

It should be pointed out that the total number of nodes in a pyramid is not much larger than the number of pixels in the original image alone. Indeed, if the sizes of the successive levels are $2^n \times 2^n$, $2^{n-1} \times 2^{n-1}$, ..., then the total number of nodes is

$$(1 + \frac{1}{4} + \frac{1}{16} + \ldots) < 2^n \times 2^n \times 1\frac{1}{3}.$$

3.2. Feature detection and extraction in pyramids

Uhr [35] proposed a class of pyramid-like structures called "recognition cones", as a model for feature extraction in biological visual systems. The general idea is that each level extracts features from the level below it, and represents them at reduced resolution. Another early use of multiple resolutions for feature detection was the work of Kelly [36], in which edges detected in a low-resolution image were used to guide the search for edges in the full-resolution image. During the past decade, several groups have used pyramids, both nonoverlapped and overlapped, in the initial stages of processing in computer vision systems [37–42].

Sets of feature detectors, e.g., edge detectors, whose sizes grow exponentially have been studied by a number of investigators [43–44]. An economical way of computing sets of such detectors is to build a pyramid and apply local edge detection operators at each level of the pyramid. The resulting values are differences of block average gray levels, rather than of single-pixel gray levels; thus they are the same as the results of applying scaled-up difference operators to the original image. If several different types of operators are needed, e.g., operators having different orientations, the pyramid does the block averaging once and for all, and we can compute each operator using only a few arithmetic operations on these block averages. Note that the larger the operator, the fewer the positions in which it is computed; but this is reasonable, since if we used large operators in every position, the values at nearby positions would be very redundant since they would be based on blocks that overlapped substantially.

Using feature detectors whose sizes and positions are powers of 2 should be adequate to detect features of all sizes in an image, particularly if an overlapped pyramid is used. For any feature, there will be an operator of about the right size (within a factor of $\sqrt{2}$) to detect it, and in fact there will be such operators that overlap the feature's position by at least 50%, so that they will respond to its presence. Note that if we did not use overlap, features in some positions would be very hard to detect; for example, a small spot exactly in the center of the image would not be "seen" by any spot detector, since no (sub . . .)-quadrant of the image would contain more than a quarter of the spot.

When a feature has been detected at some level of the pyramid, local thresholding in the appropriate part of the image can be used to extract it. For example, if we have detected a spot using a center/surround operator, let a_1 and a_2 be the average gray levels in the center and surround, respectively; then we should be able to extract the spot (from the full-resolution image) by applying the threshold $(a_1 + a_2)/2$ to the part of the image underlying the operator [45]. A similar method can be used to extract streaks [46].

3.3. Segmentation using pyramids

In an overlapped pyramid, each node at a given level contributes to four "parent" nodes on the level above. We can define a tree structure on the nodes by linking each node to exactly one of its parents, say the one having value closest to its own. If we carry out this process up to the 2×2 level, where there are just four nodes, we have partitioned the nodes into four trees, each rooted at the 2×2 level. Let us now recompute the values of the nodes, giving each node the average of the values of those nodes on the level below that are linked to it. Based on these new values, we may need to change some of the links; this in turn yields new values again; and we can repeat the process until there is no further change, which typically takes only a few iterations. It can be shown [47] that this iterative linking process is a special case of the one-dimensional ISODATA clustering algorithm, and so is guaranteed to converge. The result is (typically) a partition of the image into four distinctive pixel subpopulations, viz. the sets of leaves belonging to the four trees, with small regions (e.g., noise pixels) merged into their backgrounds if they are sufficiently isolated [48–50].

More generally, let us define link weights between a node and each of its parents, based on their similarity in value, and recompute the node values by weighted averaging. If we require that the link weights to the four parents must sum to 1, we obtain a result very similar to that using forced choice linking; the weights converge to 0's and 1's, and we again get a partition of the nodes into four trees rooted at the 2×2 level [49]. If we do not require the weights to sum to 1, many of the weights still converge to 0, and we get a partition of the nodes into a set of trees, where the leaves of each tree constitute a compact, homogeneous region in the image, and the root of the tree is at a level corresponding to the size of this region [51].

Another way of using pyramid node linking to segment an image is to combine it with the splitting process defined in Section 2.1. We can regard splitting as top-down creation of links, such that whenever a block is not split, we link it to its quadrants, them in turn to their subquadrants, and so on down to the pixel level. After this has been done, we can apply the bottom-up linking process defined above to those nodes that are not yet linked; this allows small blocks of the image to "merge with" neighboring larger blocks that they resemble [52].

3.4. Edge and curve pyramids

Multiresolution representations also appear to be useful for encoding and processing information about edges, lines, or curves in an image. If we detect edges at each level of a pyramid, we can establish links between edge elements at adjacent levels, e.g., based on similarity of orientation [53]. To extract the major edges from the image, we can select edges detected at high levels of the pyramid, follow down the links until we reach the base of the pyramid, and thereby display full-resolution representations of these edges [54]. Detecting edges at each level of a pyramid also allows us to detect higher-level features such as antiparallel pairs of edges, or points surrounded by edges, using local operations only, since at some level the edge pair or surrounding set will be only a few pixels apart [55]. Similar remarks apply to streak-like (i.e., thick) linear features, since these can be detected using local (thin) line-detection operators at each level of the pyramid.

For thin linear features we must use a different approach, since the ordinary pyramid construction process by averaging would obliterate them. Rather, we can use an encoding scheme in which, e.g., a node stores straight line approximations to the pieces of curve (if any) that pass through its image block, and its parent(s) create such approximations by combining the approximations provided by their children. Under this type of scheme, long straight line segments can be compactly encoded by single nodes high in the pyramid (the longer, the higher), and smooth curves can be encoded by small sets of nodes (the straighter, the fewer). Several variations on this type of coding scheme are currently under investigation [56]. On a related type of hierarchical curve representation, not based on subdivision of the image into quadrants, see [57].

An important feature of pyramid-based methods is that by operating at many resolutions, they convert certain types of global information about an image into local information, e.g., parallel-sided strips and compact spots become locally detectable. Using cooperation between levels, it thus becomes possible for such information to have an influence on pixel-level processes; in other words, it becomes possible for certain types of geometrical information to directly influence these processes. By linking compact objects (or straight lines, etc.) into trees, pyramids provide a natural transition from pixel arrays to more abstract data structures in which objects are represented by single nodes (here: the roots of the trees). Moreover, the root nodes and the object pixels (the leaf nodes) are closely interconnected, since the tree height grows logarithmically with object size.

3.5. Some other uses of pyramids

We have seen that pyramids can be used for image segmentation by defining various types of linking processes between nodes at successive levels. Note that low-level nodes are relatively sensitive to noise, but are less likely to overlap more than one region of the image; while the reverse is true for high-level nodes. Through cooperation across the levels, we can combine the advantages of both high resolution and low noise sensitivity. An example of image segmentation by relaxation at multiple resolutions can be found in [58].

The pyramid-based methods of image segmentation described in this paper can also be extended to segmentation based on color or texture, using color or texture inhomogeneity as a splitting criterion [59], or color or texture node/parent similarity as a linking criterion [60]. On some uses of gray-level pyramids as aids in texture analysis see [61]. On the use of multiresolution Markov models for texture synthesis see [62].

Pyramid-based methods can also be applied to arrays of dimensionality other than two. On the use of one-dimensional pyramid node linking for waveform and contour segmentation see [63]. One could also use binary-tree (analogous to quadtree) representations of contours and waveforms, or of the rows of an image, but this does not seem to have been done in practice. The pyramid concept can also be extended to three or more dimensions, e.g., one could imagine a "hyperpyramid" of three-dimensional arrays of sizes $2^n \times 2^n \times 2^n$, $2^{n-1} \times 2^{n-1} \times 2^{n-1}$, ... (so that the total number of nodes is less than $1\frac{1}{7}$ times that in the full-resolution array), and one could define segmentation processes based on splitting or node linking in such a hyperpyramid; this approach may be useful in the analysis of various types of three-dimensional data arrays.

ACKNOWLEDGMENTS

The support of the National Science Foundation under grant MCS–79–23422 is gratefully acknowledged, as is the help of Janet Salzman in preparing this paper. A slightly different version of this paper appears in *Pictorial Data Analysis*, edited by R. M. Haralick and S. Levialdi (Proceedings of a NATO Advanced Study Institute, Bonas (Gers), France, August 2–13, 1982).

REFERENCES

1. T. Pavlidis, Structural Pattern Recognition (Springer, New York, 1977).
2. A. Klinger, Data structures and pattern recognition, Proc. 1st Intl. Joint Conf. on Pattern Recognition (1973) 497–498.
3. A. Klinger and C. R. Dyer, Experiments in picture representation using regular decomposition, *Computer Graphics Image Processing* 5 (1976) 68–105.
4. C. R. Dyer, The space efficiency of quadtrees, *Computer Graphics Image Processing* 19 (1982) 335–340.

5. L. Jones and S. S. Iyengar, Representation of a region as a forest of quad trees, Proc. IEEE Conf. Pattern Recognition Image Processing (1981) 57–59.

6. M. Li, W. I. Grosky, and R. Jain, Normalized quadtrees with respect to translations, *Computer Graphics Image Processing* **20** (1982) 72–81.

7. H. Samet, Region representation: quadtrees from binary arrays, *Computer Graphics Image Processing* **13** (1980) 88–93.

8. J. Burt, Hierarchically derived piecewise polynomial approximations to waveforms and images, Techn. Rept. 838, University of Maryland, College Park, MD (1979).

9. H. Samet, An algorithm for converting rasters to quadtrees, *IEEE Trans. Pattern Analysis Machine Intelligence* **3** (1981) 93–95.

10. H. Samet, Algorithms for the conversion of quadtrees to rasters, *Computer Graphics Image Processing*, to appear.

11. H. Samet, Region representation: quadtrees from boundary codes, *Comm. ACM* **23** (1980) 163–170.

12. C. R. Dyer, A. Rosenfeld, and H. Samet, Region representation: boundary codes from quadtrees, *Comm. ACM* **23** (1980) 171–179.

13. H. Samet, A distance transform for images represented by quadtrees, *IEEE Trans. Pattern Analysis Machine Intelligence* **4** (1982) 298–303.

14. M. Shneier, Path-length distances for quadtrees, *Information Sciences* **23** (1981) 49–67.

15. H. Samet, A quadtree medial axis transformation, *Comm. ACM*, to appear.

16. M. Shneier, Calculations of geometric properties using quadtrees, *Computer Graphics Image Processing* **16** (1981) 296–302.

17. H. Samet, Computing perimeters of images represented by quadtrees, *IEEE Trans. Pattern Analysis Machine Intelligence* **3** (1981) 683–687.

18. H. Samet, Connected component labelling using quadtrees, *J. ACM* **28** (1981) 487–501.

19. C. R. Dyer, Computing the Euler number of an image from its quadtree, *Computer Graphics Image Processing* **8** (1978) 43–77.

20. N. Alexandridis and A. Klinger, Picture decomposition, tree data-structures, and identifying directional symmetries as node combinations, *Computer Graphics Image Processing* **8** (1978) 270–276.

21. H. Samet, Neighbor finding techniques for images represented by quadtrees, *Computer Graphics Image Processing* **18** (1982) 37–57.

22. G. M. Hunter and K. Steiglitz, Operations on images using quad trees, *IEEE Trans. Pattern Analysis Machine Intelligence* **1** (1979) 145–153.

23. G. M. Hunter and K. Steiglitz, Linear transformation of pictures represented by quad trees, *Computer Graphics Image Processing* **10** (1979) 289–296.

24. N. Ahuja, On approaches to polygonal decomposition for hierarchical image representation, *Computer Graphics Image Processing*, to appear.

25. P. J. Burt, Tree and pyramid structures for coding hexagonally sampled binary images, *Computer Graphics Image Processing* **14** (1980) 271–280.

26. D. R. Reddy and S. M. Rubin, Representation of three-dimensional objects, Tech. Rept. CS-78-113, Carnegie-Mellon University, Pittsburgh, PA (1978).

27. C. L. Jackins and S. L. Tanimoto, Oct-trees and their use in representing three-dimensional objects, *Computer Graphics Image Processing* **14** (1980) 249–270.

28. D. Meagher, Geometric modeling using octree encoding, *Computer Graphics Image Processing* **19** (1982) 129–147.

29. L. J. Doctor and J. G. Torborg, Display techniques for octree-encoded objects, *Computer Graphics Applications* **1**(3) (1981) 29–38.

30. M. M. Yau and S. N. Srihari, Recursive generation of hierarchical data structures for multidimensional digital images, Proc. IEEE Conf. Pattern Recognition Image Processing (1981) 42–44.

31. K. R. Sloan, Jr., Dynamically quantized pyramids, Proc. 7th Intl. Conf. Artificial Intelligence (1981) 714–736.

32. J. O'Rourke, Dynamically quantized spaces for focusing the Hough transform, Proc. 7th Intl. Joint Conf. Artificial Intelligence (1981) 737–739.

33. P. J. Burt, Fast filter transforms for image processing, *Computer Graphics Image Processing* **16** (1981) 28–51.

34. E. H. Adelson and P. J. Burt, Image data compression with the Laplacian pyramid, Proc. IEEE Conf. Pattern Recognition Image Processing (1981) 218–223.

35. L. Uhr, Layered "recognition cone" networks that preprocess, classify, and describe, *IEEE Trans. Computers* **21** (1972) 758–768.

36. M. D. Kelly, Edge detection in pictures by computer using planning, *Machine Intelligence* **6** (1971) 397–409.

37. S. Tanimoto and T. Pavlidis, A hierarchical data structure for picture processing, *Computer Graphics Image Processing* **4** (1975) 104–119.

38. S. Tanimoto, Pictorial feature distortion in a pyramid, *Computer Graphics Image Processing* **5** (1976) 333–352.

39. A. R. Hanson and E. M. Riseman, Segmentation of natural scenes, in: A. R. Hanson and E. M. Riseman (eds), *Computer Vision Systems* (Academic Press, New York (1978)) 129–163.

40. S. L. Tanimoto, Regular hierarchical image and processing structures in machine vision, in A. R. Hanson and E. M. Riseman (eds), *Computer Vision Systems* (Academic Press, New York (1978)) 165–174.

41. M. D. Levine, A knowledge-based computer vision system, in A. R. Hanson and E. M. Riseman (eds), *Computer Vision Systems* (Academic Press, New York (1978)) 335–352.

42. S. Tanimoto and A. Klinger (eds), *Structured Computer Vision* (Academic Press, New York, 1980). [See especially the papers by L. Uhr, Psychological motivation and underlying concepts, 1–30; S. L. Tanimoto, Image data structures, 31–55; M. D. Levine, Region analysis using a pyramid data structure, 57–100; A. R. Hanson and E. M. Riseman, Processing cones: a computational structure for image analysis, 102–131; R. Bajcsy and D. A. Rosenthal, Visual and conceptual focus of attention, 133–149.]

43. A. Rosenfeld and M. Thurston, Edge and curve detection for visual scene analysis, *IEEE Trans. Computers* **20** (1971) 562–569.

44. D. Marr and E. Hildreth, Theory of edge detection, *Proc. Royal Soc.* **B207** (1980) 187–217.

45. M. Shneier, Using pyramids to define local thesholds for blob detection, *IEEE Trans. Pattern Analysis Machine Intelligence*, to appear.

46. M. Shneier, Extracting linear features from images using pyramids, *IEEE Trans. Systems, Man, Cybernetics* **12** (1982) 569–572.

47. S. Kasif and A. Rosenfeld, Pyramid linking is a special case of ISODATA, *IEEE Trans. Systems, Man, Cybernetics*, to appear.

48. P. Burt, T. H. Hong, and A. Rosenfeld, Segmentation and estimation of image

region properties through cooperative hierarchical computation, *IEEE Trans. Systems, Man, Cybernetics* **11** (1981) 802–809.

49. T. H. Hong, K. A. Narayanan, S. Peleg, A. Rosenfeld, and T. Silberberg, Image smoothing and segmentation by multiresolution pixel linking: further experiments and extensions, *IEEE Trans. Systems, Man, Cybernetics* **12** (1982) 611–622.

50. H. J. Antonisse, Image segmentation in pyramids, *Computer Graphics Image Processing* **19** (1982) 367–383.

51. T. H. Hong and A. Rosenfeld, Unforced image partitioning by weighted pyramid linking, TR–1137, University of Maryland, College Park, MD (1982).

52. M. Pietikäinen, A. Rosenfeld, and I. Walter, Split-and-link algorithms for image segmentation, *Pattern Recognition* **15** (1982) 287–298.

53. M. Shneier, Two hierarchical linear feature representations: edge pyramids and edge quadtrees, *Computer Graphics Image Processing* **17** (1981) 211–224.

54. T. H. Hong, M. Shneier, and A. Rosenfeld, Border extraction using linked edge pyramids, *IEEE Trans. Systems, Man, Cybernetics* **12** (1982) 660–668.

55. T. H. Hong and M. Shneier, Extracting compact objects using linked pyramids, TR–1123, University of Maryland, College Park, MD (1981).

56. T. H. Hong, M. Shneier, R. Hartley, and A. Rosenfeld, Using pyramids to detect good continuation, TR–1185, University of Maryland, College Park, MD (1982).

57. D. Ballard, Strip trees: a hierarchical representation for curves, *Comm. ACM* **24** (1981) 310–321.

58. K. A. Narayanan, D. P. O'Leary, and A. Rosenfeld, Multiresolution relaxation, *Pattern Recognition*, in press.

59. P. C. Chen and T. Pavlidis, Segmentation by texture using a co-occurrence matrix and a split-and-merge algorithm, *Computer Graphics Image Processing* **10** (1979) 172–182.

60. M. Pietikäinen and A. Rosenfeld, Image segmentation by texture using pyramid node linking, *IEEE Trans. Systems, Man, Cybernetics* **11** (1981) 822–825.

61. M. Pietikäinen and A. Rosenfeld, Gray level pyramid linking as an aid in texture analysis, *IEEE Trans. Systems, Man, Cybernetics* **12** (1982) 422–429.

62. D. D. Garber, Computational models for texture analysis and texture synthesis, IPI–TR 1000, University of Southern California, Los Angeles, CA (1981).

63. K. A. Narayanan and A. Rosenfeld, Approximation of waveforms and contours by one-dimensional pyramid linking, *Pattern Recognition* **15** (1981) 389–396.

Azriel Rosenfeld
Computer Science Center
University of Maryland

3 Three-dimensional description of objects and dynamic scene analysis

J K Aggarwal

1. INTRODUCTION

There are fundamentally two distinct ways of describing three-dimensional scenes. These descriptions are based on: (i) viewer-centered methods and (ii) object-centered methods. An intensity image is a viewer-centered description of the scene whereas, an edge-vertex description specifying relative distances between adjacent vertices is an object-centered description of the scene. The following incomplete list presents the variety of available descriptions:

Viewer-centered descriptions
Intensity Images
Sequences of images including stereo pairs
Range Images
Perspective and orthographic line drawings

Object-centered descriptions
Solid-geometric methods
Generalized cones
Medial axis transformation
Surface descriptions

There are descriptions which, strictly speaking, do not belong to either of the above classes. For example, the tomographic description of a volume of tissue. A more detailed discussion of the methods for the description of three-dimensional objects as well as for the acquisition of data is found in Aggarwal *et al*. [1].

The subject of the present paper is the integrating of information contained in a sequence of intensity images with the twofold objective of obtaining the three-dimensional description and the motion of the object. This should be distinguished from another class of efforts which combine several distinct viewer-centered descriptions to obtain a single description of the scene. For example, one may combine an intensity image and a range image to obtain an edge map for the image (see Gil *et al*. [2]).

The research for obtaining three-dimensional structure was initially pursued as stereoscopic scene analysis (see Duda and Hart [3]). Usually, the scene was stationary whereas the camera moved (or there were two cameras) to obtain the two images. More recently, several researchers have considered the problem of several views. The following sections review the works of Ullman [4], Roach and Aggarwal [5], Nagel [6], and Tsai and Huang [7], after presenting briefly the perspective equations and the stereoscopic vision problem as discussed in [3]. This is followed by a discussion of a method based upon silhouettes developed by Martin and Aggarwal [8] together with recent results. The last section of the paper discusses the possible directions of future research. The present paper emphasizes the three-dimensional character of dynamic scene analysis in contrast to the earlier reviews by Martin and Aggarwal [9, 10] where the entire area of dynamic scene analysis was discussed.

2. PERSPECTIVE EQUATIONS

For the case of the simple geometry where the image plane is in front of the lens center, the global coordinate system coincides with that of the image plane and the optical axis aligns with the y-axis, the image plane coordinates of the perspective projection of a point in space are related as follows:

$$\frac{X}{F} = \frac{x}{F+y}$$

$$Y = 0 \tag{1}$$

$$\frac{Z}{F} = \frac{z}{F+y}$$

Here, (x, y, z) is the point in space, $(X, 0, Z)$ is its projection on the image plane as shown in Figure 1. It is clear from the geometry of the figure, and the equations of projection (1), that the position of the image point is uniquely determined from the point in space. However, given the image point, one is able to fix only the projecting ray. This becomes evident if the projection equations are rewritten as

$$x = \frac{X(F+y)}{F} = \frac{Xz}{Z}, \tag{2}$$

or better still using a homogeneous coordinate system as

$$\begin{vmatrix} x \\ y \\ z \end{vmatrix} = \frac{F}{F-Y} \begin{vmatrix} X \\ Y \\ Z \end{vmatrix} \tag{3}$$

where Y is a parameter; and for various values of Y, the point (x, y, z) traces out the projecting ray.

For the case where (x_0, y_0, z_0) represents the gimbal center, and the optical axis is panned through an angle θ and tilted through an angle ϕ, and $(\ell_1, \ell_2+F, \ell_3)$ represents the constant offset of the image plane center (relative to the gimbal center), the relationship between the image plane coordinates and spatial coordinates is considerably more complicated as given in [3]. The inverse perspective equations are also given there. The above simpler equations (1), (2), and (3) may easily be derived from the complicated equations by simply assuming $\theta=0=\phi$, $x_0=y_0=z_0$, $\ell_1=0=\ell_3$, and $\ell_2=-F$. The derivation of the above equations together with an excellent discussion of perspective equations is presented by Duda and Hart [3].

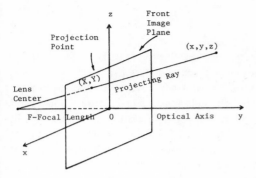

Figure 1. Lens center and image plane geometry.

In order to locate the spatial coordinates of a point in space from its image plane coordinates, one may employ a method called stereoscopy, based upon two image plane views. Each image plane point will give rise to a projecting ray and given that the point lies on both projecting rays, it lies on the intersection of the two projecting rays. This assumes that one has established the correspondence of points in the two images. In its generality, the correspondence problem is difficult and it is not discussed in the following. The interested reader is referred to [11, 12].

The stereoscopy arrangement consists of two image planes with two lens centers. Let the lens centers be given by $\underline{L}_1$ and $\underline{L}_2$, and $\Delta=\underline{L}_2-\underline{L}_1$ gives the base-line vector as shown in Figure 2. If $\underline{U}_1$ and $\underline{U}_2$ denote the unit vectors along the two projecting rays then the equation

$$a\underline{U}_1 = \Delta + b\underline{U}_2 \tag{4}$$

determines the point of intersection of the two projecting rays, and thus the point in space, where a and b are suitable scalar constants. In a noisy environment such a and b will not exist since the two rays may not intersect. The problem may be reformulated as a minimization problem with the penalty function

$$J(a, b) = \| a\underline{U}_2 - (\Delta + b\underline{U}_2) \| \tag{5}$$

with the minimum being reached for the values of parameters a_0, b_0. The coordinates of

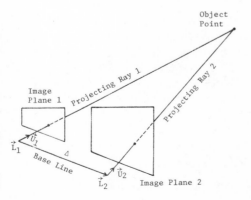

Figure 2. Stereoscopy arrangement with two image planes.

the points in space are approximated by

$$\underline{v} = (1/2)(a_0\underline{U}_1 + (\Delta + b_0\underline{U}_2)) + \underline{L}_1 \tag{6}$$

the midway point of the shortest distance between the two projecting rays. The solution to the above problem is simple and may be found in Duda and Hart [3].

This early work forms the basis of the more recent work where several researchers have considered the problem of computing the three-dimensional structure from several views together with the motion parameters of the object. The works of Ullman, Roach and Aggarwal, Nagel, and Tsai and Huang are discussed. It may be emphasized that notations used by various authors are different, and at times they are considering slightly different problems. The purpose of the review is to present a coherent view of the problem and its solutions.

3. EQUATIONS FOR DYNAMIC SCENE ANALYSIS

Several researchers have considered the equations necessary for dynamic scene analysis of images obtained by perspective transformations of scenes containing three-dimensional objects. The present section reviews the works of Ullman [4], Roach and Aggarwal [5], Nagel [6], and Tsai and Huang [7]. The underlying assumption in all these works are: (i) the objects are rigid and (ii) the correspondence of points between images has already been established.

In the above presentation of section 2 the y-axis is used as the optical axis. However, this convention is not followed universally. In particular, for the works discussed in the present section, the original notation of the author is used. For example, in sections 3.2 and 3.4, the z-axis is the optical axis, whereas in sections 3.1 and 3.3 the y-axis is the optical axis. The reader is forewarned of the obvious inconsistency. In addition, it may be noted that lower case triplets like (x, y, z) denote points in three-dimensional space in terms of the global coordinate system, whereas upper case pairs (X, Z) or (X, Y) denote image plane points in the image plane coordinate system.

3.1. Polar equations

Ullman considers the constraints imposed by the perspective projection and combines them with the constraints imposed by motion of a rigid object to derive the structure of the object. Let (x_i, y_i, z_i), $i=1, 2, 3$ represent the spatial coordinates of points on an object undergoing a translation $(\Delta x, \Delta y, \Delta z)$ and a rotation θ about the z-axis. Let the coordinates of corresponding points be given by (x'_i, y'_i, z'_i), $i=1, 2, 3$. The two sets of coordinates of the points are related by the relationships:

$$x'_i = x_i \cos \theta - y_i \sin \theta + \Delta x$$
$$y'_i = x_i \sin \theta + y_i \cos \theta + \Delta y \tag{7}$$
$$z'_i = z_i + \Delta z$$

for $i=1, 2, 3$. The perspective transformation imposes the constraint that

$$\phi_i = \frac{x_i}{y_i} = \frac{X_i}{F}$$

$$\eta_i = \frac{z_i}{y_i} = \frac{Z_i}{F} \tag{8}$$

where (X_i, Z_i) are the image plane coordinates of the point (x_i, y_i, z_i) and F is the focal length of the perspective projection.

In the derivation of the above relationships, it is assumed that the optical axis is aligned with the y-axis, and the rotation of the object takes place about the z-axis. The geometry of the above relationships is illustrated in Figure 3.

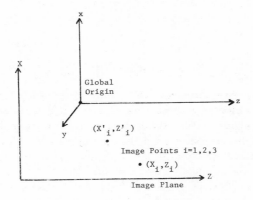

Figure 3. Geometry for Ullman's polar equation.

On combining the perspective and motion constraints, one obtains

$$\phi'_i = \frac{x'_i}{y'_i} = \frac{x_i \cos\theta - y_i \sin\theta + \Delta x}{x_i \sin\theta + y_i \cos\theta + \Delta y}$$

$$\eta'_i = \frac{z'_i}{y'_i} = \frac{z_i + \Delta z}{x_i \sin\theta + y_i \cos\theta + \Delta z} \tag{9}$$

and on substituting

$$x_i = y_i \phi_i, \qquad z_i = y_i \eta_i \tag{10}$$

the above equations reduce to:

$$\phi'_i = \frac{y_i \phi_i \cos \theta - y_i \sin \theta + \Delta x}{y_i \phi_i \sin \theta + y_i \cos \theta + \Delta y}$$

$$\eta'_i = \frac{y_i \eta_i + \Delta z}{y_i \phi_i \sin \theta + y_i \cos \theta + \Delta y}$$

(11)

The following observations may be made about these equations:

(1) There are six equations in seven unknowns Δx, Δy, Δz, θ, y_1, y_2, y_3.

(2) Ullman provides a procedure for reducing the six equations to a single equation in θ.

(3) One variable is used to provide the scale.

(4) The derived equation, called the polar equation, is of the form

$$A \sin^2 \theta + B \cos^2 \theta + C \cos \theta \sin \theta + D \sin \theta + E \cos \theta = 0$$

(5) In general the polar equation has four roots and consequently there is ambiguity in the choice of θ.

Ullman suggests a possible strategy for settling on the unique solution by introducing redundancy. In particular, by considering four points and taking the common solution corresponding to various triplets, the ambiguity may be resolved. The results of experimental examples show that the correct answer can usually be found from as little as two views of four points.

Solution of the polar equation and the correct choice of the root leads to the computation of (x_i, y_i, z_i), (x'_i, y'_i, z'_i), $i=1, 2, 3$ and Δx, Δy, Δz and θ in terms of the scaling parameter. The assumptions are centered around the rigidity of the object and the motion constraint.

3.2. A generalization of the stereo problem

Roach and Aggarwal consider the problem as illustrated in Figure 4. The object is assumed to be stationary, whereas the camera undergoes an unknown motion. In order to reconstruct the motion and the three-dimensional coordinates of object points from two-dimensional images, two views of five points are needed as shown in the following analysis.

The camera position model considered by Roach and Aggarwal is slightly more general than that of Duda and Hart [3]. In addition to the angles θ and ϕ for pan and tilt, they consider the angle κ for the rotation of the image plane coordinates axes relative to the global coordinate axes. This leads to three position coordinates (x_0, y_0, z_0) and three angle coordinates (θ, ϕ, κ) for specifying the camera position. The equations relating the image plane coordinates (X, Y) with the spatial coordinates (x, y, z), focal length F, and the camera parameters (x_0, y_0, z_0), (θ, ϕ, κ) are given in [5].

In Figure 4, there are 27 unknowns: 15 for the three-dimensional coordinates of five points and 12 for the two camera positions (for each position, three for the lens center and three for the optical axis). However, from two views of five points, there are only 20 equations (two for each point in each view). If the six parameters for the first camera position are known (it is convenient to assume them to be zero), and z-coordinates of

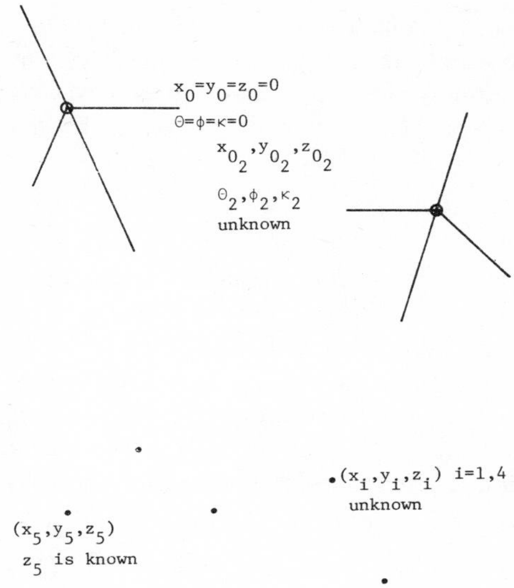

$$x_0 = y_0 = z_0 = 0$$
$$\Theta = \phi = \kappa = 0$$
$$x_{0_2}, y_{0_2}, z_{0_2}$$
$$\Theta_2, \phi_2, \kappa_2$$
unknown

$\bullet (x_i, y_i, z_i)$ i=1,4
unknown

(x_5, y_5, z_5)
z_5 is known

Figure 4. Roach's camera configuration and point coordinates.

one of the points is chosen as the scaling factor, the number of unknowns and the number of equations is exactly the same. It is convenient to compute the x- and y-coordinates of the point, whose z-coordinate is selected as the scaling factor, from the equations $X = Fx/z$ and $Y = Fy/z$. Therefore, there are 18 equations with 18 unknown parameters. In general, these 18 equations present a formidable task for solution. The difficulty of obtaining the solution is considerably aggravated by the presence of noise.

The system of nonlinear projection equations explained above can be solved by using a modified finite difference Levenberg-Marquardt algorithm due to Brown [13–15] without strict descent that minimizes the least-squared error of the 18 equations. The method employed is iterative and requires an initial guess for each unknown parameter.

This work is somewhat like the camera calibration systems of Sobel [16] and Yakimovsky and Cunningham [17]. In their work multiple images of points together with a central projection model and numerical methods are used to determine camera parameters such as focal length, position, and orientation. These studies, however, have considerably more information about the three-dimensional positions of points than we are assuming. Thus, the problems being solved and the information given for the calibration systems are different from the work described in this section.

Implicit in this work are two very important assumptions: that the objects being observed are rigid and that the images of the object are noise free and thus completely accurate. To test the effect of the second assumption on the numerical method described above, from 1 to 4 pixels were randomly added to or subtracted from the exact photo-coordinate data for a moving object. This perturbation of the data causes extreme instability in the numerical solutions. However, one of the main reasons for using a least-squared error technique to solve a problem is to make adjustments to observations that contain error (noise). Adjustment is only possible, however, when there are more equations than unknowns. Two views of five points are therefore inadequate for noisy data since there are the same number of equations as unknowns. Two views of six points or three views of four points produce 22 equations in 21 unknowns using the same problem model discussed above. Examination of experimental runs using overdetermined systems of equations shows that minimal overdetermination is not very accurate. It is only with considerable overdetermination (two views of 12 or even 15 points; three views of seven or eight points) that the results become accurate.

3.3 A generalization of the polar equation

Nagel uses compact vector notation to formulate the problem of structure and motion. The camera is assumed to be stationary, whereas the object is moving. There are two coordinate systems. One is attached to the camera and the second one to the object. In the camera coordinate system, each object point is expressed as $C_{mi}\underline{c}_{mi}$, where the subscripts i and m refer to the view and point, respectively, and the vector $\underline{c}_{mi}$ is a unit vector. The same point in the object coordinate system is denoted by $\underline{A}_m$. The object undergoes a translation $\underline{T}$ followed by a rotation R. The relationships for the two views (as shown in Figure 5) are given by

$$C_{m1}\underline{c}_{m1} = \underline{A}_m \qquad (12)$$
$$C_{m2}\underline{c}_{m2} = (\underline{A}_m + \underline{T})R \qquad (13)$$

In these equations $\underline{c}_{m1}$ and $\underline{c}_{m2}$ are known from the image plane coordinates and lens center location, whereas C_{mi}, $\underline{A}_m$, $\underline{T}$ and R are unknown. It may be noted that in this geometry the lens center is at the origin.

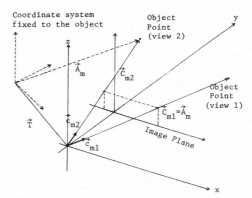

Figure 5. Nagel's geometry of the two coordinate systems.

The matrix R can be expressed in terms of the direction cosines of the axis of rotation and the rotation angle. Nagel outlines a procedure by which various unknown parameters may be computed in terms of the scale parameter C_{11}. Nagel makes extensive use of vector algebra in deriving the above results and uses five points in two views to derive the motion and structure parameters. Also, for the special case similar to Ullman, the equation reduces to the polar equation derived earlier. Thus, Nagel's results substantiate earlier work of Ullman and Roach and Aggarwal. Again, the assumption of rigidity of the · object and the existence of correspondence between the image points in two views are used to derive the above results.

3.4. The planar patch

Tsai and Huang consider an approach significantly different from the above approaches. Instead of considering the motion of individual points, they consider the motion of a planar patch characterized by eight points. The configuration of the coordinate system, camera and object points are shown in Figure 6. The perspective constraints are expressed as:

$$\text{View 1:} \quad X = F\frac{x}{z}$$

$$Y = F\frac{y}{z} \tag{14}$$

$$\text{View 2:} \quad X' = F\frac{x'}{z'}$$

$$Y' = F\frac{y'}{z'} \tag{15}$$

The motion of a point is expressed as

$$\begin{matrix} x' \\ y' \\ z' \end{matrix} = R \begin{matrix} x \\ y \\ z \end{matrix} + \underline{T}$$

where R is the rotation matrix and $\underline{T}$ is the translation vector. Tsai and Huang assume that there is a rigid planar patch in the object space characterized by eight points and

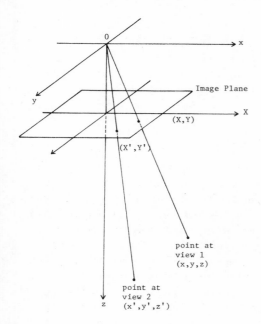

Figure 6. Tsai and Huang geometry of the image plane.

given by the equation

$$ax + by + cz = 1 \tag{16}$$

Further, they express the change of position of points in the image plane as the transformation

$$X' = \frac{a_1 X + a_2 Y + a_3}{a_7 X + a_8 Y + 1}$$

$$Y' = \frac{a_4 X + a_5 Y + a_6}{a_7 X + a_8 Y + 1} \tag{17}$$

where a_i, i=1, ..., 8 are called the pure parameters. Using the motion, rigidity, and perspective constraints, the pure parameters may be expressed in terms of the planar patch and motion parameters. After a tedious algebraic manipulation a sixth order polynomial may be formulated such that its coefficients are expressed in terms of pure parameters and its solution yields (together with additional manipulation) the motion and structure parameters in terms of scale. The sixth order polynomial appears to have only two real roots, however, no analytical demonstration of this fact is possible at this time.

3.5. A coherent view

The commonality in the underlying assumptions and the analyses of the four papers reviewed above may be summarized as follows:
(1) Rigidity of the object.
(2) Perspective transformation.
(3) Combining perspective and motion constraints.
(4) Solution being expressed in terms of a scale factor.
(5) Need to establish correspondence of points between views.
The solution of the equations, in general, is complex but manageable. However, it is the need to establish correspondence of points between views that has led us to the work described in the next section.

4. OCCLUDING CONTOURS IN DYNAMIC SCENES

In this section we will describe the results of work [8] done in the pursuit of two major goals. The first goal is the development of a dynamic scene analysis system that does not depend completely on feature point measurements. The second goal is the development of a scheme for representing three-dimensional objects that is descriptive of surface detail, yet remains functional in the context of structure from motion in dynamic scenes.

To lessen the dependency on feature point detection, occluding contours with viewpoint specifications are used. The term "occluding contour" means the boundary in the image plane of the silhouette generated by an orthogonal projection. Silhouettes can most often be formed by a simple thresholding of the intensity values. A connected component analysis [18, pp. 336–347] of the resulting binary valued image yields the boundary of the object silhouette. An ordered list of the image plane coordinates of the resulting boundary constitutes the initial representation of the occluding contour. Throughout the analysis of the dynamic image, however, another representation, referred to as the raster-ized area [19], of the contour will also be used. For its use here, the most significant attribute of this representation is that given an area so represented and an arbitrary segment on a line parallel to the "raster direction" it is a simple process to determine what portions of the given segment intersect the area, i.e., to clip that segment to the represented area.

The three-dimensional structure to be derived from the sequence of occluding contours is a bounding volume approximation to the actual object. For this reason the representation incorporated in this system is based on volume specification through a "volume segment" data structure. The volume segment representation is a generalization to three dimensions of the rasterized area description. For the rasterized area, each of the segments denoted a rectangular area. The generalization to three dimensions is to have each segment represent a volume, i.e., a rectilinear parallelepiped with edges parallel to the coordinate axes. In addition to grouping collinear segments into lists, the set of segment lists is partitioned so that the subsets contain lists having coplanar segments. The primary dimension of the parallelepiped specified by a segment is the length of the segment. The second dimension is given by the inter-segment spacing within the plane of the segment, while the third dimension is the inter-plane distance. The latter two dimen-sions are specified to be uniform throughout the volume segment representation.

The primary advantage of this structure in general situations is that the process of determining whether an arbitrary point is within the surface boundary consists of a simple search of three ordered lists: select a "plane" by z-coordinate; select a "line" by x-coordinate; and, finally, check for inclusion of the y-coordinate in a segment.

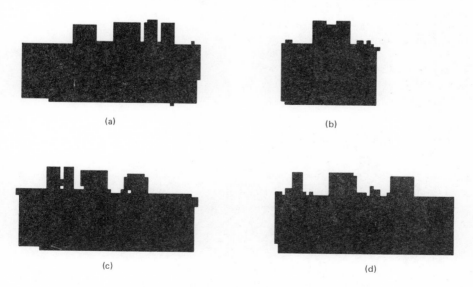

(a)

(b)

(c)

(d)

Figure 7. Silhouettes for a box from
four different viewpoints.

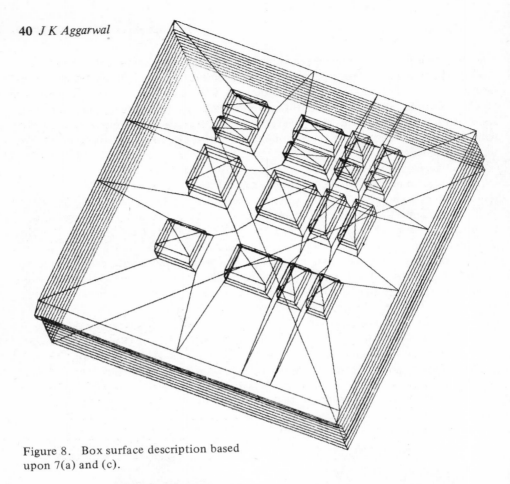

Figure 8. Box surface description based
upon 7(a) and (c).

Figure 9. Box surface description based
upon 7(a), (c), and (b).

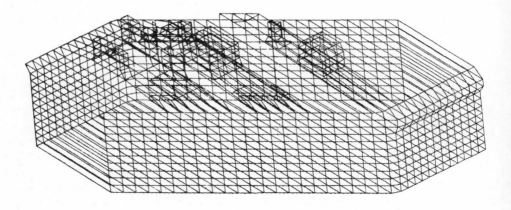

This volume segment representation is created from a dynamic image by two processes. The first process combines information from frames 1 and 2 of the dynamic image to form an initial volume segment representation. The second process then accepts each succeeding frame in order to refine the approximation represented by the volume segment structure. Thus these processes analyze the occluding contours with their view orientations to initially construct and to continually refine the volume segment representation of the object generating the contours. Algorithm summaries of the two processes are given in more detail in [20].

Two examples illustrate the results possible by this method.

Example 1. Four silhouettes of a box with knobs are shown in Figure 7. On combining occluding contours, labelled (a) and (c); (a), (c) and (b); and (a), (c), (b) and (d), one successively obtains the volumes shown in Figures 8, 9, and 10. The continuing refinement of the process is fairly apparent.

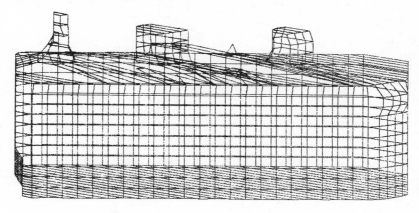

Figure 10. Box surface description based
upon 7(a), (c), (b), and (d).

Example 2. Figures 11(a), (b), and (c) give the occluding contours for a rectangular parallelepiped with a hole. Figures 12 and 13 give the volume constructed from (a) and (b); and (a), (b), and (c), respectively.

5. FUTURE DIRECTIONS FOR RESEARCH

The review of previous sections has focused on the computation of three-dimensional structure and the motion vector of objects moving in space from two-dimensional images. The need for efficient methods for inversion of perspective equations and for storage of three-dimensional description of objects becomes fairly obvious. The problems related to the three-dimensional description of objects have been around for a long time and they have received the attention of some more creative researchers. However, there is still a need for additional research in this area. The properties of an "ideal" three-dimensional description of objects are fairly easy to enumerate but this ideal is far from being achieved.

Figure 11(b). Occluding contours for rectangular parallelepiped with a hole.

Figure 11(a). Occluding contours for rectangular parallelepiped with a hole.

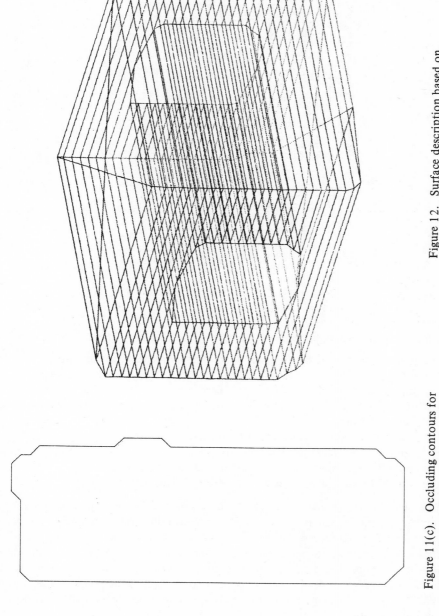

Figure 12. Surface description based on 11(a) and (b).

Figure 11(c). Occluding contours for rectangular parallelpiped with a hole.

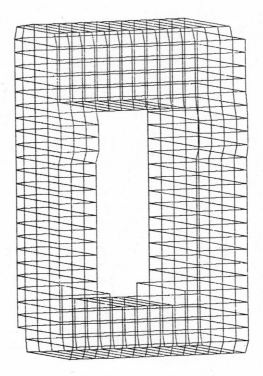

Figure 13. Surface description based on
11(a), (b), and (c).

In particular, an ideal description should have the following properties:

(1) The description should be compact so that it is amenable to easy storage and fast transmission.
(2) The description should be easily transformable to viewer-centered description.
(3) The description should be readily convertible to other object-centered three-dimensional descriptions.
(4) The description should be amenable for partial description of objects and easy to update when additional information is available.
(5) The description should be able to accommodate holes and concavities.
(6) The description should be extendable to include deformation of objects.

The intensive computational needs for inversion of perspective equations in a noisy environment are rather severe. If one adds the constraints for real time processing of video images, the computational needs are indeed astronomical. Not only must one provide for a super computer, but one must come up with rather creative and innovative solutions to the numerical drudgery. In particular, the real time processing of color video images of 512x512 with 8 bits intensity for each color requires the processing of approximately 2^{26} bits/sec. Parallel processing of data and its early reduction are important, necessary ingredients.

ACKNOWLEDGMENTS

It is a pleasure to acknowledge the help of Messrs C. H. Chien, B. Gil, Y. C. Kim and Dr W. N. Martin during the preparation of this paper. This research was supported by the Air Force Office of Scientific Research under Contract F49620-83-K-0013.

REFERENCES

1. J. K. Aggarwal, L. S. Davis, W. N. Martin and J. W. Roach, Survey: Representation Methods for Three-Dimensional Objects, in L. N. Kanal and A. Rosenfeld, (eds), *Progress in Pattern Recognition*, Vol. 1 (North-Holland Publishing Company, New York, 1981), 377–391.
2. B. Gil, A. Mitiche, and J. K. Aggarwal, Experiments in Combining Intensity and Range Edge Maps, to appear in *Computer Graphics and Image Processing*.
3. R. O. Duda and P. E. Hart, Pattern Classification and Scene Analysis (John Wiley and Sons, New York, 1973).
4. S. Ullman, The Interpretation of Visual Motion (MIT Press, Cambridge, MA., 1979).
5. J. W. Roach and J. K. Aggarwal, Determining the Movement of Objects from a Sequence of Images, *IEEE Trans. on Pattern Analysis and Machine Intelligence*, Vol. PAMI-2, No. 6 (Nov. 1980), 544–562.
6. H.-H. Nagel, Representation of Moving Rigid Objects Based on Visual Observations, *Computer*, Vol. 14, No. 8 (Aug. 1981), 29–39.
7. R. Y. Tsai and T. S. Huang, Estimating Three-Dimensional Motion Parameters of a Rigid Planar Patch, *IEEE Trans. on Acoustics, Speech, and Signal Processing*, Vol. ASSP-29, No. 6 (Dec. 1981), 1147–1152.
8. W. N. Martin and J. K. Aggarwal, Volumetric Descriptions of Objects from Multiple Views, to appear in *IEEE Transactions on Pattern Analysis and Machine Intelligence*.
9. W. N. Martin and J. K. Aggarwal, Dynamic Scene Analysis: A Survey, *Computer Graphics and Image Processing*, Vol. 7, No. 3 (June 1978), 356–374.
10. J. K. Aggarwal and W. N. Martin, Dynamic Scene Analysis, Technical Report 82-3, Laboratory for Image and Signal Analysis, The University of Texas, Austin, Texas (Sept. 1982).
11. J. K. Aggarwal, L. S. Davis and W. N. Martin, Correspondence Processes in Dynamic Scene Analysis, *Special Issue on Image Processing, Proceedings IEEE*, Vol. 69, No. 5 (May 1981), 499–657.
12. J. A. Webb and J. K. Aggarwal, Shape and Correspondence, to appear in *Computer Graphics and Image Processing*.
13. K. M. Brown and J. E. Dennis, Derivative Free Analogues of the Levenberg-Marquardt and Gauss Algorithms for Nonlinear Least Square, *Numer. Math.*, Vol. 18 (1971), 289–292.
14. K. Levenberg, A Method for the Solution of Certain Nonlinear Problems in Least Square, *Quart. Appl. Math.*, Vol. 2 (1944), 164–168.
15. D. W. Marquardt, An Algorithm for Least Squares Estimation of Nonlinear Parameters, *J. SIAM*, Vol. 11, No. 2 (1963).
16. I. Sobel, On Calibrating Computer Controlled Cameras for Perceiving 3-D Scenes, *Artificial Intelligence*, Vol. 5, No. 2 (1974), 185–198.
17. Y. Yakimovsky and R. Cunningham, A System for Extracting Three-Dimensional Measurements from a Stereo Pair of TV Cameras, *Computer Graphics and Image Processing*, Vol. 7, No. 2 (Apr. 1978), 195–210.

18. A. Rosenfeld and A. C. Kak, Digital Picture Processing (Academic Press, New York, 1976).
19. W. M. Newman and R. F. Sproull, Principles of Interactive Computer Graphics, second edition (McGraw-Hill, New York, 1979).
20. W. N. Martin and J. K. Aggarwal, Analyzing Dynamic Scenes, Technical Report 81–5, Laboratory for Image and Signal Analysis, The University of Texas, Austin, Texas (Dec. 1981).

J. K. Aggarwal
College of Engineering
The University of Texas at Austin

4 Vices and virtues of image parallel machines

P E Danielsson

1. INTRODUCTION

Almost all approaches to parallelism try to expand the so-called von Neumann bottle-neck: the data path between memory and processor. Image processing has its own domain of possibilities in this respect because of the simple fact that we are dealing with except-ionally large arrays of data. One attempt [1, 2] to classify parallelism is based on the following two assumptions that will be scrutinized toward the end of this paper.

(1) Image processing/image analysis is clearly a two-level system. Only the lower level, that closest to the input, needs to be considered for non-conventional computer architecture.

(2) A large, probably dominant, portion of low-level processing is the neighborhood type of operation. The neighborhood access problem is the von Neumann bottle-neck of image processing.

Neighborhood processing is illustrated in Figure 1. For all image points a neighborhood of input pixels produces an output pixel. A complete low-level task may consist of one or several of these operations and of different combinations of intermediate results.

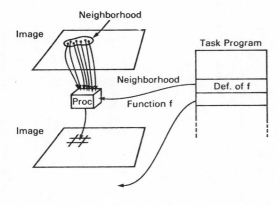

Figure 1. Neighborhood image processing.

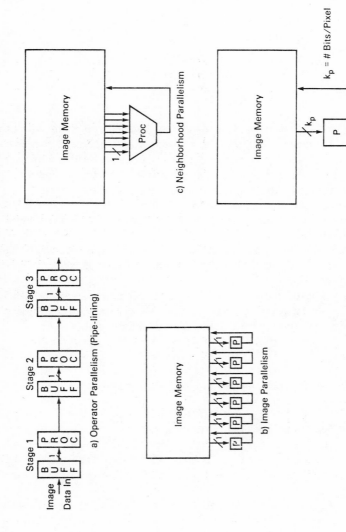

Figure 2. The four types of parallelism
for neighborhood operations.

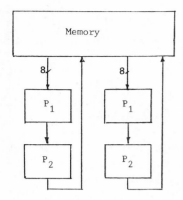

Figure 3. A mixed form of parallelism: pipe-line of depth 2, image parallelism of degree 2, 8 bit/pixel.

For a very dedicated image processing system where input data belong to a limited universe and where processing is well defined and understood, nothing can beat a pipe-lined system. All processing steps, including the image pick-up of input data, are then tuned to the same required throughput. Different processing steps may be designed differently to obtain this result.

Now, for general purpose image processing we have a much more difficult design problem. We want to exploit parallelism to increase throughput and still retain the same efficacy for big and small tasks, for images large and small, sometimes using large and scattered neighborhoods, sometimes only pixelwise operations (1×1 neighborhoods). Also, we have to deal with binary images as well as gray level or even multispectral images. "Same efficacy" means that a 1024×1024 image should require 16 times longer execution time than a 256×256 image, that a 5×5 convolution is approximately three times slower than a 3×3 convolution, and that a binary image can be stored and processed about eight times as efficiently as an 8-bit/pixel image, etc.

The four orthogonal dimensions for parallelism in neighborhood operations are
(1) the operator dimension (operator parallelism = pipe-lining),
(2) the image dimension (image parallelism),
(3) the neighborhood dimension (neighborhood parallelism), and
(4) the pixel dimension (conventional processing)
as illustrated in Figure 2. They are all independent of each other and can be used in a variety of combinations. One example is shown in Figure 3. However, as soon as we employ hardware that extends beyond what is needed for a certain task we lose efficiency. For instance, byte-wide data-paths (pixel-bit parallelism = 8) is a waste when dealing with binary images. A long pipe-line cannot be utilized for a task that consists of only one or two operations, etc.

Thus in the general purpose situation we should look for the lower size limit in all of the four dimensions to find how far we can go in parallelism. *We conclude that the lower size is one (1) in all these dimensions except the image dimension.* For this dimension the lower size limit is rather of the order of 64×64. Smaller images are either uninteresting or easy to handle with conventional processing. Consequently, only an assembly of bit-serial processing elements sequentially processing neighborhoods of the same image bears the promise of being cost effective for a wide range of problems. This is image parallelism.

2. COMMENTS ON MIMD AND SIMD

Multiprocessor systems can be of two types: SIMD (Single Instruction Multiple Data) or MIMD (Multiple Instruction Multiple Data).

MIMD machines imply that each processor carries a control unit and executes its own program. Furthermore, orchestrating the different processors into a complete task force requires some sort of master control. The complexity of this control tends to grow very fast (exponentially?) with the number of processors. Therefore, the number of processors in an MIMD machine will be rather limited.

SIMD machines have one control unit that broadcasts both memory addresses and control bits to all PEs (processing elements) (see Figure 4). The complexity of the control unit is constant and allows for a very high degree of parallelism. Figure 4 conforms with many image processing systems, such as ILLIAC III [3], ILLIAC IV, CLIP IV [4], DAP [5], and MPP [6]. Of great importance is the interconnection between PEs (not shown in Figure 4) which is discussed later.

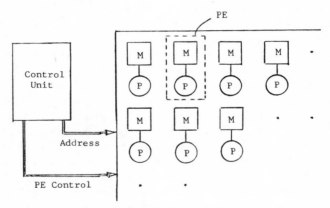

Figure 4. SIMD processing with a two-dimensional array of processing elements (PEs).

The SIMD system of Figure 4 is our main subject of study. However, before we leave the MIMD operation mode altogether let us consider a case where such image processing would be beneficial. The virtue of independent program execution of each PE appears for highly data-dependent image processing. If a neighborhood operation is very complex it might be of great advantage to execute it in a way that quickly recognizes a uniform (flat) area in the input image. Probably the output would then be zero. Then more time and sophisticated investigation could be spent on the busy parts of the input image. A SIMD system has to "wait for the worst case", it cannot short-cut the execution since one or several processors might need the whole thing. An MIMD system, on the other hand, can hopefully have a much shorter execution time, especially if flat and busy parts are equally distributed among the PEs.

Our decision to adopt the SIMD idea means that we reserve highly data-dependent processing for the higher level of our total system.

3. THE IMPORTANCE OF IMAGE-TO-ARRAY MAPPING

In all realistic cases the m × m size of the array of Figure 4 is much smaller than the n × n size of the image we want to process. Thus, we have to adopt a mapping rule between image and array or, equivalently, decide which pixels of the image should be stored in the top left-most memory module. There are two basic methods with very small variations. The most common is to store every mth pixel (in both coordinates) of the image in the same memory module. This is the method used in CLIP IV and MPP. It is illustrated in Figure 5 for a 4 × 4 array and we see that the principle is to *distribute the image of the PEs*.

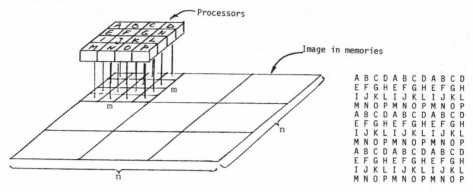

Figure 5.   Image distributed over the PEs.

Each memory module holds (n/m) pixels of the image found on intervals of m pixel units in both coordinates. It leads to a "densely packed" activity situation where all processors are simultaneously processing neighboring pixels of an m × m subimage.

If we assume eight connected PEs, 3 × 3 neighborhoods work fairly well. Larger neighborhood access requires either connections surpassing longer distances (which is probably unrealistic) or the shifting of data. The last method seems to increase the number of cycles with at least a factor of 2 to immediate access. Also, it does not allow efficient access from a more scattered or irregular neighborhood.

The most difficult problem to overcome, however, is the special edge conditions. By necessity, the single global address is correct and sufficient only as long as all processors want data from the same m × m subimage. However, for the border PEs, their respective neighborhoods extend over to other subimages, the pixels of which are found on other memory addresses. Consequently, individual address modifiers have to be introduced (+1, −1, +m, −m, etc.). Alternatively, these border pixels have to be brought into play with separate and selective access cycles. In either case complex and time-consuming overheads develop.

For the MPP system a circumvention of this problem has been suggested [7]. The full image is stored outside the array and only one subimage of 128 × 128 is brought in at a time. For a 3 × 3 operation, 126 × 126 valid results can be produced, which means that subsequent subimages have to be overlapping. The operation seems to create a considerable time penalty, however.

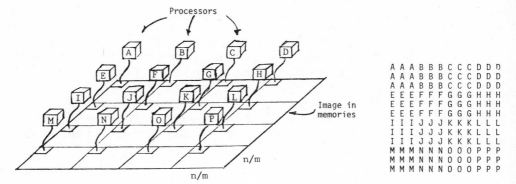

Figure 6.   Distributed processor topology.

The problem of the "densely packed" array of Figure 5 can be summarized by the fact that the processors almost stand on each other's toes when it comes to neighborhood access. The alternative is to *distribute the processors over the image* (illustrated in Figure 6). Each processor roams over its own subimage, size n/m × n/m, under control of the same global address pointer. Neighborhoods (even large ones) that overlap the adjacent subimage are reached by nearest-neighbor connection.

Steering of data from neighboring memory modules to the processors is easily controlled by the same central mechanism that delivers global addresses. Note that when one processor "reaches" out for a bit, say, to the east, and grabs something from this memory module, the neighboring processor to the west is doing the same, using the memory of the first one.

For most realistic cases, huge neighborhoods become accessible. For instance, a 512 × 512 image on a 16 × 16 array results in 32 × 32 subimages and maximum neighborhoods of 65 × 65.

One or several interconnecting schemes is shown in Figure 7. Note that we emphasize the memory-to-processor interconnection rather than the processor-to-processor interconnection. In Figure 7(a) the processors are on top of their own memory modules. A case where each processor reads information from its north-west neighbor memory is illustrated by the dashed data paths. Only four double directed lines are necessary.

It seems that distributed processor topology is drastically enhancing the potential of image parallelism. It was first suggested for DAP [5] although efficient neighborhood access and control seem to be lacking in this machine as is evident from [8].

It should be noted that so-called pyramide or cone processing [9] is easily accommodated in the topology of Figure 7. Demagnifying the n × n image in steps of 2 can go on inside the array, storing the new subimages at new memory positions. However, the top of the pyramid above the m × m level has to be handled outside the array.

The virtue of distributed processor topology might be disturbed by input problems. The normal raster scan serial format outside the array does not immediately lend itself to be stored in the format of Figure 6. Neither for that matter, does the format of Figure 5, which is one good reason for the so-called staging memories of MPP. We will come back to this problem after the next section.

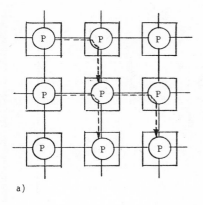

a)

Figure 7. Only four interconnecting lines are necessary for memory-to-processor interconnection.

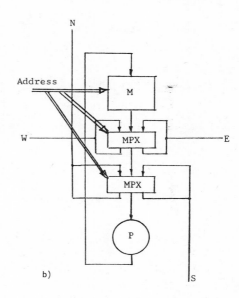

b)

4. TABLE LOOK-UP

Just as in any other kind of data processing, table look-up has been extremely common in image processing as a result of the ever decreasing cost of memory. Examples of table look-up that are hard to abstain from are the following:

(1) Arbitrary gray-scale mapping.
(2) Boolean function on a 3 × 3 binary neighborhood.
(3) Histogram functions.
(4) Multiply with a constant.

However, a large variety of other image-processing algorithms can also make good use of table look-up as shown in [2].

Implementation requires an index register in each PE. The global address is a pointer to the beginning of the table and the individual offset values in the index registers are added to the global address. This operation may seem to be a major extension of the PE hardware. However, if the index arithmetic is used for table look-up only we can spare the adder completely (Figure 8).

For simplicity assume a 16-bit address space. Let all tables occupy 2, 4, 8, or 16 bits and let a 2-bit table start at XXXO, a 4-bit table start at XXOO, etc. Then, we can shift in the offset, MSB first, in XR and reach the table entry by simple concatenation implemented as wired-OR.

Table look-up was one feature of ILLIAC IV. Quite probably it is a necessary ingredient in a competitive image parallel architecture.

Let us now make a speed estimate for *collecting the histogram* of a 512 × 512 × 8 bit image on a 16 × 16 array. The operation takes place in three steps.

(1) Local histogram collection. Each subimage is 32 × 32. Thus each bin (table entry) has to be 11 bits. To read out a pixel value to XR takes 8 cycles and to

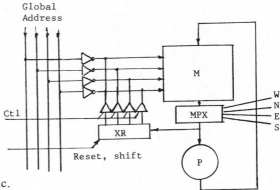

Figure 8.   Simplified index arithmetic.

update the entry value and store it back takes 22 cycles. In total 1024 (8 + 22) = 33972.

(2)   Merging of four neighboring 256-tables. Every second PE in each row is given an XR-value of 128. Half of the tables are then moved in their lower part, half in their upper part, to their neighbor. (Assume a torus connected array.) All tables now contain pairs of entries that can be merged, i.e., added. The procedure is repeated column-wise. The total number of cycles becomes 128 × 22 + 128 × 34 + 64 × 24 + 64 × 37 = 12072.

(3)   Each PE now holds a table of 64 entries of 13 bits. These are shifted out over the edge and merged into one table in 16 × 24 × 13 = 13312 cycles.

The total number of cycles becomes 59356.

Assuming a 100 ns cycle as in the MPP we arrive at a histogram collection time of 5.94 ms, equivalent to approx 43 Mpixels/s.

5.   INPUT/OUTPUT. ORTHOGONALIZATION

The most common image format is the serial raster scan data. A simple serial/parallel buffer can convert the incoming data stream to an "m-bit word" that can be moved over the edge into the m × m array. We assume that we can move the m bits vertically to any of the m rows of PEs. Thereby we can utilize a bandwidth that is m times the bandwidth of a single PE. This bandwidth it still a factor of m less than the full m^2 bandwidth of the array.

Let us number the pixels 0, 1, 2, ... from top-left-right-down in TV fashion. Also, for simplicity, assume 1 bit/pixel. (The I/O problem remains the same for other pixel formats.) Then, if the array is 4 × 4 and one line consists of, say, 16 bits, these will be stored as shown in Figure 9(a). Distributed processor topology, however, requires data to be stored in the manner of Figure 9(b). The conversion between Figure 9(a) and 9(b) is called *orthogonalization*. It is an essential technique for both I/O operations and global transforms of type FFT.

Several solutions have been suggested, among which can be included the staging memory of MPP [6], although this is a data conversion that takes place outside the array

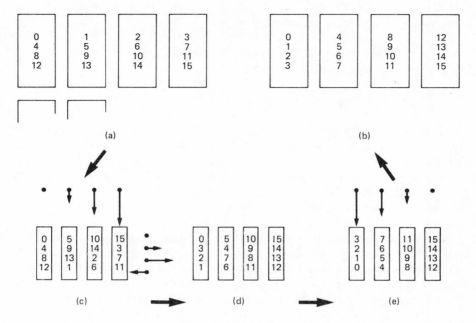

Figure 9.   Orthogonalization using shift
in a shift register in each PE ((a)→(c) and
(d) → (e)) and bit-plane shifting ((c)→(d)).

itself and does not have the distributed processor topology as its goal. Flanders [10] discusses a great variety of conversion schemes, mostly on the background of the DAP concept.

Here, we would like to point out that rather simple skewing and uniform shifts result in efficient orthogonalization inside the array without any external extra hardware (see Figure 9). From the original format we take four bits from each memory module to a shift register in each processor. These are shifted individually to obtain the result of Figure 9(c). Next we apply shift *between* horizontally adjacent PEs. Different bits are shifted a different number of steps but all PEs do the same operation at all steps. The result is shown in Figure 9(d). These bit vectors, which are still contained in the shift registers in the processors, are then individually shifted to obtain Figure 9(e), which is then finally stored back in the format of Figure 9(b).

The procedure may seem complicated but its hallmark is the utilization of the full m × m bandwidth at all operation steps. The number of cycles per bit is

$$\frac{m}{4} = \begin{array}{ll} 2 \text{ for (a)} \rightarrow \text{(c)} & \text{(load + shift)} \\ 1 \text{ for (c)} \rightarrow \text{(d)} & \\ 1 \text{ for (d)} \rightarrow \text{(e)} & \text{(shift)} \\ 1 \text{ for (e)} \rightarrow \text{(b)} & \text{(store)} \end{array}$$

In general the number of cycles will be

4 + m/4 per bit

and consequently the bandwidth is

$$\frac{m^2}{4 + m/4}$$ which equals 32 for m = 16

This is twice the bandwidth of the primary input operation that gave us Figure 9(a). For large m the orthogonalization requires only a quarter of the time spent on the primary input operation.

As a summary, distributed processor topology requires, columnwise in the processors, individual shift control. With this provision the I/O bandwidth for raster scan images is in the order of $0(4/5 \text{ m})$ times the bandwidth of a single PE. No external hardware is necessary.

Fast orthogonalization is of great value for *FFT-type transforms*. Consider a 16×16 point FFT. The first two butterflies can be performed with data in place as in Figure 9(a). The intermediate result is then redistributed to the position in Figure 9(b) which allow the last two butterflies to be executed. The number of cycles 4 + m/4 spent per bit is, by all estimates, negligible compared to the butterfly computations involved.

6. RECURSIVE OPERATIONS. PROPAGATION

Contrary to common belief, propagation-type operations such as shrinking, thinning, labelling, etc., *cannot* be performed by image parallel machines with full efficiency. This is best understood if we consider two extreme cases dealing with an $n \times n$ image, namely the uniprocessor versus the $n \times n$ processor system.

The uniprocessor may traverse the image top-left-right-down with the recursive neighborhood of Figure 10, upper left corner, followed by a second bottom-right-left-top scan using the neighborhood shown in the lower right corner. This is two n^2 operation steps but since the neighborhood is halved it is equivalent in complexity to n^2 normal (parallel, non-recursive) operations. The effect is propagation along straight lines over the entire image, i.e., over n pixel distances.

Now, consider the $n \times n$ processor array. In one time step a propagation wave moves only one pixel distance in spite of the fact that n^2 operations are performed. The efficacy is only $1/n$ and the speed-up factor relative to the uniprocessor system is n rather than n^2.

ILLIAC III and CLIP IV have implemented combinatorial "flash-through" which brings down the cycle time considerably for propagation type operations. However, only the simplest logic operations can be set up in this manner.

Now consider image parallel machines with distributed processor topology (Figure 6) having an $m \times m$ array operating on an $n \times n$ image. The processors can work with full recursive efficacy inside the n/m subimages. However, at the subimage borders the behaviour is more like a parallel system. Clearly, to propagate across a subimage we need $(n/m \times n/m)$ operations and a propagation wave that traverses the whole image requires

$$m(n/m \times n/m) = \frac{n^2}{m} \text{ time steps}$$

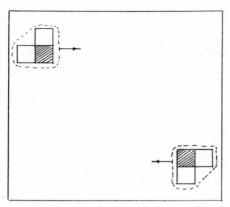

 Pixel from old image

 Pixel from new image

Figure 10. Recursive, propagating operation with a simple neighborhood operator.

Thus, the speed-up factor is m rather than m^2 which means that efficacy is $1/m$ as could be expected. It is inversely proportional to the number of processors.

However, the given numbers may be overly pessimistic. Propagation over the entire image may be an exceptional case. For reasonably large values of n/m it seems likely that the objects are of the same size as a subimage, i.e., overlapping two subimages. In such a case, the image parallel machine works close to full efficacy.

7. CONCLUSIONS

Let us now trace backward. We have found that several operation types are possible to implement with full efficacy in an image parallel machine. These are some of the virtues.

(1) Parallel neighborhood operations, thanks to distributed processor topology.
(2) Table look-up functions, thanks to the simplified index arithmetic.
(3) Input/output and FFT, thanks to a special shift register function in the processors.

We have also found that

(1) Recursive/propagation type operations work but with reduced efficacy. The reduction is proportional to the size of m of the array in one coordinate.

We also remarked (in section 2) that a SIMD system cannot implement certain operations efficiently. These are the vices.

(1) Heavily data-dependent neighborhood operations, cannot be executed because all processors have to be synchronized down to the microinstruction level.
(2) Operations that involve data-dependent addressing, e.g., contour following, have to be excluded since this would require completely different actions taken by each processor.

The last operation type has not been mentioned before.

The balance between vices and virtues is illustrated in Figure 11. In my own opinion the virtues well counterbalance the vices. Thus, image parallel machines should be given a

Figure 11. The balance in favour of image parallelism.

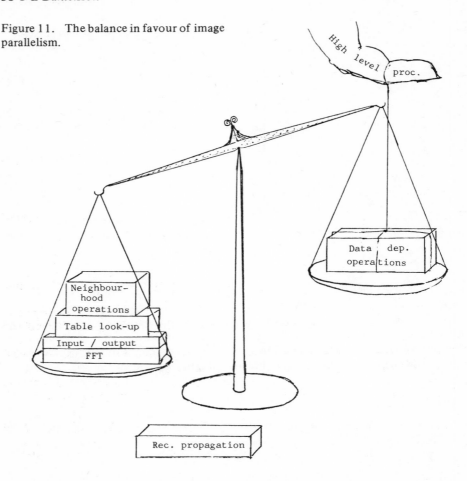

fair chance in the future, in spite of the fact that no such system has yet proved to possess a competitive performance-to-cost ratio.

But what shall we do with the vices? Well, in the first place they should be reduced in importance by inventing algorithms which substitute data-dependent operations with data-independent actions. The residue has to be taken care of by the higher level of our image processing system. Normally, this means the host.

This may seem a capitulation but need not be so. There are several indications that human vision should also be regarded as a two-level system with rather different functional characteristics. On the very low level (retina, visual cortex) everything seems to be working in image parallel mode. However, the so-called pre-attentive vision [11] is also obviously parallel in nature. Most of this activity is probably located in the right brain and it detects events in an image without conscious participation of the observer. Even very primitive animals possess this kind of vision.

The attentive vision is quite different. It has probably arrived rather late in the evolutionary arena and is a conscious behaviour guided by the linguistic and sequentially oriented left brain [12]. It is a uniprocessor system which can perform highly data-dependent scanning of the scene, at any given instant focusing on only a small part of the image. A typical attentive vision process is reading.

Machine vision does not have to imitate nature. However, when the artificial system seems to converge toward something that resembles the natural one it should be taken as a positive indication. Therefore, when we are confronting the unavoidable dichotomy it seems natural and permissible to allocate data-dependent, syntax analysis oriented processing to the higher level.

The main obstacles for the success of image parallelism are then removed.

ACKNOWLEDGMENT

This work has been supported by grant 82-3533 from the Swedish Board for Technical Development.

REFERENCES

1. P. E. Danielsson and S. Levialdi, "Computer Architecture for Pictorial Information Systems", Computer, Vol 14, pp 53–67, November 1981.
2. P. E. Danielsson and I. Ericsson, "LIPP – proposals for the design of an image processor array", in "Computing Structures for Image Processing", M. J. B. Duff ed., Associated Press, 1983.
3. B. H. McCormick, "The Illinois Pattern Recognition Computer – ILLIAC III", IEEE Trans. Computers, Vol EC–12, pp 791–813, December 1963.
4. M. J. B. Duff, "Parallel Processors for Digital Image Processing", in "Advances in Digital Image Processing", P. Stucki, ed., Plenum Press, New York, pp 265–276, 1979.
5. S. F. Reddaway, "The DAP approach", in Infotech State-of-the-Art Report on Supercomputers, Vol 2, pp 309–329, 1979.
6. K. E. Batcher, "Design of a Massively Parallel Processor", IEEE Trans. Computers, Vol C–29, pp 836–840, September 1980.
7. J. L. Potter, "Continuous Image Processing on the MPP", IEEE Computer Society Workshop on Computer Architecture for Pattern Analysis and Image Database Management, pp 51–56, 1981.
8. P. Marks, "Low-level Vision Using an Array Processor", Computer Graphics and Image Processing, Vol 14, pp 281–292, 1980.
9. S. L. Tanimoto and T. Pavlidis, "Hierarchical Data Structure for Picture Processing", Computer Graphics and Image Processing, Vol 14, pp 104–119, 1975.
10. P. M. Flanders, "A Unified Approach to a Class of Data Movements on an Array Processor", IEEE Trans. Computers, Vol C–31, pp 809–819, September 1982.
11. B. Julesz, (title unknown), in SPIE proceedings, Vol 367, "Processing and Display of Three-Dimensional Data", August 1982.
12. S. P. Springer and G. Deutsch, "Left Brain, Right Brain", Freeman, San Francisco, 1981.

Per-Erik Danielsson
Department of Electrical Engineering
Linköping University, Sweden

5 Esoteric iterative algorithms

S R Sternberg

I. INTRODUCTION

The image processing technique of *shrinking* and expanding a binary image to reduce noise or remove unwanted artifacts is very old but is still very often used. When artifacts are small a single pixel shrink and expand will suffice for image cleaning. When artifacts are large, several iterations may be required for sufficient image smoothing. In this paper we examine in close detail the operations of expand and shrink, here called dilate and erode. We will show that these operations are much more than simply a procedure for removing artifacts, rather they can be extended into a whole family of image processing algorithms. Because the algorithms combine sequences of dilations and erosions and because the results of these iterative operations are often surprising, we have called them here esoteric iterative algorithms.

A more conventional name for this approach to image processing is Mathematical Morphology. Conceived at the Ecole de Mine in Paris in the mid-1960s by G. Matheron [1] and J. Serra [2], Mathematical Morphology has grown to envelop a variety of applications and hardwares which fully demonstrate the effectiveness of the iterative neighborhood transformation process. The extensions of the morphology of neighborhood transformations from binary into greyscale processing by S. Sternberg [3] in the mid-1970s introduced a new wave of generalization into the iterative neighborhood processes. In addition to providing a solid framework for such classical greyscale processing problems as edge detection and background removal, the greyscale morphology has been used to treat binary images directly, thus unifying the approach to both binary and greyscale image processing [4].

II. BINARY MORPHOLOGY

The operations of Mathematical Morphology are based on set theoretic concepts. A binary image is viewed as a set of points X in the binary plane. We may conveniently conceive of X as being the black points constituting the objects of the image, the white points constituting the background. Besides the image X, the transformations also specify a second binary image called a structuring element. The structuring element is also a set, and its shape determines the points of X which will be affected by the transformations.

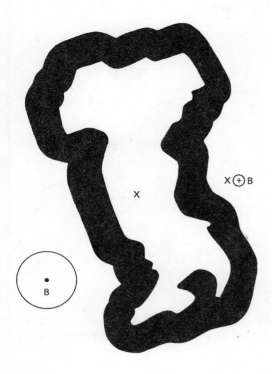

Figure 1. Dilation. The set X is shown in white, the structuring element B is a disk. Dilation increases X by the region shown in black.

For example, considering Figure 1, the set X is shown dilated by the structuring element B, a disk of unit radius. The dilation by a disk is usually considered to be an isotropic swelling of each point of X, the circular wavefront of the swelling surrounding each point of X by a circular region B. By this conception, the dilation of X by B, denoted X⊕B, can be expressed as a Minkowski sum of the sets X and B,

$$X \oplus B = \bigcup_{x \in X} B_x \tag{1}$$

where

$$B_x = \{ b + x \mid b \in B \} \tag{2}$$

is the translation of the structuring element B to the point x.

Points belonging to the dilation X⊕B either belong to X or lie within a distance of X determined by the radius of B. This leads to an alternative explanation of the dilation by disks

$$X \oplus B = \{ y \mid B_y \Uparrow X \} \tag{3}$$

the set of points y where the structuring element B can be placed such that it hits X. (The ⇑ notation is read "hits", meaning the intersection of B_y and X is non-empty.) These two notions of the dilation are dual, the swelling of X into its background X^c ($X^c = \{ y \mid y \notin X \}$), denoted by the Minkowski sum; the placing of B in the background X^c touching X denoted by the hit transform. Furthermore, the structuring element B

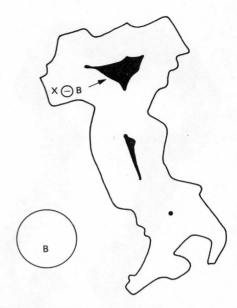

Figure 2. Erosion. The set X is reduced
to only the black points in its interior.

need not be a disk. Relationships (1) and (3) hold no matter what the shape of the
structuring element B. However, for nonsymmetric B we replace B in expression (3) by
B̌, the set B transposed about the origin

$$\check{B} = \{-b \mid b\epsilon B\} \tag{4}$$

Because dilation involves vector addition of the points of X and B it is commutative,

$$X\oplus B = B\oplus X = \bigcup_{b\epsilon B} X_b \tag{5}$$

Duality is a very necessary principle of Mathematical Morphology. Duality insists that
every transformation applied to the set X also be regarded with respect to the compli-
ment X^c. If X swells, X^c shrinks. The shrinking transformation is called erosion, and for
symmetric structuring elements B it can be written as the dual of the dilation

$$X\ominus B = (X^c\oplus B)^c \tag{6}$$

Figure 2 illustrates the erosion of X by the disk B.

Duality admits of complementary notions of the hit transformation and the Minkowski sum. Erosion of X by B can be expressed as a set containment transformation,

$$X\ominus B = \{y \mid B_y \subset X\} \tag{7}$$

The eroded set X⊖B is the set of all points y where we can position the origin of B such
that B is completely contained in X. For erosion, the Minkowski sum is exchanged for a
Minkowski difference, which for symmetric B is

$$X\ominus B = \bigcap_{b\epsilon B} X_b \tag{8}$$

Erosion is not generally commutative of course, if B ⊂ X then X ⊄ B.

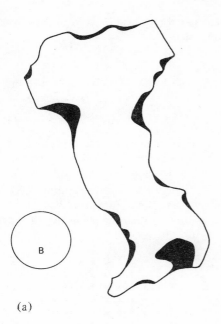

(a)

Figure 3. (a) Closing. The set X is increased by the black regions which are points that cannot be touched by the disk structuring element as it slides around the outside of X. (b) Opening. The set X is reduced by the black regions which are points that cannot be touched by the disk structuring element as it slides inside of X.

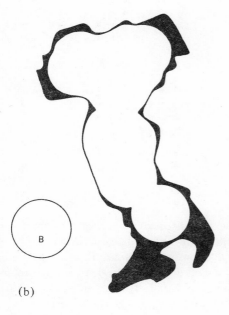

(b)

Dilations and erosions most frequently occur in pairs. A dilation of X by B followed by an erosion of X⊕B by B is called a closing and is denoted X^B,

$$X^B = (X \oplus B) \ominus B \qquad (9)$$

An erosion of X by B followed by a dilation of X⊖B by B is called an opening and is denoted X_B,

$$X_B = (X \ominus B) \oplus B \qquad (10)$$

Figures 3(a) and (b) illustrate the closing and the opening by a disk, respectively. The closing of X by B is the complement of the union of translations of B inside X^c. The opening of X by B is the union of translations of B inside X. Closing X by B is easily visualized as sliding B outside of X along the perimeter of X. Where narrow inlets appear in X, the closing by B fills them. If there are holes in X smaller than B, they are also filled. Opening X by B is visualized as sliding B inside X. Where X displays a narrow peninsula, the opening removes it. If particles of X are smaller than B, they disappear.

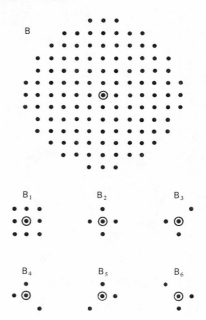

Figure 4. Digital Disk and its neighborhood structuring elements. The symbol ⊙ denotes the origin.

Dilations and erosions are usually implemented as sequences of neighborhood transformations. Because useful structuring elements might well be larger than a 3 × 3 pixel neighborhood, how then do we perform such dilations and erosions? Figure 4 illustrates a typical case. The structuring element B is a digital disk of radius 6. It can be decomposed into six smaller structuring elements, $B_1, \ldots, B_6$, all of which are subsets of the 3 × 3 neighborhood window. If we iteratively dilate B_1 by B_2, $B_1 \oplus B_2$ by B_3, etc.,

$$B = B_1 \oplus B_2 \oplus B_3 \oplus B_4 \oplus B_5 \oplus B_6 \qquad (11)$$

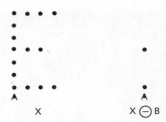

Figure 5. L-shaped structuring element.
Here, $B = B_1 \cup B_2$, where $B_1 = V \oplus V \oplus V$
and $B_2 = H \oplus H$.

Figure 6. Erosion by L-shaped
structuring element detects L-shaped
features.

(parentheses removed for convenience), then we construct the disk B. To dilate X by B, we iteratively dilate X by the neighborhoods $B_1, \ldots, B_6$,

$$X \oplus B = X \oplus B_1 \oplus B_2 \oplus B_3 \oplus B_4 \oplus B_5 \oplus B_6 \qquad (12)$$

To erode X by B, we iteratively erode X by $B_1, \ldots, B_6$,

$$X \ominus B_1 \ominus B_2 \ominus B_3 \ominus B_4 \ominus B_5 \ominus B_6 = X \ominus B \qquad (13)$$

Again the order of the operations is taken from left to right. By selecting the B_i we therefore can perform the dilations or erosions by structuring elements of a wide variety of shapes and sizes. The problem is we must know how to decompose the desired structuring element into 3×3 neighborhoods. For some structuring elements, this may be impossible.

The L-shaped structuring element of Figure 5 defies generation as an iterative sequence of neighborhood dilations. In cases such as this, the structuring element is decomposed as the union of two or more structuring elements, each of which can be generated as a dilation of neighborhood elements.

In the example of Figure 5, the L-shaped structuring element is composed as the union of a horizontal bar B_1 and a vertical bar B_2. Dilation of X by B is performed by dilating X individually by B_1 and B_2 and taking the union of the results,

$$X \oplus (B_1 \cup B_2) = (X \oplus B_1) \cup (X \oplus B_2) \qquad (14)$$

For performing the erosion, we take the intersection of the individual erosions,

$$X \ominus (B_1 \cup B_2) = (X \ominus B_1) \cap (X \ominus B_2) \qquad (15)$$

Erosions by nonsymmetric structuring elements are useful as feature detectors. For example, erosions by the L-shaped structuring element previously generated can be used for detecting the L-shaped features in the character recognition example of Figure 6.

III. GREYSCALE MORPHOLOGY

Although the Mathematical Morphology of binary images is interesting, it is only when we apply the iterative morphological transformations directly to greyscale images that the iterative neighborhood process becomes truly esoteric. In the greyscale morphology,

(a)

(b)

Figure 7. (a) Original greyscale image.
(b) Opened by a spherical structuring
element. (c) Difference image (Figure
(a)−(b)).

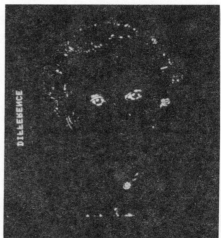

(c)

greyscale images are visualized as grey level surfaces whose relief is determined by the
grey level function $f(x, y)$ on the binary plane. The structuring elements of the greyscale
morphology are solids, like spheres or cylinders. Closing a greyscale image by a spherical
structuring element is equivalent to sliding a sphere across the grey level surface. The
closing operation fills in narrow pits and crevices in the grey level surface corresponding
to dark lines or points against a light background in the greyscale image. Opening does
the opposite: topological features which are brighter than their local backgrounds but too
small to contain the structuring element are detected.

Figure 7 illustrates the process of opening a greyscale image by a spherical structuring element. The greyscale image of the girl of Figure 7(a) is opened by a spherical

structuring element about the size of the girl's eyes. The opening removes highlights which are smaller than the structuring element (Figure 7(b)) which can be seen in the difference image (Figure 7(c)) whose grey levels are the difference between corresponding points of the original less the opening.

The extension of the Mathematical Morphology into greyscale image processing requires a new concept, that of umbra. Umbra means shadow, and the umbra of a set X includes both X and the shadow that X would cast when it is illuminated from above by a columnated light source. For greyscale images, the grey level surface is a set of points $\{x, y, f(x, y)\}$ forming a thin sheet in Euclidian three-space. The umbra of a grey level surface is formed by dilating $f(x, y)$ by the structuring element $\check{Z}$ the negative z-axis. The umbra, denoted $U(f)$, consists of points on the grey level surface and all points below the surface. It is a solid region of three-space whose surface topology reflects the intensity variations of the greyscale image.

Unions and intersections of umbrae are umbrae, the grey levels of unions and intersections of umbrae of grey level functions a(x, y) and b(x, y) being determined by the maximum and minimum of the individual grey level functions a and b, respectively,

$$U[a] \cup U[b] = U[\max(a, b)] \qquad (16)$$
$$U[a] \cap U[b] = U[\min(a, b)] \qquad (17)$$

Dilating or eroding an umbra U[a] by a structuring element B (a structuring element is a set in Euclidian three-space and not necessarily an umbra) is the same as dilating or eroding $\cup$[a] by the umbra of B,

$$U[a] \oplus B = U[a] \oplus U[b] \qquad (18)$$
$$U[a] \ominus B = U[a] \ominus U[b] \qquad (19)$$

This can be seen to be true from the definitions for dilation and erosion as Minkowski sums and differences (equations (1) and (8)). Dilation is the union of translations of umbrae by the points of the structuring element while erosion is the intersection of the translated umbrae. By equations (16) and (17), these unions and intersections are always umbrae, and their grey level surfaces are described in terms of the maxima and minima of the translated umbrae. Denoting the dilation of grey level surface a(x, y) by grey level structuring element b(x, y) as d(x, y) we can write

$$d(x, y) = \max_{i, j}[a(x-i, y-j) + b(i, j)] \qquad (20)$$

The eroded surface e(x, y) can likewise be computed as a minimum of translations of a(x, y),

$$e(x, y) = \min_{i, j}[a(x-i, y-j) - b(i, j)] \qquad (21)$$

IV. CONCLUSIONS

The familiar notions of expand and shrink are formalized in Mathematical Morphology by introducing the structuring element, a binary shape which determines the region of the expansion around a point in a binary image. Expanding and shrinking a black-and-white image by a structuring element, called dilation and erosion, respectively, are given geometric significance. The dilation denotes those points of the image where the structuring element hits the black objects, erosion signifies where the structuring element may be placed inside the black objects.

Because the morphological operations are defined on sets they are inherently dimensionless and may be applied as operations on greyscale images where the greyscale image is viewed as an umbra. This extension to greyscale images then gives rise to a convenient representation of the morphological openings and closings by spherical structuring elements, the rolling of a ball above and below grey level surfaces, the rolling ball algorithm is implemented as a sequence of neighborhood operations, these operations involving only the addition (dilation) or subtraction (erosion) of structuring element grey levels and the determination of maxima (dilation) and minima (erosion).

REFERENCES

1.   Matheron, G., *Random Sets and Integral Geometry*, Wiley, New York (1975).
2.   Serra, J., *Image Analysis and Mathematical Morphology*, Academic Press, London (1982).
3.   Sternberg, S., "Cellular Computers and Biomedical Image Processing", in Lecture Notes in Medical Informatics, vol. 17: Biomedical Images and Computers. Proceedings 1980. Edited by J. Solansky and J. C. Bisconte. Springer-Verlag, Berlin (1980), pp. 294–319.
4.   Sternberg, S., "Languages and Architectures for Parallel Image Processing", Proceedings of the Conference on Pattern Recognition in Practice, Amsterdam, May 21–23, 1980. Kanal, L. N. and Gelsema, E. S., eds., North-Holland, Netherlands (1980).

S. R. Sternberg
Cytosystems Corporation
Ann Arbor, Michigan

6 Remote sensing – a view of integration

F C Billingsley

INTRODUCTION

Remotely sensed imagery has now been available from the Landsat series of satellites for 10 years. During this period, the data use has expanded from investigations performed by a group of sponsored investigators to attempted routine use by private companies, agribusiness, local and state planners and resource managers, and federal agencies. The data are becoming a powerful tool in meeting the needs of the world in resource exploration and management. New analysis methods are being developed to take advantage of the new types of data.

The Seasat program, albeit short-lived, provided a source of synthetic aperture radar data which responds to a different set of parameters (surface angle, surface roughness, dielectric constant, which are affected by factors such as vegetative cover, soil moisture or water surface geometry) than Landsat. It also provided other instruments such as a radar altimeter and a microwave scatterometer (designed for oceanographic use, but which returned data of interest over land).

The Heat Capacity Mapping Mission returned data which has been of use in measuring the thermal inertia of surficial materials, a body property, as contrasted to the surface reflectance property measured by imaging sensors.

France and Japan are planning their own satellites using pushbroom image sensors, and the United States has future sensors on the drawing boards.

To a very large extent each satellite mission has been conceived and executed independently of the others. Each has been driven by a "technology push" – a valid approach in the days of newly available technology and in the face of unknown utility of the new data. The various potential uses of the data have developed independently of each other, and only gradually have there begun to be attempts to utilize together data from the various sources. There is still little if any use of data from divergent satellites although repeat data from a single satellite such as Landsat have proven useful. In short, during this formative period, the data sources and data usages have not been unified. Particular stumbling blocks, in addition to the natural discipline segregation during the learning period, have been the diversity of resolutions, formats, and geocoding methods, the lack of data continuity, and the effort and cost of data preparation.

During the same decade the cost of computation has dropped dramatically, and the potentials of large-scale data transmission have improved. Digital analysis techniques

Table 1.   The Many Faces of Integration

Objectives of different scenarios
   Research
   Applied, Demonstration
   Commercial, Profit-making

Political
   Why Have a Program?
   International Competition
   International Cooperation

Multiple discipline
   Land and Sea
   Cross-discipline Missions – Global Ecology
   Parameter Satisfaction

Service to multiple users
   From Single Sensor and its Archive
   From Multiple Sensors and their Archives
   Networks and Data Delivery

Multiple source data used together
   Multitemporal, Same Sensor
   Different Sensors and Data Types – Images, Polygon, SAR
   Formats and Geographic Information Systems

Processing services at various levels
   Acquisition and Archiving
   Value-added Preprocessing
   Analysis Services

---

have developed and the use of digital data has increased manyfold. Data delays are being overcome as the operational necessities of the program are recognized. It is now time to consider the current situation and to decide whether an integrated approach (to something) is warranted and/or feasible.

We will consider a number of facets of integration, as listed in Table 1. A totally integrated program might well integrate all of the factors. Present practical programs do not have any of the topics integrated, let alone integration of the set. Table 1 will form the outline of the discussions as we consider what might be integrated, assuming that there is to be a program, not just a group of isolated projects. Although in the real world sense the definition of the "program" may follow determinations of what may be practical, it may be profitable to consider a number of motivations or desiderata concerning programmatic decisions during the formative definition stage. The list in Table 2 is offered for starters. These are not necessarily exclusive, and all have ramifications and subsequent decisions/actions required for their implementation. For those who believe

Table 2. Possible Programming Desiderata

---

1 **Provide a federal data source for the common good**
  If this is to be Landsat data, GOTO 8
  If not Landsat:
    Buy what is available – SPOT? MOS?
    Sponsor other data sources

2 **Keep the country foremost in earth remote sensing**
  Sponsor a good mix of research and demonstrations
  Support nascent operations until they are self-supporting

3 **Get out/stay out of Federal operational remote sensing**
  Let other countries (e.g., France, Japan) take over
  Convince private industry to supply the products needed (by whom?)
  Don't use space products

4 **Provide Federal large scale research results**
  Do the research. Support future research programs (such as Landsat?)
  Sponsor university research

5 **Provide a Federal research facility for others to use**
  Do it or sponsor it
  Should it be made to self support? How much?

6 **Provide technology advancements in high cost areas**
  Do it. Support follow-on sensors and data technology
  Sponsor others to do it

7 **Get out of Federal remote sensing research**
  Let other countries take over
  Hope that private industry will take over

8 **Continue to obtain (how much?) Landsat data for federal agency use**
  Keep Landsat alive through direct support (i.e., GOTO 9)
  Purchase Landsat type data from private industry
    Private industry must be there to supply products as defined by
    Federal agencies. This may require subsidy

9 **Keep the Landsat program alive in the present configuration**
  Continue to provide whatever support is required
  Find a sponsor willing to take it on without Federal support

10 **Private industry to run the Landsat program**
  Find out how much to make it break even
  Don't compete
  Subsidize it – but then is it private industry?

11 **Private industry to run a remote sensing program, not necessarily Landsat**
  Get out of the way. Don't compete
  But Federal agencies may not get Landsat type data

12 **Foster private industry/Federal agency use of Landsat data**
  Demonstrate its use
  Guarantee continuity to justify users' investments. GOTO 9 or 10

---

in the ultimate good of appropriately applied technology (I am one), an earth observations program must include items 1, 2, 4, and 6, and perhaps 5. The implementation implications follow directly. For this discussion let us assume that some sort of positive program has been defined, and consider the remainder of the integration questions.

OBJECTIVES OF DIFFERENT MISSION SCENARIOS

In an ordered world, research will have been done to gain some understanding of the world, most often without intent of direct commercial gain. The understanding so gained will then be developed into demonstrations of some possible applications with the expectation that later commercial applications will develop which are suitable for private industry to undertake for profit. The real world is not so ordered, in that the fundamental research may not have been done to completion prior to the attempt at demonstration. The Landsat program is a case in point — a set of technology became available accompanied with enough understanding of potential results to warrant a technology push for the system. The central column of Table 3 illustrates the sequence of steps in the applied approach. In the Landsat case, we are somewhere between the last two steps — demonstrate application and develop user interest: potential applications certainly have been demonstrated, but user interest is not to the point of developing the groundswell required to make evident a potential profit-making market for the data.

The "research" sponsored by the Landsat program has been directed to the fourth item, determine what information can be extracted, a valid step in progressing down this column. Early research was defacto dedicated to the use of the Landsat data with little regard for other data sources as the basic knowledge was being developed. Because of this, the sensing parameters have been severely limited to those of Landsat — no off-nadir, preselected spectral bands and ground resolution, and fixed revisit interval. During the latter part of this period, data from other sources was beginning to be incorporated. This is now allowing the various data sources to take their relative places in importance — remotely sensed data is now recognized to be ancillary to other data for many applications. However, because of the (probably necessary) fixation on the use of Landsat data, there has been a hiatus in true research as outlined in column 1 of Table 3. Specifically, there has been little research in non-Landsat parameters: off-nadir viewing, other spectral bands, variable viewing angles, different times of day, different ground resolutions, and so forth. The Thematic Mapper, SPOT and MOS are relatively small (sciencewise) extensions of the MSS, although it is recognized that they are major engineering extensions. In some sense we are mired down by success and stuck in the same place parameterwise. We are now realizing that to move forward, we must go back to square two to do the research required to develop new concepts. Thus there is starting a movement toward doing the column 1 research. At this point it is not clear that this is driven for love of knowledge, but more likely the prime driver is the realization that we have not developed the knowledge base to determine parameters for future sensors.

At the same time, enough experience has been developed that some forays into column 3 are possible. As an example, a number of companies have developed analysis systems for image data, some are in the business of image analysis, and some, such as exploration companies, have purchased data for in-house analysis. It is to be recognized,

Table 3. Steps in the Development of Scenarios for Remote Sensing (expanded from Taranik, 1982, personal communication)

Research approach	Applied approach	Commercial approach
Determine current state of knowledge	Develop measurement system	Determine a range of potential businesses
Identify gaps in knowledge	Collect data	Determine potential markets for each
Formulate scientific questions	Analyze data	Set up potential operating scenarios for each
Design experiments to answer questions	Determine what information can be extracted	Determine decision criteria
Define measurement requirements	Determine how information applies	Evaluate scenarios
Evaluate measurement technology	Demonstrate potential application	Select business (what)
Develop measurement system	Create user interest	Select market segment (who)
Collect scientific data		Select scenario (how)
Analyze scientific data		Implement selected system
Develop theory and models		Develop customer interest
Understand what is measured		Operate the system and supply products
Final product is: Understanding	User interest	Physical products

however, that image analysis systems are also used for other imagery (medical, for example), and that even when Landsat data are used, images are generally ancillary to other information. Potential facets of the space imagery activity to consider are building spacecraft and sensors (either or both), launch and operations, data receipt and archiving, value-added services, data analysis equipment or services, data communications, and perhaps others. If we assume (logically, I believe) that the motivation of commercial venture is to make a profit, no amount of technology push will entice a potential private enterprise into the business unless the pull of profit is evident. Thus, if private industry is to take over the land remote sensing activity, it must discover or be shown avenues for profit. It is not evident even to those now in the business (government funded) that a profit can be made in the operation of a Landsat-type system as a whole, if spacecraft and launch costs are to be recovered.

THE POLITICAL FACE OF PROGRAM INTEGRATION

Thus, if the operation is to be continued in its present form (back to the desiderata of Table 2), other motivations than monetary profit must be sought. A data base for the common good, large scale research, and high cost technology development are valid areas for governmental motivation but not likely to be areas of private industry motivation. As a Landsat-scope program clearly falls in these categories, government sponsoring will continue to be required. Government "profit" will be in the research results, strengthened industry through the technology gained, and a data base (e.g., for land use studies, urban growth, geologic studies, hydrology studies) useful to both government and private industry (e.g., oil and mineral exploration, agribusiness). But note that it will never be possible to give a dollar-for-dollar accounting of benefit/cost in these areas; attempts at such have been unconvincing and fruitless.

It has been recognized from the beginning that Landsat data are world-wide. The United States open skies policy has been a source of wonderment in many other countries. The remote sensing program may well be the instrument which can facilitate world-wide programs in global food supplies, global pollution, global habitability, and hence foster global inter-country cooperation. Each participating country must find the balance between maintaining its technical capabilities and sharing them for the common good.

MULTIPLE DISCIPLINE

Current NASA missions have been largely dedicated to one area of investigation or another, as exemplified by three notable satellites (and their names): *Land*sat, *Sea*sat, and the *Coastal-Zone* Color Scanner. Admittedly, each has some special sensing features, optimized for its task:

Landsat

> The Multispectral Scanner and the Thematic Mapper — moderate spectral selectivity and reasonably good spatial resolution; have produced interesting ocean as well as land data.

Seasat

> SAR, scatterometer, radar altimeter — these are optimized for oceans use, but have produced interesting data over land.

CZCS

> An instrument flown on Nimbus. More, narrower spectral bands than Landsat, but poor resolution, suited to coastal zone work.

These satellites have served admirably for research in the various fields (as have others also). Independently targeted research satellites will continue to be required as new parameters or phenomena are investigated. These may serve more than the first investigator if suitable data structure is planned ahead of time (see below). However, more than incidental cross-discipline utilization of satellites can and should occur if the designers can consider multiple disciplines. For example, in land use, there are threads common to the requirements of several disciplines, as indicated in Table 4, as well as divergent requirements. These may be integrated (Figure 1) to serve the majority of needs, but there are remaining needs not satisfied in the normal designs. If these in turn are considered, it may be found that all need not be satisfied simultaneously, so that some tradeoffs become available. With this thought in mind, a multiresolution sensor has been proposed as food for thought [1]. This concept covers the divergent needs of short repeat time, high spatial resolution, specialized spectral resolution, with a not-unreasonable data

Table 4. Remote Sensing Desiderata by Discipline

Agriculture: Revisit interval 5–8 days to pick up emergence and other crop calendar-related events.
  Spatial: High resolution (10–15 m?) to pick up field boundaries; coarser resolution would suffice for field interiors.
  Spectral: Properly placed bands; four to six probably sufficient.

Mapping: Revisit interval not critical; eventually need complete cloud-free coverage.
  Spatial: 3–10 m resolution needed for 1:24,000 mapping.
  Spectral: The fine resolution may be panchromatic or principal components.

Geology: Revisit interval should be seasonal to pick up variations indicative of the geologic information desired.
  Spatial: No hard requirement; 15 m is a logical next step.
  Spectral: Perhaps seven or so bands, placed for soils and rock recognition. Bands probably different than agriculture bands.

Land use: Revisit interval perhaps seasonally or semiannually.
  Spatial: No hard requirement; 15 m is a logical next step.
  Spectral: No hard requirement; bands placed for other purposes may suffice.

rate, provided that not all are needed simultaneously (i.e., high spatial resolution with long repeat time and low resolution with short repeat time may be met simultaneously, but not high resolution and short repeat time). Figure 2 shows the concept.

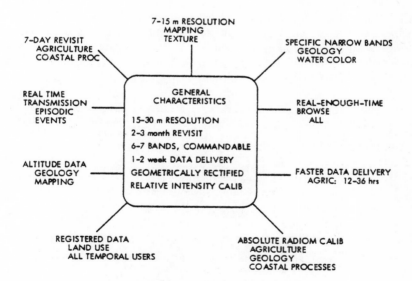

Figure 1. A general purpose design will leave many needs unfulfilled.

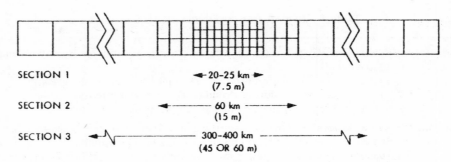

Figure 2. Pixel layout for a
multiresolution sensor.

Consideration of specific tasks, such as global ecology, will indicate the need for various sensing parameters: atmospheric constituents with very low resolution but very narrow specific spectral bands, land cover and desertification with good (say, 15–30 m) resolution and "agricultural" spectral bands, urban effects with better resolution and bands suited to anthropogenic features, ocean color sensing for, say, chlorophyll, with CZCS-type bands and low resolution sensors, probably microwave, for soil moisture. In addition, ground-based sensors for such as stream gauging will require satellite data links. If a single sensor such as the multiresolution one described cannot be devised, several sensors on one or more satellites will be required, requiring the careful integration of the data formats and processing.

An integrated system must be designed as an information system, not as a sensor with its data system. This may be modeled as shown in Figure 3 [2]. The forward model is required to convert units of information to units of required data, and answers the question "What set of measurements will best carry (i.e., allow the best derivation of) the information?" This box provides the set of measurement "requirements" to the measuring system, which responds with a set of real measurements which hopefully will be

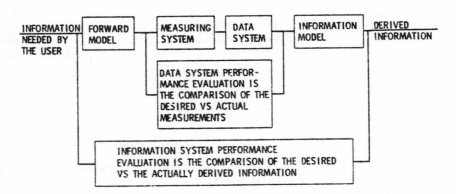

Figure 3. A total information system
design includes analysis at two levels.

somewhat near to the desired set. However, the data system may have an inaccurate or uncertain transfer function, so that the set of apparent measurements presented to the information model deviate further from reality. It is with this set that the user attempts to derive his information, using the information model.

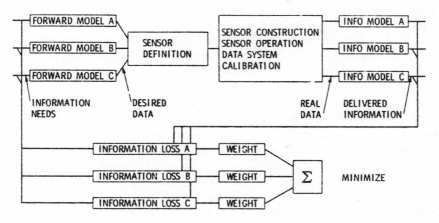

Figure 4.   Total information system
design is a linear programming problem.

The system may be analyzed on two levels, the data system level (the usual approach) or the information system level (rarely done). We will consider the latter as shown in Figure 4. It can be seen that this is a linear programming problem, with the desire to minimize the sum of the information losses caused by the system. The sleepers in the analysis are the forward and information models: in the forward model, the analyst must evaluate how well the desired data will carry the information desired (this is where he asks for certain tolerances, presumably derived from knowledge of their effect on his eventual analysis), and in the information model the analyst must state what amount of degradation in his interpretations will occur (his loss function) if the data is somewhat different than requested. The former estimate may be available, the latter usually is not. However, the design will proceed anyway, with assumed loss functions. A needed research topic is the determination of at least the methodology of determining these loss functions — their implicit assumption by the system designer, while satisfactory for "technology push" systems, is not adequate for systems designed to "do a job".

SERVICE TO MULTIPLE USERS

In the earlier days of the Landsat program, and at the present time in other programs, data preparation and delivery were to a relatively small group of experimenters, generally part of an experiment team. There was no particular need to "sell" the data to others, nor for extensive preprocessing by the data suppliers. With the success of Landsat, it was found necessary to systematically archive, catalog, and distribute the data to an ever-growing group of outside analysts. Planning of such a system for SPOT is in progress.

Some of the attributes required of the data to facilitate usage may be stated as follows:

(1) With the increasing analysis complexity, even experimental use of data requires a sizeable investment. Operational use generally implies a commitment on the part of the user to continue to use a particular data type. These investments will not come about without commitment on the part of the data suppliers to continuity of the data.

(2) For many problems, timeliness of the data is critical, either because the data must be acted upon immediately or because the data must be analyzed before the next set arrives. In addition, even for retrospective data analysis, rapid response from the system is required to deliver data while the problem is still being addressed.

(3) The data are of no use until analyzed. Thus, the data must be in a form suitable for rapid and cost-effective analysis. This typically may require that all calibrations have been applied or the data registered to a suitable reference.

(4) The data must be available at what appears to the user to be an equitable cost. It must be recognized by data suppliers that relatively few tasks will be dependent on a particular source of data. If they wish their data to be used, pricing policy must appear to be reasonable, as well as actually be reasonable.

(5) The growing diversity of users will require preprocessing of the data in many ways. To avoid reprocessing data from a certain form to that required by the user suggests that the data be stored with a minimum — perhaps zero — of preprocessing in the archive. The archive will then provide the special processing on retrieval to provide data users as close to first generation as possible. This will also avoid the preprocessing of data which is then never used.

(6) It is almost universally true that the required ancillary data to allow easy rectification and registration of imagery is not included with the images. Various standards exist in different communities for definition of some of the parameters (for instance, standard ways of denoting latitude and longitude), but there has been little effort to organize an integrated whole. Therefore, each user is left to his own devices when registration is required.

For data sets from multiple sources, the last point above is exacerbated when multiple data registration is required. This adds new desiderata:

(1) To allow the users to assemble data sets from various sources, the design of archival and retrieval systems must be coordinated to allow the data sets to be assembled with implicit registration, not requiring explicit registration in the process. This is of particular importance when data such as SAR images are to be registered with (for example) Landsat images, as the techniques for rectification are quite different and perhaps best done by central system experts. As one simple example, if each archive defines a standard data grid, they all must be the same (at least in multiples) and a common way defined for ground location (e.g., is a pixel to be located at each grid intersection, or in the square formed by the grid?) and nesting.

(2) There must be format compatibility. This is discussed more below.

Even with a system in place for archiving data which may meet most of the desiderata above, the local user still has a choice to make as to whether to use it or its data. This

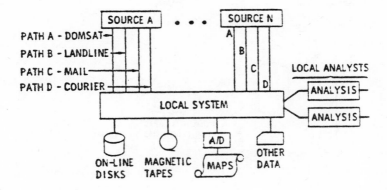

Figure 5. The local system has many
sources.

decision will normally revolve around two factors: money (front end costs or operations)
and data delays (which may be converted to money equivalent). The system design job
has been to aggregate a very large number of potential users, to estimate their needs in
terms of data traffic, and to provide a large system with low-cost, minimum potential for
failure, suitable cataloguing, browse and query capability, and reasonable queuing performance. The local user (by definition, one with a small number of analysts, within a
relatively small geographic area) is concerned with these factors to a lesser extent and
has much less flexibility in invoking the law of large numbers. He will have available to
him a number of data sources and delivery methods (Figure 5), each of which has its own
characteristic delay function (Figure 6). These must be weighed against data urgency and
cost of delays. He must also determine how much to rely on the various services of the
central system.

The central system and the local system will each be trying to optimize the parameters available to it. However, the goals will be quite different: the central system must
design a cost-effective system satisfying all of the potential design drivers, but probably
not provide many specialized services; the local user has a choice of using the central

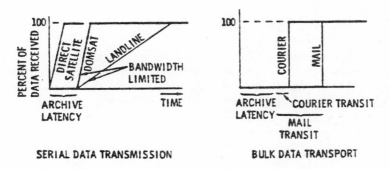

Figure 6. Each delivery route has a
characteristic delay.

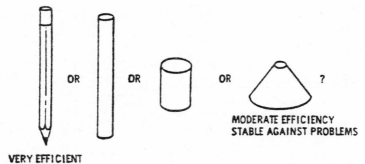

OR   OR   OR   ?

MODERATE EFFICIENCY
STABLE AGAINST PROBLEMS

VERY EFFICIENT
WHEN WORKING

EASY TO TOPPLE

Figure 7. A system may be fragile or robust depending on its design and purpose.

services (such as cataloguing or registration) or not, and may provide both generic and specialized services to its analysts. Thus the "tuning" of the two systems may be quite different (Figure 7). The central system must be stable against failures, perhaps implying lesser services provided, whereas the local system may be more critically designed. Interactive activity on the central system will probably be mostly cataloguing and data ordering, with interactive analysis being done on the local system. A discussion of a number of cost factors is given in Ref. [3].

MULTIPLE SOURCE DATA USED TOGETHER

Perhaps the most exciting development during the decade of Landsat has been the ability to use data of various types and sources together. This began with the use of multitemporal data from the same source, such as Landsat, for detection of changes in visible land cover, leading to the ability to determine changes in land use. Multitemporal data was also used throughout a growing season to monitor the phenologic stages of crops [4], which is leading to the recognition of crop types via the crop calendars. These efforts required precision registration of the images from various dates, in turn requiring that each image be reduced to the same ground grid. As the sensor does not deliver data to this requirement, extensive technology has built up for the required correlations and rectifications. It was soon apparent that Landsat data, far from being prime, was an additive factor in analyses being carried out already in various disciplines. This led to the need to register Landsat data with data from other sources, including maps. At the same time, the general improvement in computer technology has allowed the conversion and analysis of map data in polygon form. It was inevitable that the two should come together in the form of geographic information systems. Figure 8 illustrates the salient features required in systems which are image-based. Note that they will be independent of the incoming image data form (except for the required data conversion and registration). Because these systems have been developed to solve geographic problems, they will not necessarily be bent to accommodate space data, but are independent of these data. At the same time, the space data have not (with the exception of Canadian processing) been cast into optimum form for use in the geographic information systems [5].

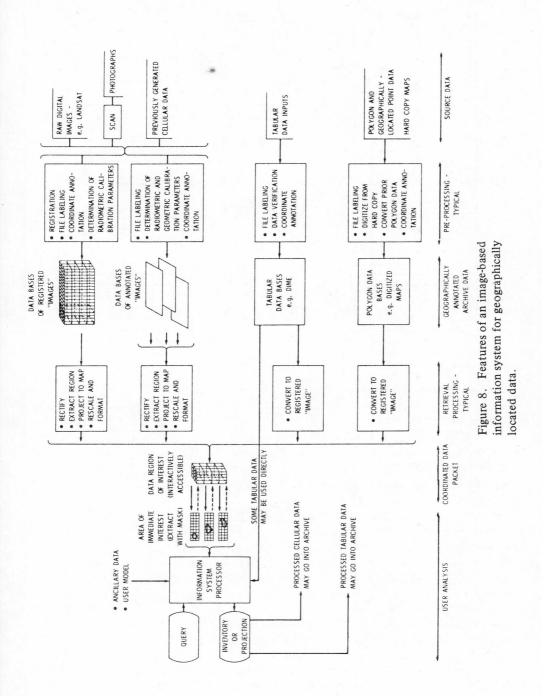

Figure 8. Features of an image-based information system for geographically located data.

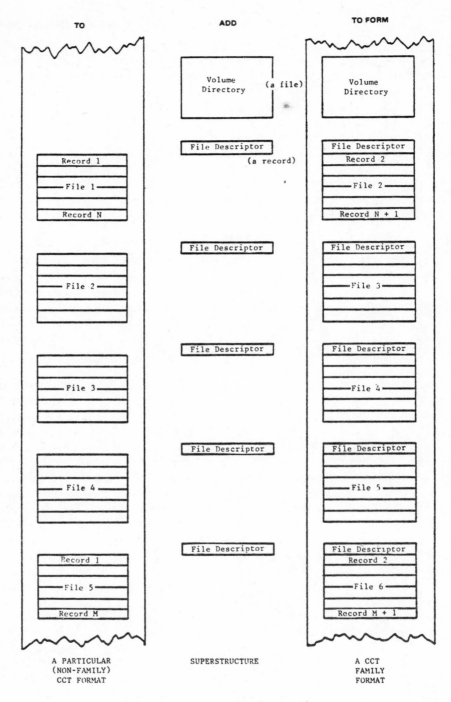

Figure 9. Example of converting a particular CCT format to a CCT family format.

A number of desiderata were stated in a meeting on the impact of geobase information systems on space image formats [6]. Unfortunately, few of these have been adopted. In general, various data are produced in various formats and scales, having been produced for the most part to match the data acquisition/production requirements for the system instead of attempting to make the downstream data handling by the user easier.

One example of the possible production of space data in user form is in the location and scaling of pixels. A clear statement was made that, for UTM users, an eminently usable form would be to have pixels on Easting lines, with spacings of, say, 5, 10, 20 m, or 25, 50 m, etc., with a scale to suit the data. Except for Canada, Landsat 4 data are being produced at the rotation angle of the orbit at the latitude of each scene center with 28.5 m pixels (TM) or 57 m pixels (MSS). Thus, users of UTM data must register each Landsat image before use to match the other data. It may be that data conversion from mission-dependent to mission-independent form is a valid task for private enterprise.

As stated by McCandless [7]: "... Ultimately, the real data combining must occur in image processing and it is here that the game is either won or lost. Processing, if properly designed as part of the system at the outset, can alleviate some demands on system complexity and expense. Processing can fuse diverse sensor technologies and spectral regions into a combination that exceeds the sum of its parts in its effectiveness. The early success or failure of multisensor application will be very dependent on how closely processed data products match user needs and a useful system design must begin and verifiably end at this point."

The remote sensing community has taken a large step forward recently with the adoption of the new Landsat International CCT Format. Hopefully, its adoption by other segments of the community will foster integration of the different data types. The major impetus for working toward a CCT format standard is the need to exchange data on the international, national, local, and system level, as well as to merge data from many different sources. Many different types of data must be handled. Obviously, the format must be well suited to the handling of multispectral image data. If the format is to be useful in geographic information systems, it must also handle other types of data such as polygons, profiles, point data, text, etc. The format has been designed to be compact, precise, and simple so that noncomputer experts can understand it readily. At the same time, it is flexible and suited to a wide variety of data. It has been designed to be self-documenting, so that proper software can read the data without prior knowledge of its source or specific format.

The essence of the structure is the use of superstructure records which, when combined with any tape format, provide access to the data [8, 9]. These are used as shown in Figure 9. The volume directory contains File Pointers which describe the structure of each file in conjunction with the File Descriptors which accompany each file. The overhead incurred is small. The format has been defined as a family, with specific embodiments to be defined for each application. To date, applications include Landsat MSS and TM, SAR images and polygon files (Canada), and some digital telemetry data. SPOT is intending to adopt the format, and hopefully the Japanese MOS will do so also. Now that the International Format is in place for these applications, the remote sensing community is in position to expand its definitions and to drive toward a truly integrated data standard for this type of data. What must be done now is to adopt data coding and referencing

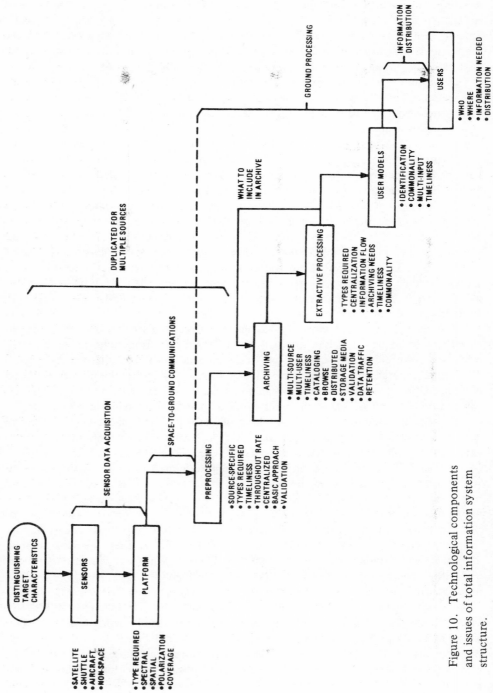

Figure 10. Technological components and issues of total information system structure.

schemes which can be adopted by geocoders, map digitizers, and others. As these procedures are already in use in these communities (e.g., lots of maps are being digitized in various places), the likely situation will be that the remote sensing community will adopt these standards and apply them to the remotely sensed imagery. The format has been designed to allow such future expansion. It is hoped that other countries will inspect these embodiments, adopt them if possible, or work toward an international standard which can be used by all.

PROCESSING SERVICES AT VARIOUS LEVELS

In the total scheme of things, the data sequence might be stated as: calibration — sensing — preprocessing — archiving — retrieval processing — analysis — modeling. Some of the considerations for these steps are indicated in Figure 10. In the beginning (of Landsat), the defined product was the photographic reproduction of the data, based on the feeling that users would not have extensive computer processing facilities or analysis knowledge to handle the digital data. This required that all images be rectified both radiometrically and geometrically before the images were converted to film, encouraging the archive to be the rectified digital data.

In actuality, the digital data became extremely popular. However, the techniques for the rectifications were only being developed, leading to the desire on the part of the users to obtain raw, uncorrected, data. This was argued on the bases that (1) the striping removal (by the system) left residuals; (2) after geometric corrections, removal of residual striping is no longer possible; (3) even if the system did geometric interpolations, many users would have to reinterpolate to make the data fit their problem, and there was worry about what the second interpolation would do to the data; (4) small features which had been inadequately sampled Nyquist-wise would be degraded. As a result, there was strong pressure to store raw, not geometrically processed, data in the archive.

Objections 1, 2, and 3 are gradually being overcome. Current feelings are swinging toward acceptance of the system-destriped data as the destriping has been improved. There has been little experience with absolute radiometry (and the atmosphere disturbs this anyway), so there has been little pressure for absolute calibration. With recognition that second interpolation does minimal harm to the data, concern for this, while still present in some quarters, is not as violent. Objection 4 will always be a problem when users are concerned with objects of pixel or subpixel size — they will always require unresampled data.

For the Landsat MSS, the official archive is the radiometrically corrected but geometrically unprocessed data; this will allow earliest-generation data to be delivered to users. For the Thematic Mapper, the official archive product is film, but raw data is being saved for future inclusion. With the buildup of geographic information systems an additional requirement for geographic location, not just intraimage geometric accuracy, has been placed on the system. This is being met by the inclusion of capabilities within the archiving system for the recognition and location of ground control points and their subsequent location in the images. When and if sufficient ground control points are available for a given scene, that scene will be geodetically located to within a fractional pixel. Lacking sufficient control points, the scenes will be geometrically corrected only.

A complication is introduced by the vibrational jitter in the Landsat 4 spacecraft, making it especially desirable that the system correct the data. The net result is that the system will produce correctly warped and (with control points) geodetically located images, a task which is tedious, if not impossible, for most users.

All of these considerations lead to a set of principles for digital handling of earth resources:

(1) If the data are acquired digitally, store them digitally.
(2) Do a minimum of processing, preferably zero, before archiving.
(3) Collect and reduce all ancillary data and relate to the archived images.
(4) Provide browsing and cataloguing for the users.
(5) Provide as a retrieval option, rectification, scaling, subarea selection, and data registration to the user's grid.

It is recognized that the complete list of services, particularly the retrieval processing, may encroach on desires of private industry to provide value-added services, and that such services may be beyond the desires of the archive organization. Canada has adopted an intermediate position: preselect the output grid to satisfy most users (UTM, with 25 m (TM) or 50 m (MSS) pixel spacing, aligned with topographic maps) and supply all products in that form.

The required analysis may be performed at the user's facility, or supplied by analysis companies. Both are being used. The keys to successful analysis are the familiarity of the analyst with the area being analyzed, and with the ability to take advantage of any serendipitous findings during the analysis. Given the expectation of data continuity and with it the expectation that the data usage will grow, it is reasonable to expect that these services could show a profit.

CONCLUSIONS

In some sense, the entire paper is a set of conclusions. As satellite data becomes integrated with other data, they form a powerful tool in meeting the growing needs of the world in resource exploration and management. The growing number of potential applications is made possible by the detailed numerical analysis of the combined data sets. New models for the use of the data in day-to-day operations are being developed. Today's experimental uses will become tomorrow's necessities. Indeed, the reduction of data use to a routine may be taken as a measure of maturity of the technique.

Geographic information systems and the integration of different data types will form the foundation of the new uses. The integration requirements will affect the designs of new sensors and place requirements on the platform orbit control and attitude stability. The digital processing must allow the assimilation of vast quantities of data as the desired resolution elements decrease in size. This in turn will require a new look at data compression to keep the data links and archives from overload.

This paper has not touched on perhaps the ultimate in integration — the development of "smart systems" which integrate not only the integrated data, but also the user's models and thinking process. Such systems are becoming possible and are being developed for medical diagnosis and mineral exploration.

In short, the concept of integration has led us far beyond the simple thought of combination of different data sets. It affects the very act of data acquisition, through archiving and processing, to the ultimate smart system.

ACKNOWLEDGMENT

This paper represents the results of one phase of research performed at the Jet Propulsion Laboratory, California Institute of Technology, sponsored by the National Aeronautics and Space Administration under Contract NAS7–918.

REFERENCES

1. Billingsley, F. C., "Concept for a Multiresolution Sensor", SPIE Paper 345-16, Washington, D.C., April 1982.
2. Billingsley, F. C., "Data vs. Information: A System Paradigm", NASA Workshop on Registration and Rectification, JPL Publication 82-23, June 1982.
3. Billingsley, F. C., "Data Base Systems for Remote Sensing, Cost and Technology Considerations", 3rd Conference on the Economics of Remote Sensing Information Systems, Incline Village, Nevada, November 1979.
4. LACIE (Large Area Crop Inventory Experiment), Proc. of the LACIE Symposium, NASA Johnson Space Center, Houston, Texas, October 1978, Report No. JSC-16015.
5. Estes, J. E., "Remote Sensing and Geographic Information Systems Coming of Age in the Eighties", Proc. of the 7th Pecora Symposium, Sioux Falls, S.D., October 1981, available through ASP, 105 N. Virginia Ave., Falls Church, VA 22046.
6. Simonett, D. S., et al., "Geobase Information Systems Impacts on Space Image Formats", Univ. California Santa Barbara Remote Sensing Unit, 1978.
7. McCandless, S. W., Jr., "MultiSensor Processing — The Importance of its Influence on System Design", Proc. of the Commission II Symposium (ISPRS), Ottawa, Canada, September 1982. Available from the Canadian Institute of Surveying, Box 5378, Station F, Ottawa, Canada, K2C 3J1.
8. "LGSOWG, The Standard CCT Family of Tape Formats", LGSOWG CCB-CCT-002D, available through ORI, Inc., c/o Ms. Lynn Buhler, 1400 Spring Street, Silver Spring, MD 20910.
9. Thomas, V. L., and Guertin, F. E., "Standardization of Computer Compatible Tape Formats for Remote Sensing Data", Proc. of the International Geoscience and Remote Sensing Symposium (IGARSS '81), IEEE Geoscience Society, Washington, D.C., June 1981.

Fred C. Billingsley
Jet Propulsion Laboratory
California Institute of Technology
Pasadena

7 A critical analysis of remote sensing technology

G Nagy

1. INTRODUCTION

Thousands of specialized systems for geographic data processing are in current use to help us cope with the resource-related social and economic problems facing us. Some digital terrain models encompass the entire globe. Many countries are developing integrated planning systems. Smaller units of government struggle with databases for transportation networks, land-use, water and mineral rights, soil-type, or recreation areas. Even small towns are rationalizing and computerizing their cadastral records to improve tax collection. The direct and only mildly distorted view of the earth afforded by satellite imaging systems provides a valuable source of up-to-date information for many of these endeavours.

Geographic analysis may be viewed primarily as a data-reduction process: a vast amount of data is reduced to a few elements that can be assimilated by human decision-makers. The bulk of the data is, in many applications, satellite image data. A single multispectral Landsat-D frame contains over 100,000,000 bytes, more than most entire nonimage geographical databases. Fortunately, an increasingly wide range of image processing technologies can be applied to geographic analysis as systems originally designed to do pattern recognition and image processing are being expanded to perform the full gamut of operations needed for data integration and spatial analysis. However, as with any new technology, there are technical, economic, and conceptual barriers to wide-scale use.

The objective of this paper is to review briefly the major current sources of remotely sensed data and to present sundry recent developments which show some potential for affecting the automatic extraction of information from remotely sensed data. Although no attempt is made at exhaustive coverage of all potentially useful items, both theoretical concepts and technological developments are mentioned. The author's impressions from a NASA-sponsored task force on suitable directions for pattern recognition research for remote sensing are also summarized.

2. DATA ACQUISITION

The number of civilian image acquisition platforms in orbit is increasing steadily as is their spatial, temporal, and spectral range, resolution, and accuracy. In the United States, responsibility for earth resource satellite operation is gradually being transferred to the

Department of Commerce, and political pressure for private operation of the data acquisition, processing, and distribution networks is growing. The expected profitability of such operations is perhaps the best indication of the progress achieved since the launching of the first earth resources satellite, ERTS-A, in 1972.

Landsat-D, launched in July 1982, carries both a four-channel Multispectral Scanner (MSS) similar to those on Landsat 2 and 3, and a Thematic Mapper (TM) with 30-m resolution and seven narrower spectral bands. Up to 133 MSS scenes per day may be available by January 1984, with 12 TM scenes per day (requiring an entirely new image processing facility) becoming available by mid-1984. These figures apply to data over the United States only; direct data acquisition by foreign receiving stations may support up to 660 MSS and 250 TM scenes per day worldwide. Acquisition of MSS data on an operational basis will be governed largely by user requests addressed to the United States Geological Survey's Earth Resource Observation Systems (EROS) Data Center.

The image information is either transmitted directly to earth or, when the satellite is out of range of the ground receiver, stored on board on a magnetic tape recorder for subsequent transmission. On-board storage has proved unreliable and will be obviated when the Tracking and Data Relay Satellite System (TDRSS) is installed. Each geosynchronous TDRSS satellite will have high bandwidth (up to 300 Mbits/s) transponders and will relay Landsat data to a single station in New Mexico. From there, the data will be transmitted to Goddard via Domsat and, after processing at Goddard, transmitted via Domsat to the EROS Data Center for product generation and archiving.

After reception at the EROS Image Processing Center, the data is archived on high-density tapes from which both laser-recorded photographic data and computer-compatible tapes (digital data) are generated for distribution. A digital browse file of selected catalog information, including cloud cover, is available through dial-up terminals.

The National Oceanic and Atmospheric Agency (NOAA) is also proposing to increase the resolution of the weather satellite scanners to render them more useful for earth resource studies. A high-quality photographic camera capable of providing 10-m resolution, and an imaging radar system will be tested on Space Shuttle flights. The Soviet METEOR satellite launched in 1980 carries low and medium resolution optical-mechanical scanners and a high-resolution solid-state multispectral scanner. The European SPOT satellite, to be operated by a commercial consortium, is scheduled for launching in 1984.

The combined Goddard and EROS facilities for processing Landsat data must surely be the highest volume civilian digital image processing facility in existence. During the 1980 fiscal year, for example, 26,000 different Landsat scenes were received and processed at the EROS Data Center. In spite of the potential value of digital image processing, however, it should be noted that current user requests at EROS run 30:1 in favor of photographic scenes against digital scenes (129,000 versus 4300 in 1981).

The digital images are system-corrected at the NASA/Goddard Image Processing Facility on the basis of ephemeris data received from the tracking stations, including position, attitude, and altitude. Radiometric correction based on on-board calibration sources removes sensor and digitizer anomalies. Geometric correction compensates for satellite altitude and attitude changes, slewing introduced by satellite motion, and the effects of the angle subtended by the satellite's field of view. The images are resampled to a Hotine Oblique Mercator projection. Resampling means that the gray value assigned to a pixel in the target image is a weighted average of the points corresponding to the target point in the source image.

Haze removal, an optional process, compensates for atmospheric scatter. Another optional process, contrast stretching, allows use of the full dynamic range for low-contrast images. Edge enhancement, which exaggerates the difference in intensity between a given pixel and a user-specified neighborhood (kernel), is also available. Users who prefer to avoid the loss of discrimination due to resampling may request only radiometrically corrected data; this results in some loss of positional accuracy [1].

3. THEORETICAL DEVELOPMENTS IN CLASSIFICATION

Most of the early work on partitioning satellite scenes into areas corresponding to pre-defined categories — for crop classification, land-cover maps, forest inventories — was based on classification according to spectral signatures. The last few years, however, have seen the development of techniques based on temporal signatures, which take into account changes in spectral reflectance between several observations obtained at different seasons of the year. Furthermore, local context, predicated on the idea that category changes between neighboring pixels are infrequent, has been used to advantage. Even more importantly, information from ancillary sources of data, such as terrain topography (elevation and slope) in forest classification, is now being taken into account. An ex-cellent review of progress in classification techniques through the last decade has been written by David Landgrebe, former director of the Purdue Laboratory for Applications of Remote Sensing (LARS) [2].

Since so much information is already available on what has been done (see also the proceedings of the Michigan Symposia on Remote Sensing, of the Purdue Symposia on Machine Processing of Remotely Sensed Data, of the William T. Pecora Symposia, to name a few, and the Journal *Remote Sensing of Environment*), this paper concentrates on what has *not* been done. Some of the items mentioned are well established in other disciplines; others are still somewhat esoteric. With a few exceptions, however, the ideas referenced below have not yet been applied to satellite remote sensing. The concepts discussed in the remainder of this section are related to classification.

3.1. Exploitation of Spatio-Temporal Context

A potentially far-reaching method of integrating the spatial, temporal, and spectral components of multiple observations of the same scene has been proposed by Haralick and Shapiro [3-5]. The consistent labeling procedure that they advocate is a hierarchic extension of decision theoretic classification. In addition to the customary conditional probabilities for each pixel given the class, a 'world model' is postulated that specifies, in either probabilistic or deterministic terms, the permissible spatial or temporal configur-ation of pixels bearing each class label.

Configurations that satisfy the category constraints of the world model are deter-mined by dynamic programming, Viterbi algorithm, relaxation, or other search tech-niques. World models may be specified by semantic networks, prediction rules, or Markov models. The process begins at the pixel level, combining pixels into homogeneous units, but these units are then assembled into higher-level units in the same manner.

The approach attempts to unify structural and statistical classification methods on the one hand, and pattern recognition and artificial intelligence techniques on the other. The transition from local to consistent global interpretation may correspond, in practice, to obtaining land use, eco-type, or habitat information from landcover in a systematic manner. Much remains to be worked out in detail, particularly with regard to the specification of higher-level constraints, before the method is likely to prove applicable to practical problems.

Levine and his colleagues [6–8] at McGill University approach scene analysis of color photographs of outdoor scenes from the point of view of computer vision and artificial intelligence. Their modular, rule-controlled systems are data directed and knowledge based. The knowledge base (rules) are in a permanent relational database, while the results of current operations, including hypotheses, are in a similarly organized short-term memory. Boundary and region analysis and relaxation labeling algorithms are used to partition the scene into interpreted components. These cooperative processes or algorithms operate at various levels in the system hierarchy. Although there is no unified theoretical formulation that encompasses the entire system, and the different components interact in such a complex manner that it may be difficult to find systematic means to improve performance, this approach represents a distinct departure from current operational methods of processing remotely sensed data.

The notion of 'spectral-temporal' trajectories is demonstrated by Wheeler and Misra [9]. A trajectory for a specific surface patch is characterized by the angles of the vectors between pairs of two-dimensional points in the brightness-greenness plane obtained at different dates. The results are, however, difficult to assess, since no ground truth measurements for the actual pixels was available.

The practical problems of integrating spatial and spectral information are discussed in an article by Landgrebe [10] that offers an unusual perspective on the development, over a 9-year period, of a set of pattern-recognition procedures ('ECHO') for image data obtained through satellite remote sensing. The alternatives available at the various stages of development are discussed and perspicacious arguments are advanced in support of the final approach. The presentation of half a dozen different criteria for the comparison of various algorithms is itself instructive, particularly considering that every criterion was evaluated in hundreds of separate experiments, each requiring the assignment of class labels to tens of thousands of observation vectors.

3.2. Imbedding Observations in Abstract Spaces

A bold attempt to unify structural and statistical pattern recognition is the abstract representation of Goldfarb [11]. Goldfarb's principal argument against the customary Euclidean representation of patterns is that, contrary to widespread misconception, Euclidean spaces cannot accommodate arbitrary measures of similarity between patterns.

Consider, for example, four patterns A, B, C, and D such that

$$d(A, B) = d(B, C) = d(A, C) = 1.00$$

and

$$d(A, D) = d(B, D) = d(C, D) = 0.55$$

The dissimilarity measure d actually satisfies the metric inequality, yet there is no Euclidean space, of any dimensionality, wherein the patterns can be embedded in such a way as to preserve the pairwise similarities.

Since the notion of vector representation has proved so important in pattern analysis, the question is whether there exists any type of vector space that preserves the original structure of the data as obtained through the notion of pairwise similarities – considered as the fundamental property – yet lends itself to the machinery of discriminant analysis, statistical decision theory, clustering, and other standard analytical tools of pattern recognition.

Goldfarb's solution is to imbed the patterns in a 'pseudo-Euclidean' vector space. A pseudo-Euclidean space is a real vector space on which is defined a symmetric bilinear form or generalized inner product. The measure of dissimilarity or 'distance' between two points in such a space can take on negative values and vectors may be orthogonal to themselves. A widely-known example is the Minkowsky space of special relativity.

Goldfarb develops the mathematical properties of this representation and demonstrates the correspondence necessary to implement the standard pattern analysis techniques on a given set of data. He argues that the approach accommodates mixed-mode observations, including structural or syntactic features, in a manner inherently impossible in Euclidean space. The appropriate pseudo-Euclidean space is defined by the relations of the patterns themselves, resulting in a parsimonious representation of minimum dimensionality. The addition of any new pattern may modify the space and affect the representation of all other patterns. If, however, the original sample size is sufficiently large, then a new pattern introduces no additional constraint and can be located in the existing space.

To the author's knowledge, these ideas have not yet been subjected to any significant experimental test. Goldfarb's dissertation does, however, include many suggestive examples.

3.3. Finite Sample Error Estimates

At the 1966 IEEE Workshop on Pattern Recognition, Louis Fein made a plea for 'Impotence Principles for Machine Intelligence', citing inspiring examples from other disciplines such as Heisenberg's Uncertainty Principle, the Postulate of Relativity, the Second Law of Thermodynamics, and Godel's Theorem [12].

A fine instance of an impotence condition in pattern recognition that may prevent some futile effort is a recent result by Devroye [13] to the effect that it is impossible to guarantee the finite-sample classification performance (i.e., the error rate based on a fixed training set) for any nonparametric discrimination rule. The rate of convergence of the probability of error to the asymptotic probability of error may be slower than that of any prespecified sequence. Thus knowledge of the asymptotic Bayes risk – which can be calculated for nearest-neighbor and k-nearest-neighbor classification in terms of the asymptotic Bayes risk, regardless of the distribution – does not tell us anything about finite-sample performance. Furthermore, putting restrictions on the distribution of the *unlabeled* samples is insufficient; some information about the class-conditional pattern distributions is required to estimate the finite-sample error rate.

There are, nevertheless, a number of methods for estimating the expected probability of error on new data using only the samples on the training data. Glick [14] compares a number of estimators, including the popular leave-one-out, bootstrap, and posterior probability estimators, with regard to bias (optimistic prediction because the classifier is tuned to the training set), variance (uncertainty of the estimate), robustness (influence of 'abnormal' samples), and computational cost. He argues that smoothing the data reduces both the bias and the variance of the estimator and advocates a smoothed modification of the sample success proportion. A still valuable bibliography on the estimation of error probabilities is that of Toussaint [15].

3.4. Efficient Nearest-Neighbors Classification

In 'difficult' classification problems, the optimal discriminant function between classes is generally neither a hyperplane nor a quadratic surface. An adaptive algorithm for piece-wise linear boundaries has been demonstrated long ago by Duda and Fossum, and there have been numerous attempts to cluster the patterns in each class in order to obtain more suitable discriminants. Nearest-neighbor and k-nearest-neighbor techniques have agreeable asymptotic properties and have been shown to yield excellent practical results, but require storage and computation proportional to the number of patterns in the training set [16].

It is clear, however, that the 'optimal' discriminant or boundary is influenced most heavily by the patterns close to the boundary; portions of the boundary that are far from any pattern are not critical to classification performance. This is precisely the weakness of clustering techniques, where patterns within each category are grouped without regard to their relations to patterns in other categories.

Toussaint, Bhattacharya and Poulsen summarize previous methods for thinning or editing the training set and propose a method based on the Gabriel graph (subgraph of the Voronoi graph or Thiessen diagram) for retaining only precisely those patterns that affect the nearest-neighbor decision boundary. A nearest-neighbor classification based on the retained points will thus yield exactly the same error rate on new data as nearest-neighbor classification based on the original training data. The preprocessing required is of order N log N, where N is the number of patterns in the training data [17].

This result is an example of possible contributions to statistical decision theory from studies in computational geometry, a current and exciting field of research with much to offer to remote sensing and geographic data processing.

3.5. Clustering

Our understanding of the manifold aspects of clustering or unsupervised classification — surely one of the most fundamental of cognitive operations — continues to grow. More than a dozen books specifically dedicated to clustering have appeared (see, for instance, Hartigan [18]), and new contributions continue to appear from diverse disciplines such as graph theory, statistics, linear algebra, and computational geometry. A number of global objective criteria have been advanced for comparing algorithms, but cluster validity remains an elusive concept that seems difficult to define except in terms of a specific application.

3.6. Size of Experiments

The size of experiments, in terms of pixels or acreage, has continued to grow throughout the decade, but the machinery to process entire Landsat frames is still available only in very few laboratories. The importance of large experiments lies in the fact that they render it increasingly difficult to hand-tailor the analysis procedure to the pecularities of the data set. Multitemporal experiments have also become more practicable with improvements in registration procedures. The verification of the results against independently obtained 'ground truth' remains a major difficulty.

4. RESEARCH GOALS IN MATHEMATICAL PATTERN RECOGNITION

Pattern recognition research currently sponsored by NASA is based on the recommendations of a task force commissioned by R. B. MacDonald, Chief Scientist for Earth Resources Programs. The task force was part of a larger study intended to define objectives for 'basic' research on Scene Radiation and Atmospheric Effects, Electromagnetic Radiation and Data Handling, and Information Utilization and Evaluation. The pattern recognition group was chaired by Professor L. F. Guseman, Jr., of Texas A and M University, and included ten scientists with extensive experience in remote sensing and formal training in mathematics, statistics, electrical engineering, physics, photogrammetry, or geography. After protracted discussion and presentations by a dozen invited specialists, recommendations were offered for research in three related areas: image registration, image representation, and classification.

4.1. Registration and Rectification

Image registration and rectification was considered by the participants to be a much neglected area of research where significant new results were necessary. Topics singled out for attention include the registration (overlay) of digital image arrays obtained at different times and possibly from different sensors (satellite MSS and RBV, radar, airphoto); the precise earth-location of images including consideration of topographic effects; the definition and automatic extraction of control points; the exploitation of USGS-prepared digital terrain models; the development of meaningful measures of accuracy for both registration and rectification; the relation between scale, orientation, photometric quantization, spatial sampling and coordinate system; and the need for increased understanding of the resampling process.

4.2. Image Representation

It was agreed that the central problem of image representation is the relation of observed features to the class of objects or items defined by an application-dependent taxonomy. Topics selected for further investigation are: texture information from mutliple (multispectral, multitemporal, multisource) images; the relation between spatial (shape, texture, topology) and spectral features; image segmentation techniques; the role and appropriate representation of ancillary (nonimage) information; syntactic ('structural') techniques, spatial context and temporal context; and the integration of nonimage information (atmospheric, illumination, and sensor correction) into the generation and definition of primitives in application-independent scene models. These topics are often included under the heading of image restoration. Also sought is the extension of spectral dimensionality-reduction techniques to spatial and temporal dependencies.

4.3. Classification

The incorporation of previously defined digital image representations into systematic methods of determining the required attributes of object scenes is the subject of classification. The primary objectives of the classification process are considered to be mapping, inventory, and monitoring of natural resources. 'Mapping' shows the location of classes, objects, items, or other types of interest; it includes both hardcopy and interactive display. 'Inventory' is concerned with counting, aggregation, census, or planimetry of items without specific retention of spatial coordinate information. 'Monitoring' refers to change detection, discovery of unusual conditions, and other operations of limited spatial and temporal scope.

Classification includes concepts such as categorization, identification, recognition, clustering, partitioning, taxonomy, and segmentation. Of concern are supervised and unsupervised learning, teaching, or training; estimation of parameters, distributions, and error rates; the assignment of identities, labels, or symbols by either automatic or interactive means; and the general evaluation of the accuracy, dependability, and robustness of the entire process. Of particular interest is the role of the human and the contributions extracted from ancillary data. Techniques based on statistical as well as structural, syntactic, relational, and other deterministic approaches are germane. Algorithms need to be developed for multisource data, including multisensor observations, multitemporal observations, and combinations of multi-image data with nonimage data.

When classification is not performed in a single step, the intermediate variables are called features, components, signatures, dimensions, transformations, factors, primitives, characteristics, or measurements. Methods must be developed for obtaining these intermediate variables from the digital images and ancillary data and for incorporating them into the classification process.

The information sought by the end-user may be in a form different from the simple, nonoverlapping, mutually exclusive and totally exhaustive model provided by standard pattern-recognition texts. An example of the fuzzy taxonomies of possible interest is the class of grizzly-bear habitats, for which a complete specification may not even conceptually exist.

Considerable attention was devoted to mathematical techniques of proportion estimation. Among approaches deemed worth pursuing are enumeration through classification, stratified area estimators, regression estimators, and direct estimators. Further progress is dependent on the development of algorithms which require only a small number of training samples, can deal with a large number of object classes, are responsive to non-stationary distributions, and can account for 'mixed-pixel' measurements resulting from finite sensor resolution.

4.4. Other Tools

In contradistinction to map displays or statistical inventory information which forms the final product of the classification process and benefits the 'end-user', data displays are intermediate products intended to improve the classification process itself. Specifically, they provide the opportunity for human interaction. The scope of the displays ranges from simple histograms, which allow judgment of the overlap between statistical distributions, to digital images providing photointerpreters a means of assigning labels to representative samples.

Data structures, data compression techniques, and special parallel computer architectures are of interest to the extent that they impact the classification process. Among data structures to be investigated are pixel-by-pixel storage, bit-plane structures, vector (polygon) methods, chain encoding, contour coding, hierarchical pyramid and quadtree structures, and various two-dimensional polynomial approximations. Architectures to be investigated (though at a relatively low priority) are pipeline, multiple-instruction single-data-stream, and multiple-instruction multiple-data-stream machines. Given the relative economies of special-purpose VLSI chip development and the rapidly decreasing cost of general purpose processors (particularly bit-slice architectures), it is expected that networks of processors, possibly with common memory access, will predominate. The development of adequate operating systems for these configurations is, however, a monumental task that cannot be borne by the remote-sensing community alone. In the expectation that parallel machines will be available in the next decade, however, increased attention to models of computation not based on the single-instruction, single-data-stream model is recommended.

4.5. Caveat

This summary represents only the perceptions of one of the participants of the Working Group on Mathematical Pattern Recognition and may differ in substance and emphasis from the final report of the Group. The official report of the Task Force (a document of about 200 pp.) is available from the NASA Earth Observation Directorate, Johnson Space Center, Houston, Texas.

5. OTHER DEVELOPMENTS

In contrast to the above, the items mentioned in this section are broadly applicable to various aspects of digital data processing in remote sensing applications, not just classification.

5.1. Computational Complexity

Computational complexity is a central conception in computer science and some notions may be worth summarizing here. The basic ideas are derived from the theory of formal languages where one wishes to count the number of elementary operations required by an automaton for accepting or rejecting a sentence depending on whether or not it is part of a predefined language. To apply this notion to an optimization problem, it must first be transformed to a decision problem with a 'yes' or 'no' answer. This is always possible. Important classes of applications in pattern recognition where computational complexity is important are clustering and unsupervised learning.

A major result of the past decade is the recognition that there is a class of significant problems which can be solved using a nondeterministic model of computation in a time proportional to the number of data points raised to a fixed power, but which 'almost certainly' cannot be so solved using a deterministic model. This class of problems is called NP-complete. If a solution based on a deterministic model of computation were found for one member of the class, then a solution would be known for all such problems. Thus the preferred tool for showing that a given problem is NP-complete is to demonstrate that it is computationally equivalent to a known member of the class. Examples of NP-complete problems are the travelling salesman problem and the knapsack problem [19].

More recently, it has been shown that even if the worst-case situation requires an exponentially large number of operations for solution, for an arbitrary set of data points the correct solution may be obtained in polynomial time except for a very small number of specific cases. The probability of achieving the correct solution for a randomly selected example is thus arbitrarily close to 1. Furthermore, in some situations even if the optimal solution cannot be obtained in reasonable time, an almost-optimal solution can be obtained.

The notion of a model of computation has been refined to allow accurate comparison between algorithms developed in different notations, but the choice of model does not affect the question of NP-completeness or other asymptotic results. Models of parallel computation are playing an increasingly important role as processes which were once considered to be of an essentially sequential nature, such as bounded searches of game trees, are implemented on networks of processors.

In computational geometry, three kinds of 'concrete' complexity are usually defined: storage complexity, run-time complexity, and preprocessing complexity. In supervised pattern classification, 'storage complexity' would refer to the storage requirements of the classification algorithm rather than of the training algorithm; 'preprocessing complexity' is the number of operations required to design the classifier.

5.2. Software Engineering

Software engineering is the methodical specification, design, development, testing, and documentation of large programs. It extends over the entire life-cycle of a program, from conception to last use. Useful developments include management techniques (programming teams, chief programmer), notational techniques (pseudo-code, HYPO-charts, display layouts), computer aids (the Programmer's Workbench, PSL-PSA), and stylistic programming rules (modular programs, indentations, naming conventions). Programs developed under such discipline are, it is widely agreed, easier to understand, debug, modify, and transport than programs written in a highly individualistic manner concerned mainly with machine efficiency. For the majority of current computer-science graduates such techniques are second nature and good programming techniques will eventually permeate the remote sensing community as they already have permeated much of the business world. 'New' languages, such as PASCAL and ADA, will help in this endeavour [20].

Software Physics, developed largely by Kolence and his disciples, is the calculus of computer performance evaluation. It deals with concepts such as capacity, power, and work related by equations analogous to those of elementary physics and applied to the central and peripheral (channel) processing units and to the input/output devices. In view of the extraordinarily high data volumes necessary for processing satellite image data, the continued monitoring of efficient device utilization and the consideration of alternative computer configurations is essential.

Software Science, on the other hand, is an attempt at the quantification of the world of computer programs. Many experiments based on the seminal work of Halstead demonstrate its predictive value in relating 'language level', counts of operators and operands in a program, and program length, to the time and effort necessary to develop a program and to the number of residual bugs after different stages of testing. As image processing and pattern recognition increasingly draw on ancillary information, the complexity of the necessary software is likely to increase immensely. Tools drawn from software science will consequently assume increasing importance.

5.3. Small is Beautiful

In the late 1960s some image processing groups shifted to minicomputers, and in the late 1970s they began shifting to microcomputers. It is increasingly recognized that intelligent input/output devices, such as displays, plotters, and optical scanners, can substantially reduce the load on the central processor; in fact, the central processor has lost its centrality. Distributed processing systems, where the processing power can be more flexibly balanced against the load, are gradually appearing.

There are, however, more than 40 major manufacturers of microprocessors in the United States alone. Because of the high cost of software development, most processors support only a very limited number of languages, and image input/output operations must still often be programmed in a low-level language. Consequently, software portability remains as much of a problem as ever. Furthermore, in mix-and-match academic installations depending largely on student maintenance, reliability is almost impossible to ensure.

5.4. Interactive Methods and Human Factors

Software psychology is the study of the factors that affect the understanding, creation, debugging, and modification of computer programs. It also includes display and dialog design. At the Human Factors in Computing Conference in Gaithersburg in April 1982, attendance exceeded the expected 200 by more than fourfold. The development of experimental techniques and a body of observations on man-machine interaction is occurring simultaneously with increased interest in interactive techniques in the analysis and utilization of remotely sensed data. High-quality image display and image entry devices are still extremely expensive, but perhaps by the time they become generally available we will develop a consensus on how to use them effectively.

5.5. Image Databases

The last 5 years have seen a number of attempts to extend database techniques to images. In some projects the images are stored as integral entities that can be inspected or retrieved but not modified. Indeed, tools for image modification are still essentially nonexistent, except for overlays. Nor has there been much software developed to combine images and geographical information from other sources; it is, at best, a laborious effort [21].

It is this author's conjecture that real progress will come only when commercial database systems can be expanded to two-dimensional applications. The overhead costs of developing communications protocols, back-up and archiving facilities, privacy and security measures, query languages, I/O interfaces, programming language interfaces, adequate maintenance and user documentation, and so on, are just too high in proportion to the restricted volume of geographic applications. Furthermore, even if the bulk of the storage available must be reserved for the image components, for most applications it will be essential to provide appropriate facilities for textual and formatted alpha-numeric data.

At this time most installations can keep only a very limited amount of image data 'on line'. Video disk technology may eventually provide a more economical storage medium than magnetic disk.

5.6. New Computer Architectures

The race between general purpose sequential computers and special purpose parallel architectures for image processing has been on for more than 20 years. Because of the cost of manufacturing processors using discrete components, the early machines exhibited only a relatively low degree of parallelism. Current VLSI designs may process thousands of pixels simultaneously. The recent resurgence of experimental parallel systems for image analysis is cogently reviewed by Danielsson and Levialdi [22]. In spite of the abundance of ideas and even of commercially available processors, most installations still depend largely on old-fashioned uniprocessors whose cost per operation continues to decrease dramatically. Although image-processing operations have been embedded in extensions of standard high-level programming languages, lagging software development and lack of uniform interfaces to standard processors appear to continue to retard the wide-scale acceptance of specialized systems.

6. CONCLUSION

Present successes and expected advances in remote sensing, telecommunication, data storage, and information extraction technologies lead us to believe that the next 30 years hold an exciting promise: no less than a complete inventory of the world's major resources. Assessment of food and fiber crops, forage biomass, forests, wetlands, littorals, deserts, oceans, climate, weather, and demography will all be affected by this vast enterprise. The inventory process will not be a simple operation of counting pixels in satellite images. It will, instead, draw on increasingly sophisticated automated techniques to enhance the methods and models in present use by discipline scientists through the incorporation of remotely sensed data. Between the promise and its fulfilment, however, there remains plenty of hard work for us all.

ACKNOWLEDGMENT

The support of the University of Nebraska Conservation and Survey Division through NASA Grant No. 28-004-020 is gratefully acknowledged. Part of this material has appeared in References 23 and 24.

REFERENCES

1. *Landsat Data Users NOTES*, U.S. Geological Survey, EROS Data Center, Sioux Falls, SD 57198.
2. Landgrebe, David A., 'Analysis Technology for Land Remote Sensing' *Proc. IEEE 69*, No. 5, May 1981, pp. 628–642.
3. Haralick, Robert and L. Shapiro, 'The Consistent Labeling Problem I' *IEEE Trans. Pattern Analysis and Machine Intelligence 1, #2*, March 1979, pp. 173–184.
4. Haralick, Robert and L. Shapiro, 'The Consistent Labeling Problem II' *IEEE Trans. Pattern Analysis and Machine Intelligence 2, #3*, May 1980, pp. 263–313.
5. Haralick, Robert M., 'Decision Making in Context' *IEEE Trans. on Pattern Analysis and Machine Intelligence*, to appear, 1982.
6. Levine, Martin and S. Shaheen, 'A Modular Computer Vision System for Picture Segmentation and Interpretation' *IEEE Trans. Pattern Analysis and Machine Intelligence 3, #5*, September 1981, pp. 540–556.
7. Zucker, S. W., E. V. Krishnamurthy and R. L. Haar, 'Relaxation Processes for Scene Labeling: Convergence, Speed, and Stability' *IEEE Trans. on Systems, Man and Cybernetics 8*, 1978, pp. 41–48.
8. Shaheen, S. I. and M. D. Levine, 'Experiments with a modular computer vision system' *Pattern Recognition*, to appear.
9. Wheeler, S. and Misra, P. N., 'Crop Classification with LANDSAT Multispectral Scanner Data II' *Pattern Recognition 12, #4*, 1980, pp. 219–228.
10. Landgrebe, David, 'The Development of a Spectral-Spatial Classifier for Earth Observational Data' *Pattern Recognition 12, #3*, 1980, pp. 165–176.
11. Goldfarb, Lev, *A New Approach to Data Analysis*, Ph.D. Dissertation, University of Waterloo, Department of Systems Design, 1980 (to be published as a monograph); see also Wong, A. K. C., and L. Goldfarb, 'Pattern Recognition of Relational Structures', in *Pattern Recognition Theory and Applications* (J. Kittler, K. S. Fu, and L. F. Pau, eds.), D. Reidel Publishing Company, Dordrecht, Holland, 1981, pp. 157–176.
12. Fein, Louis, 'Impotence Principles for Machine Intelligence' in *Pattern Recognition* (L. Kanal, ed.), Thompson Book Company, Washington, 1968, pp. 443–447.
13. Devroye, Luc, 'Any Discrimination Rule Can Have an Arbitrarily Bad Probability of Error' *IEEE Trans. Pattern Analysis and Machine Intelligence 4*, March 1982, pp. 154–157.
14. Glick, Ned, 'Additive Estimators for Probabilities of Correct Classification' *Pattern Recognition 10*, 1978, pp. 211–222.
15. Toussaint, Godfried, 'Bibliography on Estimation of Misclassification' *IEEE Trans. Information Theory 20*, 1974, pp. 472–479.
16. Cover, Thomas and P. E. Hart, 'Nearest Neighbor Pattern Classification' *IEEE Trans. Information Theory 13*, January 1967, pp. 21–27.
17. Toussaint, Godfried, 'Pattern Recognition and Geometrical Complexity' *Proc. Fifth Int'l Conf. on Pattern Recognition*, Miami Beach, December 1980, pp. 1324–1346.

18. Hartigan, John, *Clustering Algorithms*, John Wiley and Sons, New York, 1975.

19. Horowitz, E. and S. Sahni, *Fundamentals of Computer Algorithms*, Computer Science Press, 1978.

20. Wasserman, Anthony and S. Gutz, 'The Future of Programming' *Comm. ACM 25,* #3, March 1982, pp. 196–206.

21. Zobrist, Al and G. Nagy, 'Pictorial Information Processing of Landsat Data' *IEEE Computer 14,* #11, November 1981, pp. 34–42.

22. Danielsson, Per-Erik and S. Levialdi, 'Computer Architectures for Pictorial Information Systems' *IEEE Computer 14,* #1, November 1981, pp. 53–67.

23. Nagy, George and S. Wagle, 'Geographic Data Processing' *ACM Computing Surveys 11,* #2, June 1979, pp. 139–181.

24. Nagy, George, 'Advances in Information Extraction Techniques' *1982 Machine Processing of Remotely Sensed Data Symposium Proceedings*, pp. 550–556.

G. Nagy
Department of Computer Science
University of Nebraska
Lincoln

8 Current problems in astronomical image processing

G Sedmak

1. INTRODUCTION

Astronomy is an observational science. The basic stages of astronomical image processing include removal of the instrumental signature, processing aimed at visual inspection of the observations, and application of oriented processing aimed at any specific set of scientific tasks. The fine "astronomical" images shown in Figs. 1 and 2 are typical examples of images that astronomers do not usually process. The typical astronomical image that is used in practice is shown in Fig. 3 with all its problems.

To achieve optimum information recovery, in general the basic processing stages are not separated owing to the characteristic of astronomy to use any available methodology at its technological limit, since it is this limit that will define the current astronomical horizon. In time, this has resulted in an enormous amount of scattered, highly dedicated, poorly flexible and sometimes unprofessional solutions to contingent astronomical problems, implemented through parallel efforts by astronomers to interface and use available techniques developed for different image processing applications. Such was the situation, with a few brilliant exceptions, in the 1970s and early 1980s, when the increasing quantity of high-quality data, generated by remote ground-based and space-borne telescopes, forced the astronomical community to adopt a more integrated and cooperative approach. Astronomers are now calling for a new generation of high-standard astronomy-oriented image processing to be realized on the basis of advanced image processing science.

The present review gives a very concise summary of the current problems in astronomical image processing as seen by the astronomer in terms of observations, instruments, and methods. It is clear that any review of this type cannot be other than a partial, biased cross-section of astronomy and image processing science. The main purpose of this review is to foster the identification of selected requirements in astronomical image processing that could obtain the greatest advantage from cross-fertilization with image processing science.

2. ASTRONOMICAL IMAGES

The astronomical images presently employed show a tremendous variety of types, formats, supports, and standards. It is possible to identify a few basic types of images forming an approximate standard set for astronomical applications by cross-reference

Fig. 1. Saturn observed by US Voyager space probe. Typical astronomical image not used by astronomers.

Fig. 2. Mars soil observed by US Viking space lander. Another type of astronomical image not used by astronomers but of high interest to planetary scientists.

Fig. 3. The Large Magellanic Cloud. This image can be taken as representative of the images processed currently by astronomers. It includes all typical problems.

with astronomical observations and available technologies. The basic astronomical observations are the following:

Zero dimensional: Point source astrometry and photometry.
One dimensional: Point source spectrometry.
Two dimensional: Objective prism spectrometry.
　　　　　　　　　Astrometry and field photometry.
　　　　　　　　　Surface photometry.
Three dimensional: Spectrophotometry.

The observations are carried out in a spectral range extending from gamma-rays to metric radio wavelengths including exotic tests on neutrinos and gravitational waves. The constraints set by the levels of sources and background and by the technologies available

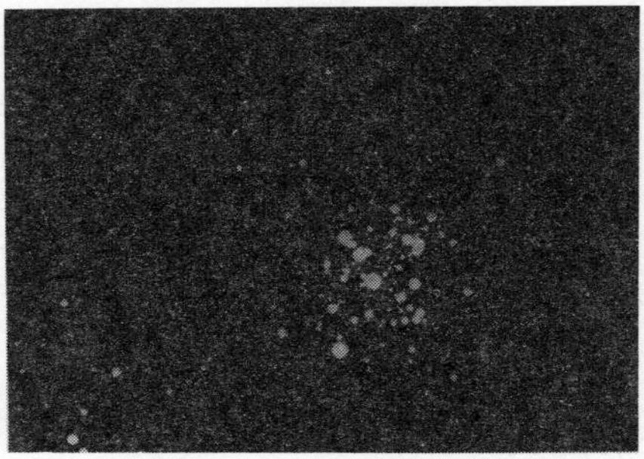

Fig. 4. Open star cluster NGC3293.
This sparse set of stellar point sources is
dominated by the observational PSF. The
sky background is clearly visible around
the well separated stellar sources.

Fig. 5. Globular star cluster NGC6205.
A dense set of stellar point sources that
overlap at the centre of the image into
a strong continuum-like background.

Fig. 6. Ring Nebula NGC6720. An
example of simply-structured extended
object not dominated by the observational
PSF. The segmentation of such an image
is relatively easy.

Fig. 7. Orion Nebula NGC1976. This
image is representative of extended
objects with an irregular shape that does
not show any evidence of geometrically
regular patterns but a rough radial
concentration.

for astronomical observations lead to a rather consolidated scene in the UV to near-IR and radio range, while the technological situation is rapidly evolving in gamma-ray, X-ray, and IR astronomy.

The typical observational scene in astronomy is summarized by the images shown in Figs. 4–20. The two main classes of objects — those limited and those non-limited by the observational point spread function (PSF) — are shown as is the great variety of shapes of all extended objects.

The technologies available show up the following basic types of astronomical images listed according to increasing wavelength of observation:

Small-to-medium format, up to 1000×1000 pixels at 8 bit/pixel, two- and three-dimensional X-ray images from coded mask and grazing incidence X-ray telescopes and spectrometers. Gamma-ray images are included many times in this format [1, 2].

Fig. 8. Centaurus A source. A very complex and well-known astronomical source. The many field stars superimposed to the object are able to distort completely the faint features of the halo in far regions.

Fig. 9. Monoceros region. This image can be taken as a summary of all astronomical image processing problems: instrumental aberrations, strong field sources, varying background, and very irregular shapes.

Fig. 10. Elliptical galaxy NGC4406. A typical extended object of elliptical shape in a clean region of the sky. The useful intensity range of the galaxy image can exceed four decades.

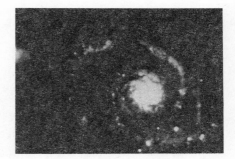

Fig. 11. Spiral galaxy M51. The spirals of the arms are evident but show very poor continuity in structural sense. An example of the problems presented by astronomy to image processing science.

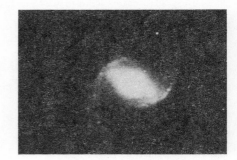

Fig. 12. Spiral galaxy NGC4548. The measurement of the complex nucleus of this image calls for an accurate interpolation of the background into the nuclear region. This problem is presently not fully solved.

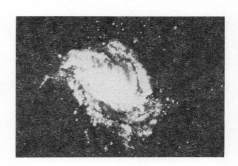

Fig. 13. Complex spiral galaxy NGC5236. A typical example of the reasons why two-dimensional segmentation can be very difficult in astronomy: strong, varying background, irregular shapes, and morphological discontinuities.

Fig. 14. Encountering galaxies NGC4567–8. Two relatively simple objects interact and generate an image very difficult to be analyzed in terms of the two original images.

Fig. 15. Cluster of galaxies in Coma. Many galaxies comparable in size to stellar objects are present in the field: a typical input for an astronomy-oriented discrimination and classification procedure.

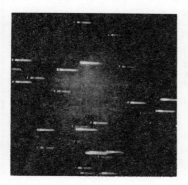

Fig. 16. Objective prism spectral image of an uncrowded stellar field. Each star contributes one single spectrogram. The regular y-shape of the spectra was obtained by suitable control of the telescope guiding system.

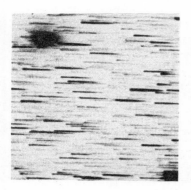

Fig. 17. Objective prism spectral image of a crowded stellar field. The individual spectrograms overlap together in all axes. The shape of the spectra is dominated by the observational PSF.

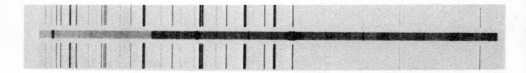

Fig. 18. High-resolution grating
spectrogram of the star CH Cygni in
H-Beta region. Original dispersion 12
Å/mm. The upper and lower spectra are
used for wavelength calibration.

Large format, up to 80,000×80,000 pixels at 8 bit/pixel, two-dimensional photo-graphic plates from high resolution spectrometers, objective prism and standard imaging telescopes in the UV to near-IR spectral range. The plates are digitized to numerical arrays by a digital microdensitometer [3, 4].

Medium format, up to 2000×2000 pixels at up to 16 bit/pixel, two-dimensional photoelectric detectors for the same applications as the photographic plate plus three-dimensional spectrophotometry and high time resolution applications such as speckle analysis [5–7].

Small format, up to 128×128 pixels at 8 bit/pixel, IR to far-IR images by coded mask telescopes and spectrometers and solid-state IR arrays [8, 9].

Small-to-medium format, up to 512×512 pixels at 8 bit/pixel, radioimages from synthesis radiointerferometers (which do not uniformly sample the Fourier con-jugated radio sky) [10, 11].

The basic image types listed above are reduced in practice to an average format on the order of 1 Mpixel because the large photographic plates of 10,000 Mpixels can generally be considered as collections of several smaller images.

The hardware and software for astronomical image processing should, therefore, be committed to managing an astronomy standard format of typically 1000×1000 pixels at 8–16 bit/pixel possibly extracted from images of a larger format taken as archives.

What cannot be standardized at all is, of course, the set of requirements presented by the technology of each particular detector for the image processing aimed at the removal of the instrumental signature. This matter lies beyond the scope and limits of the present review. However, it is useful to identify the standard calibration procedures methodologically common to all technologies, which are:

Removal of fixed background patterns.
Correction of defective pixels.
Calibration of geometrical transfer function.
Calibration of intensity transfer function.
Calibration of spectral response.

The list of calibrations given above is substantially application-independent. Taking into account the detection and noise models and the instrumental PSF is generally considered

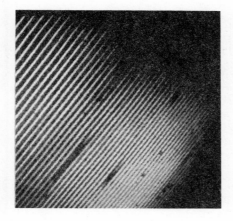

Fig. 19. Ultraviolet echelle spectrogram from IUE satellite. A subimage of the original 768×768 pixels spectrum is shown on a 512×512 pixels display. The spectral orders are partially overlapped and the recovery of the information is background-dependent.

application-dependent processing, which is determined quantitatively by parameters estimated through the calibration procedures.

The definition and operation of the calibration procedures is now going on more and more under the responsibility of the team that provided the design, realization, and maintenance of the instrumentation. This situation is practically unavoidable, owing to the increasing complexity of astronomical instrumentation, but recent experience shows that a high degree of feedback from users is necessary to achieve a suitable calibration quality and minimize operational errors.

3. HARDWARE AND SOFTWARE FOR ASTRONOMICAL IMAGE PROCESSING

The development of hardware and software systems for astronomical image processing started in the 1970s and continued with low convergence until the present availability of low-cost virtual-memory 32-bit computers, large solid-state memories, and very large disc mass memories. Now the computer hardware scene looks mature and this situation should remain stable until computer systems and programming languages with new architectures are available at a competitive cost to performance ratio and marketing price.

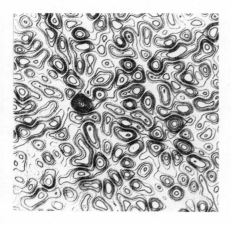

Fig. 20. Raw radioimage of the double radio source 3C 244.1 from synthesis two-dimensional radiointerferometry. The two components are very confused owing to the artifacts generated by restoring an undersampled image.

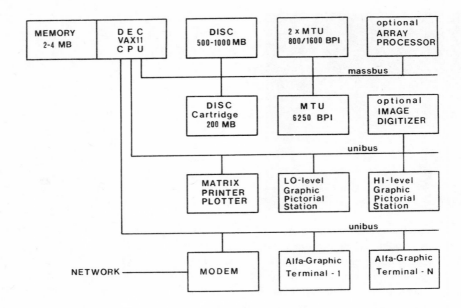

Fig. 21. Astronomy-standard image
processing hardware based on 32-bit
computer with virtual memory capability
and graphic and pictorial interactive
environment.

The situation is still not fully mature in the field of graphic and pictorial subsystems in spite of the very high number of different devices available on the market. The raster scan technology is now a consolidated standard for relatively large formats of up to 2000×2000 pixels and more, but there is no clearly identified approach on how and where, and at what level of accuracy, the high-speed operations needed in interactive image processing for display have to be executed. Trade-off studies are now going on to investigate the pros and cons of intelligent versus non-intelligent stations and the use of special hardware, such as an array processor, shared between many stations.

The core of the problem consists in the image processing rate expected in astronomy in the 1990s: the increasing rate of astronomical data processing averaged over the whole astronomical community looks to be more due to the increase in the number of users rather than to the processing rate per user, at least on medium-sized astronomical systems. If this trend is confirmed, the flexibility of a distributed multiuser environment will possibly outrun the performance of single-user graphic and pictorial stations of even outstanding performance.

The configuration of a typical astronomical image processing hardware system built in line with this philosophy is shown in Fig. 21. Such a system may be considered as the astronomy standard for image processing hardware for the 1980s, at least at the European level.

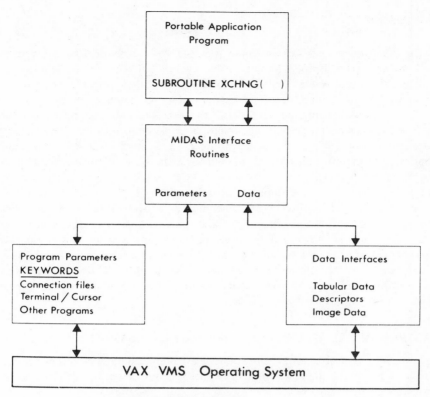

MIDAS User Interface

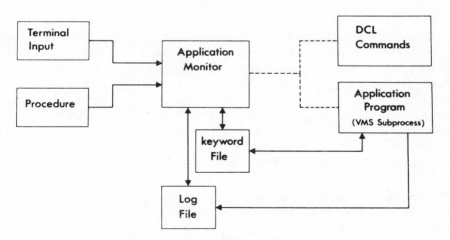

Command line structure : command/modifier   parm 1, ... n

Fig. 22. ESO MIDAS: an astronomy-standard modern software environment for imagery.

On the software side an emerging standard for astronomical image processing is nearly mature and will probably be available by 1983. It consists of a standard software environment hosting several application modules. The environment is based on a structured control language with user-definable interactive and batch capabilities, an optimized data structure emphasizing the potential of the virtual memory mapping, and a substantially device-independent graphic and pictorial subsystem with interactive, visual and hard-copy facilities. The software environment is such as to allow self-documenting operations and modifications and to present a standard set of interface requirements for the application modules, generally written in FORTRAN. The typical software environment is shown in Fig. 22. Complete astronomical image processing systems consistent with the hardware and software guidelines quoted above are, or are going to be, made available by 1983 by NASA, ESA, ESO, and some national institutions such as STAR-LINK in England and ASTRONET in Italy [12–16]. The general form of these systems should be that of a network of nodes at the main local concentrations of users. The nodes are connected by medium-speed data links so as to allow distributed circulation and administration of documentation and low volumes of data while maintaining software standardization in a growing environment.

4. ASTRONOMICAL APPLICATIONS OF IMAGE PROCESSING

A cross-section by problems of the list of basic astronomical observations and image types reported in Section 2 allows us to identify the basic set of operations involved in astronomical image processing, which are:

Presentation of data for visual inspection.

Deconvolution of observational PSF.

Recovery and segmentation of the spectrum of one source from the associated raw spectral image.

Detection and segmentation of PSF-limited objects in the presence of non-stationary background and noise and in crowded fields.

Detection, segmentation, and classification of extended objects blurred by observational PSF in the presence of non-stationary background and noise and superimposed field objects.

Detection and restoration of images from non-equisampled two- and one-dimensional interferograms.

The operational environment common to these basic tasks is characterized by large to very large volumes of multidimensional data containing sparse to very sparse subsets of relevant data. The practical management of astronomical data thus calls for two further basic tasks, which are:

Data compression.
Experimental databases.

The tasks given above are implemented through so large a variety of methods, algorithms, software, and hardware facilities that it is clearly impossible to list them in detail here.

However, all operations are based on a limited number of fundamental techniques concerned with the physical nature of the problem under consideration.

The standard image presentation functions are the following:

Display.
Rototranslation.
Scaling.
Histogram manipulation.
Arithmetics.
Filtering for enhancement.

These functions are implemented by well-known and consolidated methods that do not need any further review here.

The fundamental problems are the following:

Image deblurring.
Transversal mapping.
Background estimation.
Data compression.
Statistical classification.
Pattern recognition.
Restoration of undersampled images.

These fundamental problems have been faced by many sciences and several solutions, with very different degrees of final quality, have been worked out up to now. The next section gives a concise and unavoidably partial summary of the best-known methods used in astronomy.

5. IMAGE PROCESSING METHODS USED IN ASTRONOMY

5.1. Deconvolution of Observational PSF

Deconvolution of observational PSF should be considered necessary for any astronomical image processing owing to unavoidable seeing and/or instrumental effects, but it leads to an independent problem in all cases where the PSF dominates the information to be recovered. Typical problems are the measurement of stellar radii, the separation of multiple stellar systems, the detection of external planets, the recovery of blended spectral lines, and so on.

While the outstanding importance of deconvolution of observational PSF is clearly recognized in all fields of astronomy, the application of deconvolution is limited in practice on account of the difficulty of obtaining algorithms that give stable results at the signal-to-noise ratio and morphological situations typical of astronomy. The techniques used include classical Fourier deconvolution [17], sometimes implemented by least-squares-fit methods [18], maximum entropy restoration [19], statistical deconvolution by clustering and Bayesian estimation [20, 21], and speckle imaging [22, 23].

Fourier deconvolution is well known, and is the most critical of all with regard to the accuracy required in the definition of the PSF and the signal-to-noise ratio. It shows

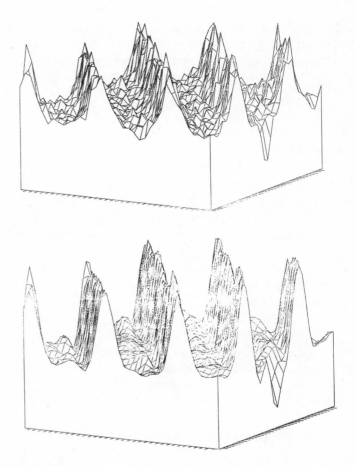

Fig. 23. Sample Fourier deconvolution
of part of an IUE echelle spectrogram.
Upper, original image. Lower,
deconvoluted image. The algorithm
included a low pass filter to ensure
stability at high frequencies.

Fig. 24. Sample maximum entropy
restoration. The object is the galaxy M87
with its jet. Left, original image. Right,
restored image. The effects of the
restoration are evident in the shrinking
of the nucleus of the galaxy and in the
increase of the signal-to-noise ratio of
the jet feature against the sky background.
Notice also the strong decrease of the
halo around the nucleus.

the maximum rate of spurious artifacts in the deconvolved image. However, careful use of this technique yields fairly good results, as shown in the sample deconvolution of a subimage of an IUE echelle spectrogram in Fig. 23.

Maximum entropy restoration is comparable in results to Fourier deconvolution in practical cases. The real advantage of the maximum entropy algorithm as against Fourier deconvolution consists in the lower rate of artifacts, obtained at the expense of a longer computing time. An example of maximum entropy restoration is shown in Fig. 24.

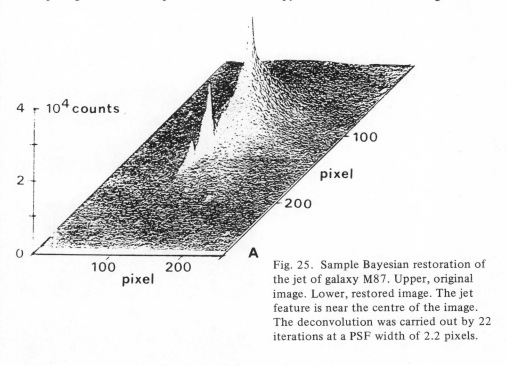

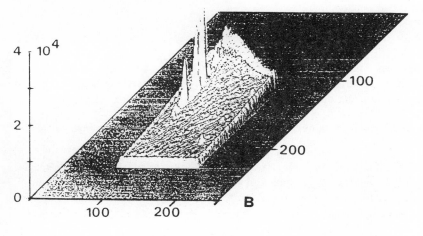

Fig. 25. Sample Bayesian restoration of the jet of galaxy M87. Upper, original image. Lower, restored image. The jet feature is near the centre of the image. The deconvolution was carried out by 22 iterations at a PSF width of 2.2 pixels.

Statistical deconvolution techniques, well known in the field of physics, now look promising for classical astronomy also. The great advantage of statistical deconvolution consists in the stability of the algorithm at low signal-to-noise ratio. The clustering approach has been used successfully to process gamma-ray images, while Bayesian deconvolution has been tested successfully on optical images, as shown in Fig. 25.

Speckle imaging is the last technique applied to extreme astronomical cases. The classical approach by Labeyrié made it possible to separate a binary star by the auto-correlation of the average of several two-dimensional Fourier spectra of individual speckle interferograms. Present numerical techniques on digital images make possible the implementation of speckle holography by isoplanatic patch cross-correlation of the raget and template images with results as shown in Fig. 26.

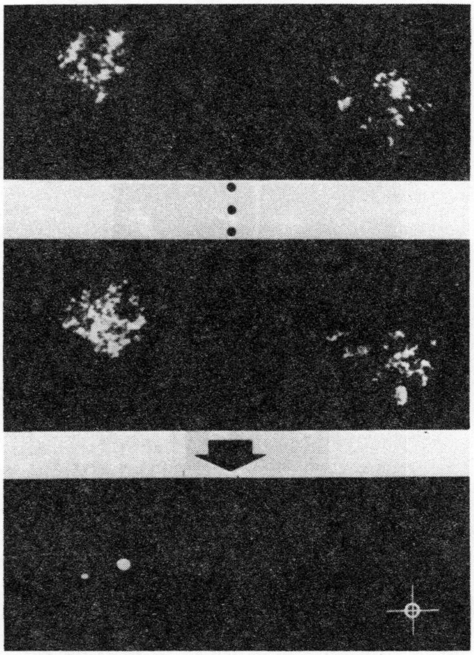

Fig. 26. Sample speckle holography of a binary star. Upper and centre, speckle of the binary star, at the left, and of a star used as template, at the right. Lower, the images of the binary star and of the template after restoration. The final image used 600 speckle interferograms. The separation of the binary star, Z Cancri, is about 0.8 arcsec, comparable to seeing figure.

These techniques have been proposed for application to space telescope images with an expected spatial resolution of 0.0015 arcsec at 1400 Å wavelength, one order of magnitude below the diffraction figure of the 2.4 m aperture of the space telescope.

5.2. Recovery and segmentation of one-dimensional spectra from two-dimensional raw spectral images

The recovery of one-dimensional spectra from two-dimensional raw spectral images, generally of optical sources, is implemented by very straightforward techniques in all cases where single source spectrometry is performed using standard prism or grating spectrographs at moderate resolution.

Fig. 27. Recovery of a one-dimensional spectral segment from an echelle spectrogram by direct wavelength mapping. Upper, the original echelle image of 3000×3000 pixels format scaled to a display format of 512×512 pixels. Centre, a sample of the order selected on the original image, showing the reference spectrum in its upper part and the stellar spectrum in its lower part. Lower, the processed spectrum. Notice that the y-axis of centre and lower images are the same, while the x-axis of the final image is wavelength instead of length.

The recovery of one-dimensional spectra from echelle spectrograms, ultrahigh resolution grating spectrometers, and all types of distorted spectrograms, like those obtained through image intensifiers, requires special techniques for mapping the two-dimensional raw spectral image into a one-dimensional calibrated spectrum. An independent problem is, finally, the recovery of individual one-dimensional spectra from an objective prism image containing many individual and possibly overlapping spectra.

The best approach to recovery of a one-dimensional spectrum from its raw image is the direct wavelength mapping of the raw domain $R(x, y|\lambda)$ onto the calibrated domain $I(\lambda)$. The parameters of this non-linear transversal mapping are usually inferred from the information contained in the raw spectral image of the reference spectrum of the source used for wavelength calibration. The direct wavelength mapping operates as shown in Fig. 27. It should be emphasized that the simple-looking result shown in Fig. 27 follows from a very complicated determination of the set of parameters that characterizes the mapping. These parameters depend on a variety of observational and instrumental figures, such as source alignment, detector temperature, and so on [24, 25].

The recovery of single one-dimensional spectra from a two-dimensional raw image containing many individual spectra is limited to the case of objective prism spectra of low to very low spectral resolution. The astrophysical importance of objective prism

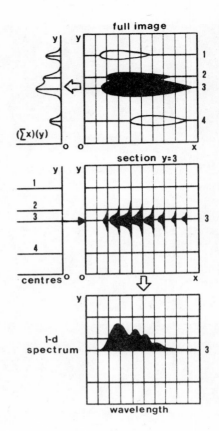

Fig. 28. Scheme of the procedure for the reduction of a crowded objective prism image. The first stage yields the centres of the spectra, the second the spectral intensities.

spectra consists in the great number of spectra contained in a single image. This number is generally sufficient to implement a statistically valid approach to a given astrophysical problem which is consistent with the available spectral resolution. The recovery of isolated spectra of the type shown in Fig. 16 is relatively simple, even though not trivial, owing to the dominating effect of the observational PSF on the cross-section perpendicular to the dispersion axis, unless this effect is corrected for by suitable image moving techniques as shown only in Fig. 16. The overlap of the individual spectra, as shown in Fig. 17, renders impossible a straightforward determination of the background, as well as the definition of a unique transversal section of the image parallel to the dispersion axis. The general approach to recovery of crowded objective prism spectra consists of a two-stage procedure [26]. The first stage performs a transversal fit, perpendicular to the dispersion axis, of a linear combination of PSF-like profiles centred around the projections, made parallel to the dispersion axis, of the positions of all objects detected on the image. The second stage consists in the extraction, through a transversal cross-section of the image output from the first stage, of each individual one-dimensional spectrum. The scheme of this procedure is shown in Fig. 28. The accuracy in the recovered spectral intensities depends on the accuracy of the estimation and removal of the background. The procedure described for the first stage should be able to remove the background with quite good efficiency up to medium image crowding. A special

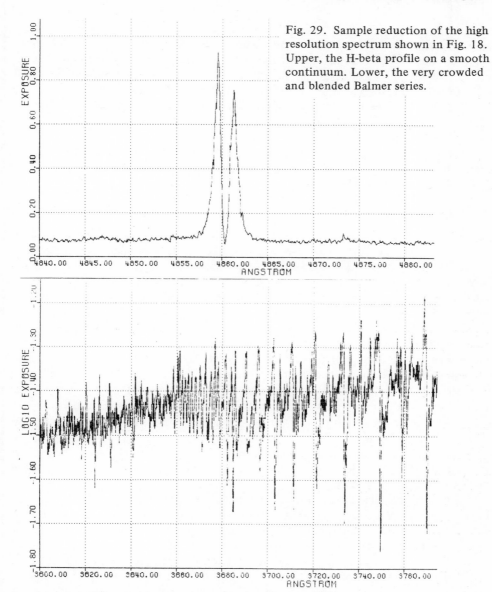

Fig. 29. Sample reduction of the high resolution spectrum shown in Fig. 18. Upper, the H-beta profile on a smooth continuum. Lower, the very crowded and blended Balmer series.

procedure is clearly needed for very crowded objective prism spectral images.

The problem of the segmentation of one-dimensional spectral images for the estimation of parameters of astrophysical importance is, in general, application-dependent. The measurement of spectral line parameters like central wavelength, central depth, bandwidth, symmetry, equivalent width, and so on is relatively straightforward in the case of isolated lines on a gently varying continuum, but most astronomical spectra do not show such pleasant features. Typical sample spectra are shown in Fig. 29. Consequently, the segmentation problem is more difficult in low-resolution prism and crowded objective prism spectra rather than in high-resolution and uncrowded spectra. One current approach is to compare the observed spectrum to a grid of theoretical patterns computed from a model [27], but this approach is very sensitive to spectral resolution, image blurring, and model hypotheses, and in any case is not suitable for application to a large number of spectra, at least given the present quality of spectra and models.

The final problem of the classification of the objects which generated the observed spectra should, of course, be faced within a more comprehensive environment integrated by multispectral analysis methodology.

The present average state of astronomy is still far from such a level, even if some tentative applications of automatic spectral analysis have recently shown positive results [28]. This work is in progress.

5.3. Background estimation and detection of two-dimensional objects

The detection of two-dimensional objects on a two-dimensional astronomical image is limited by the local levels of background and noise, which are non-stationary in most astronomical cases. The non-stationarity of the background depends on the non-uniformity of the distribution of background sources on the observed projected sky and on the non-uniform characteristics of available two-dimensional detectors. The non-stationarity of the noise depends on the statistics of the light carrier, the propagation path, and the detector. The photon noise is substantially Poisson-distributed. The propagation path generally contributes a multiplicative noise, the seeing, that is dominant in strong sources below magnitude 10 for ground-based telescopes. The seeing noise is, of course, negligible for space-borne telescopes. The detector generally contributes a signal-dependent noise plus an additive normal noise. The signal-dependent component may be small, as in photoelectric detectors, or relatively large, as in photographic plates. This scene sets the constraints on any detection scheme for astronomical objects. The general case could be illustrated by the example in Fig. 30.

Two basic detection schemes can be defined, one for PSF-limited objects and another for extended blurred objects. The determination of the background is preliminary and integrated to any detection scheme.

Several methods have been proposed and tested for the estimation of the background in two-dimensional astronomical images and for the interpolation and/or extrapolation of the background nearby and/or below the target objects. The techniques most used are based on semi-interactive determinations of "background" or "non-background" segments followed by polynomial or spline fit to interpolate the background surface [29–31] and on the Bayes estimation of the background histogram [32, 33]. These techniques work reasonably well, but may introduce unpredictable artifacts in unfavourable cases. A possibly safer approach based on Rosenfeld's Scene Labelling by Relaxation Operators was proposed by Brosch [34] and is to be adapted for practical applications.

The detection of PSF-limited objects follows as a secondary yield of the procedure for background estimation by simply thresholding the difference of the orignal image and its estimated background. The threshold level can be assigned in terms of a given probability of detection above background noise. One typical output of such a procedure is shown in Fig. 31.

The detection of extended blurred objects could in principle be made by the same

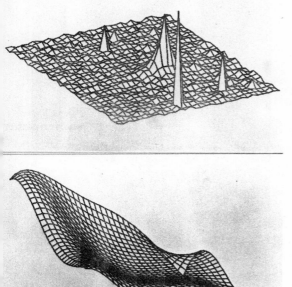

Fig. 30. Detail of an image containing one extended object, an elliptical galaxy, and several point source objects, stars. Upper, full image. The galaxy differs from the stars by its different profile, which is larger than the PSF profile dominating the stellar images. Lower, the background estimated for the image shown above. Notice that the density range of upper image is about 4 while the density range of the background is about 0.2. This value is, however, sufficient to completely distort the morphology of the outer parts of the galaxy if not accurately removed from the raw image. One critical problem is given by the outer wings of the PSF propagated from strong stellar sources.

techniques used for PSF-limited objects. However, the great decrease in the surface brightness of astronomical objects as a function of the distance from the centre of the object results in an exceedingly conservative detection scheme for the outer parts of the extended objects, which are the most important ones for astrophysics.

The accurate estimation of the background and its interpolation over the full image remains, perhaps, the most important problem presently open in astronomical image processing.

5.4. Image segmentation and classification

The segmentation and classification of two-dimensional images are straightforward for PSF-limited objects, while difficult or very difficult for practical astronomical images of extended blurred objects.

The segmentation of PSF-limited objects is reduced to the estimation of the first moments of the detected object, usually the centre and the volume. The estimation is commonly implemented by a two-dimensional fit of the observed image to an analytical or empirical model of the observational PSF. The centre and the volume of the image are used in astrometry and field photometry to determine the coordinates and the magnitude of the object, as shown in Fig. 32 with regard to the magnitudes. The problem becomes non-trivial in crowded fields, owing to the difficulty of estimating a realistic value of the background. This problem is particularly important in the photometry of globular clusters such as that shown in Fig. 5. The accuracy of the HR diagram of the cluster is, in such

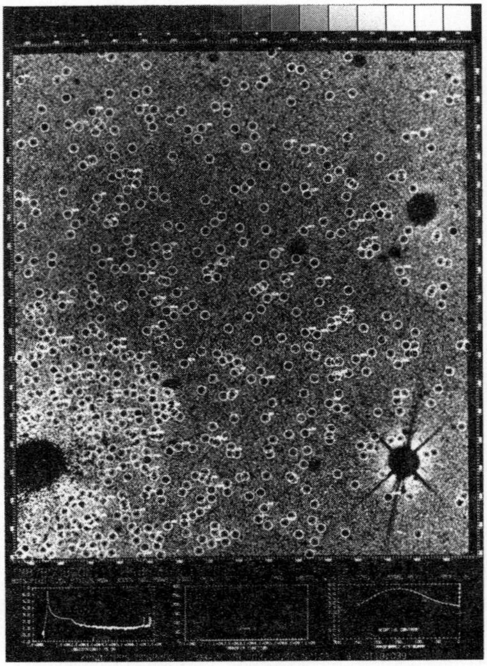

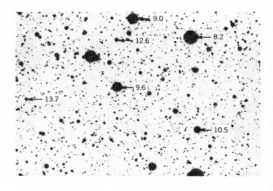

Fig. 31. Sample detection of PSF-limited
stellar objects near the galaxy M87. The
image shown is the difference of the
original and background images
thresholded to detect the stars. The
background was computed using a
running median filter. The circles identify
the detected stellar objects above the
threshold.

case, substantially determined by the errors in the estimate of the background [35, 36].
The problem of background estimation in crowded fields is still open.

The segmentation of extended blurred objects is generally aimed at visual inspection
and discrimination for classification. Visual inspection is fundamental in astronomy,
owing to the complexity of the images and to the practical impossibility of building a
fixed scheme of standard objects. Nearly all techniques of image processing for inspection
are also used in astronomy: linear filtering, histogram processing, slicing, colour coding,
dynamic video techniques, and so on. However, all these fundamental techniques make it
possible to look and perhaps understand an image in terms of a given model, but cannot
be used for really quantitative discrimination and classification of fainter objects. The
problem is given by the fact that in astronomy the fainter objects are also the smaller
ones and those more affected by blurring due to observational PSF.

Fig. 32. Example of field stellar
photometry. The magnitude of the stars
is determined by the volume of the PSF
image centred at the location of the star
examined. The apparent diameter of the
images is thus proportional to the signal
from the stars.

Discrimination for classification is presently used in astronomy to select stellar and non-stellar objects on a large-format image. Very limited work is being done for the discrimination and classification of non-stellar images in classes of increasing complexity, as in galaxy classification.

The simple approach of measuring the ellipticity of the isophotes of the images to select stellar objects of circular shape from non-stellar objects of elliptical or more complex shape is still used as a working approximation of classification by models or moments in high-speed special machines like COSMOS [37], which, however, are also able to perform a more refined moments classification at a lower operational speed.

The really effective methods of segmentation for classification are, however, those based on the statistical approach, the only one capable of a quantitative confidence level for large samples of objects of varying characteristics.

Statistical classification methods are generally divided into two stages, a data compression stage and a classification stage. The data compression stage maps the original image onto a smaller data volume by means of a given criterion of data compression. The techniques most used are reduction by splines, models, moments, and reduction by the co-occurrence matrix [30, 37–42]. Reduction by moments is the one most widely used, while reduction by the co-occurrence matrix is the most effective for high compression figures and at the same time leaves the user free to select his own classification parameters. After performing data compression, the classification is implemented by selecting the compressed data by means of a suitable discriminator. The classification may be non-supervised, or supervised using a training set calibrated by the user on a specially selected sample of "known" objects. The supervised methods are, of course, less impersonal in principle than a non-supervised method, but could prove to be superior if the bias introduced by the personal generation of the training set is, after all, lower than the bias introduced by the model necessary for any non-supervised classification. An example of typical objects selected as a training set is shown in Fig. 33, together with the result of the classification on the whole training set. The images shown in Fig. 33 are also interesting because they could be considered typical of the standard level of images available to astronomers. The fine images shown in Figs. 4–20 are, generally speaking, of a tutorial level.

It should be emphasized that in principle many segmentation techniques well known in image processing science could be applied to astronomical images to classify, in quantitative terms, complex images such as spiral galaxies, and the even more complex objects shown in Figs. 4–20. The problem consists in the severe blurring usually found in most astronomical images, particularly at faint levels. New instruments like the space telescope will certainly improve the situation by an increase of the spatial resolution, but it is clear that in astronomy it is necessary to make an effort to find special classification techniques that are able to perform complex classification schemes under severe blurring conditions. This is another field open to the image processing scientist who is willing to cooperate with astronomers on a stimulating problem.

Also, the quantitative segmentation of well-defined extended objects is very limited in the field of astronomical applications. In practice, consolidated techniques are available only for elliptical galaxies, using methods that are generalizations of the methods used for stellar objects [43, 44]. All the information available is connected morphologically to

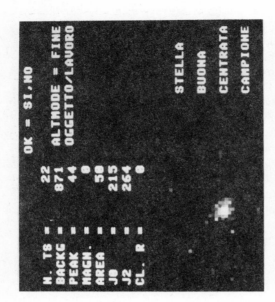

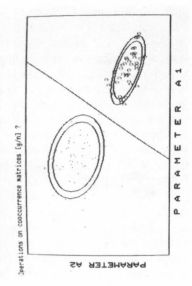

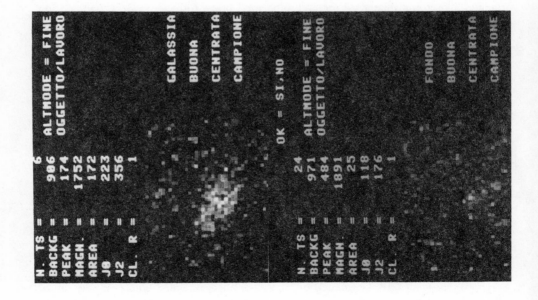

Fig. 33 (facing page). Sample of a training set selected for a supervised discrimination and classification procedure based on co-occurrence matrix statistics. Upper left, a galaxy. Upper right, a star. Lower left, background level. Lower right, the discrimination space selected for the classification on the base of the selected training set. The results shown include all the data of the training set from which the sample images given here were selected. Notice that the galaxy and the star are fine examples of astronomical images in spite of the very low number of pixels involved. Notice also that the background contains objects, but below the threshold.

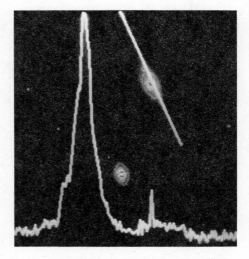

Fig. 34. Sample exposure cross-section of a galaxy image through its centre after the removal of the background. The spike at left is the PSF-limited profile of the star at the lower-left end of the section mark.

the cross-section of the object through its centre, as shown in Fig. 34. Some work is also being done on spiral galaxies [45], but this field remains substantially open for all complex objects of non-stellar shape.

It should be emphasized that practically all the work on two-dimensional image processing of extended astronomical objects is done on single spectral bands one at a time, even when images taken in several bands are available. The application of true multispectral analysis to astronomy is, in practice, a fully open field for astronomers as well as for image processing scientists. A useful tool for describing astronomical objects has been suggested in [46].

5.5. Three-dimensional image processing

All astronomical images of sources beyond the solar system are two-dimensional projections on the focal plane of the observer. However, three-dimensional images can be obtained by ordering a series of two-dimensional images by a third variable such as time, wavelength, polarization, radial velocity, and so on, as shown in Fig. 35.

The great requirements of three-dimensional processing have limited this technique to only a few astronomical problems, i.e., substantially solar research, radioastronomical survey of the H-21 line, and Fabry-Perot imaging spectrophotometry at variable wavelengths [47, 48].

The practical use of three-dimensional image processing in astronomy is now at a primitive stage that is highly dependent on the hardware, software, and display device available for the application concerned. Tentative dynamic data presentation techniques

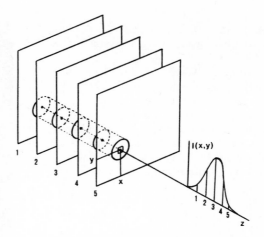

Fig. 35. Scheme of three-dimensional image processing.

are being tested successfully to enhance the visual inspection of the multidimensional data array. Probably three-dimensional image processing in astronomy, as well as in other sciences, will be pushed by the planned future availability of three-dimensional visual displays based on advanced technologies like holographic devices and three-dimensional plasma devices.

5.6. Image restoration from non-equisampled interferograms

The realization and use of large-base radiointerferometers constitutes one major observational improvement in current astronomy on account of the high spatial, temporal, and spectral resolution figures of such instruments. The basic problem in using large-base radiointerferometers consists in the practical impossibility of covering the Fourier-conjugated radio-sky by a uniform grid of samples.

Restoration from non-equisampled two-dimensional interferograms leads unavoidably to an image full of artifacts which make the interpretation of the image very confusing. An example of this effect is shown in Fig. 36.

Several techniques have been proposed and tested to improve the restoration [10, 11, 49]. The best compromise presently available seems to be the clean and restore technique (CAR). An example of the performance of this elegant technique is also shown in Fig. 36. Apparently, the statistical restoration approaches based, for example, on maximum entropy give results inferior in quality to CAR technique, even if other techniques very close to statistical ones, like the Maximum Sharpness one [50], appear to give promising results.

The problem clearly requires further study directed at the general problem of the restoration of undersampled images, a field in which astronomy expects great advantages from advanced image processing science.

6. CONCLUSION

The survey of basic problems and techniques of astronomical image processing given above allows us to identify a few selected lines of methodological development which

Fig. 36. Sample restoration of an undersampled two-dimensional radiointerferogram of a double source (3C 244.1). Upper, the non-uniform grid of samples in the Fourier-conjugated radio-sky. Centre, the raw restored image. Lower, the image restored by CAR technique smoothed to improve the signal-to-noise ratio. The two sources are clearly visible.

show growing importance for astronomy and astrophysics; these are:

Estimation and interpolation of non-stationary background.
Detection, segmentation, and classification of complex, multispectral, blurred images.
Restoration of undersampled images and interferograms.

Substantial improvement in the methodologies concerned with these lines of development is expected to set off an explosive expansion of astronomical image processing and a parallel tremendous extension of the present astronomical horizon.

A cooperative effort of image processing scientists and astronomers is the only visible means of fostering this cross-fertilizing venture.

ACKNOWLEDGMENTS

This work has been supported by the Group of Astronomy of the National Research Council. I wish to thank S. Levialdi and M. L. Malagnini for useful discussion and criticism.

The original sources of the figures shown in this paper are acknowledged here for all the images not specifically produced by author for this paper:
Figs. 1 and 2, Southern European Regional Planetary Image Facility. Figs. 3–18 and 32, Image Archive of Trieste Astronomical Observatory. Fig. 19, VILSPA IUE SWP4616 13-03-79 image observed by M. Hack. Figs. 20 and 36, reproduced from Ref. [10]. Fig. 22, reproduced from Ref. [14]. Fig. 23, courtesy of C. Morossi, C. Allocchio, M. Ramella of Trieste Astronomical Observatory. Fig. 24, reproduced from Ref. [19]. Fig. 25, reproduced from Ref. [21]. Fig. 26, reproduced from Ref. [23]. Fig. 31, courtesy of D. C. Wells of AURA, U.S.A. Fig. 33, courtesy of M. L. Malagnini, M. Pucillo, and P. Santin of Trieste Astronomical Observatory.

REFERENCES

1. R. Giacconi et al., The EINSTEIN (HEAO 2) X-Ray Observatory, *The Astrophysical Journal* 230 (1979) 540.
2. V. Schonfelder, U. Graser, R. Diehl, Properties and performance of the MPI balloon borne Compton telescope, *Astronomy and Astrophysics* 110 (1982) 138.
3. Eastman Kodak Company, Kodak Plates and Films for Scientific Photography (Eastman Kodak Company USA, P-315 1973).
4. J. C. Dainty, R. Shaw, Image Science (Academic Press Inc., London, 1974).
5. A. Boksenberg, University College London Image Photon Counting System, in: M. Duchesne and G. Lelievre (eds.), *Applications astronomiques des recepteurs d'images a reponse lineaire* (Observatoire de Paris Meudon (1976) 13).
6. J. L. Lowrance, P. Zucchino, T. B. Williams, SEC Tube development for Space Astronomy, in: M. Duchesne and G. Lelievre (eds.), *Applications astronomiques des recepteurs d'images a reponse lineaire* (Observatoire de Paris Meudon (1976) 18).
7. D. G. Currie, A photon counting array photometer using an integrated charge couples device, in: M. Duchesne and G. Lelievre (eds.), *Applications astronomiques de recepteurs d'images a reponse lineaire* (Observatoire de Paris Meudon (1976) 30).
8. P. Hansen, J. Strong, High resolution Hadamard transform spectrometer, *Applied Optics* 11 (1972) 502.

9. C. T. Elliott, D. Day, D. J. Williams, An integrating detector for serial scan thermal imaging, *Infrared Physics* 22 (1982) 31.

10. J. A. Hogbom, Aperture synthesis with a non-regular distribution of interferometer baselines, *Astronomy and Astrophysics Supplement* 15 (1974) 417.

11. J. G. Ables, Maximum entropy spectral analysis, *Astronomy and Astrophysics Supplement* 15 (1974) 387.

12. NASA, Science operations with the Space Telescope, Current Concepts January 1979 (NASA Washington D.C., USA 1979).

13. F. Macchetto, T. Westrup, FIPS: The Faint Object Camera Image Processing System, in: G. Sedmak (ed.), *ASTRONET 1982* (Memorie SAIt 53 (1982) 31).

14. P. Crane, K. Banse, The Munich Image Data Analysis System, in: G. Sedmak (ed.), *ASTRONET 1982* (Memorie SAIt 53 (1982) 19).

15. K. P. Tritton, Starlink, in: G. Sedmak (ed.), *ASTRONET 1982* (Memorie SAIt 53 (1982) 55).

16. G. Sedmak, Il Progetto ASTRONET, in: G. Sedmak (ed.), *ASTRONET 1982* (Memorie SAIt 53 (1982) 9).

17. J. W. Brault, O. R. White, The analysis and restoration of astronomical data via the Fast Fourier Transform, *Astronomy and Astrophysics* 13 (1971) 169.

18. C. R. Subrahmanya, A new method of deconvolution and its application to lunar occultations, *Astronomy and Astrophysics* 89 (1980) 132.

19. J. Lorre, Applications of digital image processing techniques to astronomical imagery 1980, *JPL Publication 81-8*, NASA JPL, Caltech, Pasadena, CA USA 1981.

20. V. Di Gesù, B. Sacco, G. Tobia, A clustering method applied to the analysis of sky maps in gamma-ray astronomy, in: G. A. De Biase and G. Sedmak (eds.), *3rd Meeting on Information Processing in astronomy* (Memorie SAIt 51 (1980) 517).

21. V. Di Gesù, M. C. Maccarone, A direct deconvolution method for two-dimensional image analysis, in: G. Sedmak (ed.), *ASTRONET 1982* (Memorie SAIt 53 (1982) 279).

22. A. Labeyrie, Attainment of diffraction limited resolution in large telescopes by Fourier analysis of speckle patterns in star images, *Astronomy and Astrophysics* 6 (1970) 85.

23. G. Weigelt, Restoration of images degraded by the atmosphere or telescope abberations, in: G. Sedmak, M. Capaccioli and R. J. Allen, *Image Processing in Astronomy*, Osservatorio Astronomico di Trieste (1979) 422.

24. R. W. Thompson, B. E. Turnrose, R. C. Bohlin, IUE data reduction, *Astronomy and Astrophysics* 107 (1982) 11.

25. W. W. Weiss, R. Albrecht, H. M. Maitzen, E. Mondre, A. Schnell, H. Jenkner, K. Rakos, IUE Data Reduction (Universitat Wien, Institut fur Astronomie, Wien, 1980).

26. R. Coluzzi, C. E. Corsi, G. A. De Biase, F. Smriglio, Automatic reduction of low dispersion objective prism stellar spectra, in: G. A. De Biase and G. Sedmak (eds.), *2nd Meeting on Information processing in astronomy* (Memorie SAIt 50 (1979) 437).

27. M. L. Malagnini, Una procedura automatica e veloce per il confronto tra distribuzioni spettrali calcolate e osservate, in: G. Sedmak (ed.), *ASTRONET 1982* (Memorie SAIt 53 (1982) 22) 7.

28. H. Zekl, An outline of a computer program for two-dimensional spectral classification, *Astronomy and Astrophysics* 108 (1982) 380.

29. L. Benacchio, M. Capaccioli, G. A. De Biase, P. Santin, G. Sedmak, Surface photometry of extended sources by an interactive procedure, in: G. Sedmak, M. Capaccioli and R. J. Allen (eds.), *Image Processing in Astronomy* (Osservatorio Astronomico di Trieste (1979) 196).

30. J. J. Renes, B-Spline approximation. Background and algorithms, in: G. Sedmak, M. Capaccioli and R. J. Allen (eds.), *Image Processing in Astronomy* (Osservatorio Astronomico di Trieste (1979) 329).

31. R. Buonanno, G. A. De Biase, I. Ferraro, R. Lombardi, Interactive procedure for astronomical plates background fitting, in: G. A. De Biase and G. Sedmak (eds.), *2nd Meeting on information processing in astronomy* (Memorie SAIt 50 (1979) 467).

32. A. Bijaoui, Automatic treatment of large scale astronomical images, in: G. Sedmak, M. Capaccioli and R. J. Allen (eds.), *Image Processing in Astronomy* (Osservatorio Astronomico di Trieste (1979) 173).

33. R. Martin, R. K. Lutz, Background following on Schmidt plates using a digital filtering technique, in: G. Sedmak, M. Capaccioli and R. J. Allen (eds.), *Image Processing in Astronomy* (Osservatorio Astronomico di Trieste (1979) 211).

34. N. Brosch, Improved detectivity of faint objects via an iterative probabilistic approach, in: G. Sedmak, M. Capaccioli and R. J. Allen (eds.), *Image Processing in Astronomy* (Osservatorio Astronomico di Trieste (1979) 252).

35. R. Buonanno, G. Buscema, G. A. De Biase, G. Iannicola, R. Lombardi, Interactive approach to integrated photometry of globular clusters, in: G. A. De Biase and G. Sedmak (eds.), *3rd Meeting on information processing in astronomy* (Memorie SAIt 51 (1980) 473).

36. R. Buonanno, G. Buscema, C. Corsi, G. Iannicola, Automatic stellar photometry on photographic plates, in: G. A. De Biase and G. Sedmak (eds.), *3rd Meeting on information processing in astronomy* (Memorie SAIt 51 (1980) 483).

37. R. S. Stobie, G. M. Smith, R. K. Lutz, R. Martin, The performance of the COSMOS machine, in: G. Sedmak, M. Capaccioli and R. J. Allen (eds.), *Image Processing in Astronomy* (Osservatorio Astronomico di Trieste (1979) 48).

38. J. A. Tyson, J. F. Jarvis, Evolution of galaxies: automated faint object counts to 24th magnitude, *The Astrophysical Journal* 230 (1979) L153.

39. W. L. Sebok, Optimal classification of images into stars or galaxies – a Bayesian approach, *The Astronomical Journal* 84 (1979) 1526.

40. G. Sedmak, M. L. Trujillo Lamas, Automatic classification of galaxy images by Fourier structural analysis, *Astronomy and Astrophysics* 104 (1981) 93.

41. M. L. Malagnini, G. L. Sicuranza, Classificazione e discriminazione di immagini astronomiche: proposta per lo sviluppo del software applicativo, in: G. Sedmak (ed.), *ASTRONET 1982* (Memorie SAIt 53 (1982) 267).

42. S. A. Butchins, Automatic image classification, *Astronomy and Astrophysics* 109 (1982) 360.

43. E. B. Newell, The reduction of panoramic photometry. II. Automatic reduction of imperfect data, in: G. Sedmak, M. Capaccioli and R. J. Allen (eds.), *Image Processing in Astronomy* (Osservatorio Astronomico di Trieste (1979) 100).

44. W. B. Jones, D. L. Gallet, K. M. Obitts, G. De Vaucouleurs, Astronomical surface photometry by numerical mapping techniques (University of Texas at Austin, Publ. Astronomy Department Ser. II, Vol. 1, N. 8 (1967)).

45. M. F. Duval, Etude photometrique et cinematique du systeme double NGC4485-4490, *Astronomy and Astrophysics* 98 (1981) 352.

46. M. Balestreri, A. Della Ventura, G. Fresta, P. Mussio, A proposed pictorial language for describing astronomical objects, in: G. Sedmak, M. Capaccioli and R. J. Allen (eds.), *Image Processing in Astronomy* (Osservatorio Astronomico di Trieste (1979) 268).

47. G. De Vaucouleurs, W. D. Pence, Velocity fields in late type galaxies from H-alfa Fabry-Perot interferometry, *The Astrophysical Journal* 242 (1980) 18.

48. E. Recillas-Cruz, P. Pismis, Fabry-Perot radial velocities of S274: a planetary nebula, *Astronomy and Astrophysics* 97 (1981) 398.
49. J. V. Pendrel, D. E. Smylie, The maximum entropy principle in two-dimensional spectral analysis, *Astronomy and Astrophysics* 112 (1982) 181.
50. P. L. Baker, On the recovery of images from incomplete interferometric measurements, *Astronomy and Astrophysics* 94 (1981) 85.

G. Sedmak
Osservatorio Astronomico Trieste

APPENDIX. ASTRONOMICAL IMAGE PROCESSING IN ITALY

Research activities in astronomy and astrophysics are carried out at several institutions in Italy, including astronomical observatories, university institutes of astronomy and university astronomical observatories of the Ministry of Education (MPI), and institutes of the National Research Council (CNR). Many of these institutions are committed to astronomical image processing, as are a few other institutes, mainly of the CNR, which are not specifically concerned with astronomy and astrophysics.

This appendix will summarize the 1982 Italian scene in astronomical image processing as gathered from the proceedings of the 1st, 2nd, and 3rd national meetings on information processing in astronomy, held in 1978, 1979, and 1980 [1–3], the international workshop on image processing in astronomy held in 1979 [4], and the ASTRONET 1982 national meeting [5].

The astronomical observatories of the MPI are located in Catania, Florence, Milan, Naples, Padua, Rome, Turin, and Trieste. The university astronomical observatories are located in Bologna, Cagliari, and Palermo. There is substantial integration between research made in the astronomical observatories and the institutes of astronomy of the universities of the centres listed above.

The institutes of the CNR devoted to astronomy and astrophysics are located in Bologna and Frascati (Rome). Other institutes of the CNR devoted to cosmic physics and related techniques are located in Milan and Palermo and in some of the centres already named. The main institute of the CNR devoted to image processing, but not specifically to astronomy, is located at Pisa.

The list of the addresses of the major institutions concerned with astronomy and astrophysics and image processing with possible astronomical applications is given in the references.

Astronomical image processing was carried out in a nearly incoherent environment until the definition of the ASTRONET project in 1980. In the 1970s, fine results on the configuration of hardware for astronomical image processing were obtained in Pisa, Rome, and Trieste. In the same period, software systems and application modules for specialized astronomical and general purpose image processing were successfully written in Bologna, Milan, Naples, Padua, Pisa, Rome, and Trieste [1–4]. The typical Italian hardware for pre-ASTRONET astronomical image processing was based mainly on Digital Equipment PDP11 and some Hewlett Packard HP2100 series of 16-bit computers. The graphic hardware consisted of Tektronix 4000 series of storage-type interactive video terminals and Versatec electrographic printer-plotters. Integrated software systems for astronomical image processing were proposed and realized in Milan and Pisa [4], but not extensively used for practical applications, with some exceptions for solar image processing [2, 5].

The ASTRONET project, started in 1981 by the MPI (University National Research Programme) and the CNR (National Group of Astronomy and National Space Plan), marked a milestone for astronomical image processing in Italy [5]. The project coordinates the efforts in this field of the Italian astronomical community through a network of standard hardware and software image processing systems specialized for astronomical applications. The scope and structure of ASTRONET is very similar to the British STARLINK network for astronomy and astrophysics [4, 5].

The ASTRONET nodes are located in Bari, Bologna, Catania, Florence, Naples, Padua, Palermo, Rome, and Trieste. Each node is equipped with a 32-bit virtual memory Digital Equipment VAX computer with 2–4 Mbytes of central memory, 500–1000 Mbytes of disc mass memory, and raster graphic-pictorial video and hard-copy facilities. The network will use the X.25 packet switching protocol as soon as it is available in Italy – by autumn 1983. At present, seven of the nine nodes have been completed and the last two completely financed. ASTRONET is expected to be fully operative by the end of 1983 and able to host remote users as secondary nodes from 1984 on.

The ASTRONET standard software is being written by selected special interest groups coordinated by a national committee. The greatest efforts are devoted to the basic software environment, graphics, pictorial standard software, documentation, and databases. The ASTRONET environment will include at optimum compatibility the NASA VICAR, ESA FIPS, ESO MIDAS, and STARLINK SCL image processing software systems and application modules as secondary ASTRONET standards. The startup of ASTRONET project in 1981 generated a great increase in the production of image processing software for astronomical applications. The greatest efforts are devoted to spectra processing, two-dimensional field photometry, surface photometry, image discrimination and classification by statistical methods, and restoration of synthesis radiointerferograms.

ASTRONET now allows Italian astronomers to focus on the astronomy-oriented application software in a standardized, internationally compatible environment. This activity should gain maximum advantages from the cooperation of astronomers and image processing people. The purpose is definitively that to enhance the performance of the astronomical image processing software presently available in Italy up to the figures predicted for effective processing of the data and images expected from the advanced ground-based and space-borne instrumentation planned for astronomy and astrophysics in the 1980s.

REFERENCES

1. lst Meeting on Information processing in astronomy, G. A. De Biase and G. Sedmak (eds.) Mem. SAIt 49 (1978).
2. 2nd Meeting on Information processing in astronomy, G. A. De Biase and G. Sedmak (eds.) Mem. SAIt 50 (1979).
3. 3rd Meeting on Information processing in astronomy, G. A. De Biase and G. Sedmak (eds.) Mem. SAIt 51 (1980).
4. International workshop on Image processing in astronomy, G. Sedmak, M. Capaccioli and R. J. Allen (eds.) Osservatorio Astronomico di Trieste (1979).
5. ASTRONET 1982 national meeting, G. Sedmak (ed.) Mem. SAIt 53 (1982).

Addresses:

Bari
Istituto di Fisica, Università di Bari,
Via Amendola 173, Bari
Tel: 080-331044

Bologna
Osservatorio Astronomico, Università di
Bologna, Via Zamboni 33, Bologna
Tel: 051-222956
Istituto Radioastronomia CNR
Via Irnerio 46, Bologna
Tel: 051-262804

Catania
Osservatorio Astrofisico, Città Universi-
taria, Via Andrea Doria, Catania
Tel: 095-330734

Firenze
Osservatorio Astrofisico di Arcetri,
Largo Enrico Fermi 5, Arcetri, Firenze
Tel: 055-220034

Milano
Istituto Fisica Cosmica e Tecnologie
Relative CNR, Via Bassini 15/A, Milano
Tel: 02-299653

Napoli
Osservatorio Astronomico di Capodi-
monte, Via Moiariello 16, Capodimonte,
Napoli
Tel: 081-440101

Padova
Osservatorio Astronomico di Padova,
Vicolo dell'Osservatorio 5, Padova
Tel: 049-661499

Palermo
Osservatorio Astronomico di Palermo
Palazzo dei Normanni, Palermo
Tel: 091-422588
Istituto Fisica Cosmica e Informatica CNR
Via Archirafi 36, Palermo
Tel: 091-231936

Pisa
Istituto Elaborazione Informazione CNR
Via S. Maria 46, Pisa
Tel: 050-500159

Roma
Osservatorio Astronomico di Monte Mario,
Via Parco Mellini 84, Roma
Tel: 06-3452794
Istituto Astronomico, Università di Roma 1,
Piazzale A. Moro, Roma
Tel: 06-4976519
Istituto Astrofisica Spaziale CNR
Frascati (Roma)
Tel: 06-9421483

Trieste
Osservatorio Astronomico di Trieste,
Via G. B. Tiepolo 11, Trieste
Tel: 040-793921

PART TWO
CONFERENCE PAPERS

Chairman's opening address

The presence at this second international conference on image analysis and processing of so many prominent scientists is evidence of the growing interest in these meetings, at which new ideas on fundamental theory and on techniques for the solution of problems of application can be examined and properly evaluated.

In fact, as in most scientific disciplines, so in the field of digital image analysis and processing, theoretical development and experimental applications have always been deeply interwoven. The marked interest aroused by the many important applications has certainly been fundamental to the development of both theoretical studies and experimental work. Certainly, a special role in this development has been played by image-processing projects for space applications and, similarly, by projects for digital image transmission via space telecommunication systems, image reconstruction from planetary probes exploring the solar system, and evaluation of the Earth's image in connection with resource and environmental observations from orbiting spacecraft.

Space telecommunication systems are rapidly evolving from applications in intercontinental linkage to more general use in continental and national coverage. The new spaceborne TLC systems are speeding up the transformation of existing ground-based networks by the introduction of new flexible networks which allow experimental integration of all information exchanges in digital form. Voices, images, and computer data will be treated together in the new telecommunications systems, intermixed on the same digital support.

In this framework, in the development of commercially competitive teleconference systems various projects for band reduction in TV frame transmission have led to important advances in the theory and in the algorithms for digital image compression.

The commercial application stage is also coming closer for spaceborne remote sensing for Earth resources and environmental monitoring. Digital images from observation satellites such as Landsat, Meteosat, Tyros, Nimbus, Seasat and HCMM have triggered worldwide activity with important developments in the theoretical aspects of image processing, in the design of new algorithms, and in the implementation of specialized hardware/software systems. Many applications have already been developed at an experimental or preoperational level. New observation satellites have clearly been designed with a view to their operational use as prototypes for future commercial systems. Particularly relevant in this respect are the Landsat D, recently launched by NASA, with improved geometric, radiometric, and spectral resolution, and the two future European satellites —

SPOT with its high geometric resolution, multiangle viewing ability for multitemporal and stereo scene analysis, and ERS-1 with its imaging radar for all-weather observation.

Major developments are thus under way in the form of new algorithms and new digital system architectures to handle the enormous flow of data from the new imaging systems with proper precision and adequate throughput.

New image-processing systems are being developed in industrial projects which will benefit from the theoretical and experimental work that is going on, mostly in academic and research establishments; in turn, these projects will certainly be the source of new ideas for research in the field of digital image processing and analysis. The present time thus seems particularly interesting and favourable for advances in this field of research, in view of the enormous economic and social commitment to the applications of space-borne remote-sensing technologies for Earth resource evaluation and environmental monitoring. As administrator of the Italian National Space Program, I am convinced that the theoretical and experimental results presented to this conference and the ensuing debate will constitute vital input for our future decisionmaking. The Italian Space Program, approved in 1980 by the Italian government, is developing several major industrial projects in satellite telecommunication systems and shuttle-based advanced space systems. In addition to these programmes, a great effort is being made to develop new application systems for spaceborne remote sensing.

One project is devoted to the development of a synthetic aperture radar (SAR), capable of producing images of the Earth's surface by means of variations in local radar reflectivity in the target. Microwave signals in the frequency range 5−10 GHz can penetrate clouds and heavy rain and can be used as an all-weather observation system. Each observed pixel is illuminated by about 300 radar pulses during the satellite overflight and thus reflects 300 echoes, each characterized by four quantities: amplitude, phaseshift, time of arrival, and Doppler shift, the last being caused by satellite motion and the Earth's rotation. Since all signals generated by the radar are coherent, it is possible to add coherently all 300 echoes from the same pixel, thus creating a synthetic aperture antenna more than 1 km wide; this corresponds to a final geometric resolution of a few metres, comparable with that of the optical sensor. Time delays and Doppler shifts are used in image processing to extract the signal corresponding to a single pixel.

Range compression and azimuth compression are performed by two one-dimensional filters. The filter for range compression is calculated only once, using the known geometry, whereas the filter for the azimuth compression is range-dependent and must be calculated first for each range and then for each processing block in the azimuth direction. This filter is computed using the raw data directly with an autofocus technique to obtain the satellite motion parameters that give maximum contrast in the processed image.

For an SAR processing facility, capable of quick-look processing of images at the rate of one frame (75 × 75 km) for every 3 min, about 20 million complex multiplications and 30 million complex additions have to be performed every second. This indicates that the development of an SAR facility also requires the development of an adequate processing system based on new architecture for parallel processing.

In the Italian Space Program, a second important project in the remote-sensing area is also related to image processing. The aim of the project is the systematic determination of the most relevant physical parameters by integration of multitemporal and multiplatform spaceborne data, digital terrain models from available cartography, and actual ground data. By appropriate modelling of electromagnetic radiation interactions with the atmosphere and with the Earth's surface materials, it is intended to define procedures and algorithms for periodic determination of major physical parameters to be used as input in application models for agriculture, hydrology, geology, and oceanography. Digital tapes describing, in the form of images, physical parameters such as soil temperature and humidity, sea surface temperature, wind and wave field distributions, etc., seem to be relevant to a wide class of applications and likely to meet the requirements and the capability of the application user. In fact, experience indicates that the integration of multisource data for each individual problem is highly impractical for the final user.

The system under development requires precise radiometric and geometric correction in order that the multisource image registration will allow adequate exploitation of traditional MSS data together with the new data from Landsat D Thematic Mapper, the data from SPOT, and the imaging radar data from ERS-1. This programme also requires a processing system with a very high throughput capacity.

A third project is therefore being developed as part of the Italian Space Program, both for SAR data processing and for systematic production of physical parameter digital tapes. This concerns the design and implementation of a new architecture for parallel image processing. The solution being studied is based on a multiprocessor system success- fully developed for problems of realtime pattern recognition. The system has a parallel multilevel multibus architecture with a distributed operating system supporting communi- cations and synchronization between processes and with automatic reconfiguration capability. It will be able to perform precision processing of the 94 Gbits produced daily by Landsat D Thematic Mapper and MSS in only 9 hours.

The subjects covered by this conference are thus of particular relevance to the remote-sensing projects of the Italian Space Program, which will certainly benefit from the new ideas and the scientific and technical contributions presented here.

Finally, I should like to express sincere appreciation to all the scientific institutions and industrial corporations that promptly recognized the relevance of this second confer- ence on image analysis and processing and assured their sponsorship and financial support. In particular, I should like to thank the University of Bari, the National Science Council, the CSATA Study Centre, the Digital Equipment Corporation, IBM, and the Società Tele- spazio. Thanks are also due to the Scientific Committee of the conference, together with our colleagues and technical personnel of the Physics Department and of the Institute of Information Science who had the hard task of the conference organization.

L. Guerriero
University of Bari

9 Impossible figures: illusion of spatial interpretation of pictures

Z Kulpa

The so-called "impossible figures" (see Figure 1) are interesting to psychologists as a new type of visual illusion (providing a source of additional information about our spatial interpretation mechanisms), and, for similar reasons, to artificial intelligence researchers (providing cues for organization of algorithms modelling human abilities to see a three-dimensional world in flat pictures). It should be mentioned also that these effects are of growing interest to the theory and practice of visual arts and graphic design (including computer graphics).

The notions of impossible figures and of some other sorts of figures, closely related to the impossible ones (so-called "likely" and "unlikely" figures), are precisely defined here.

An *impossible figure* is a drawing making an impression of some three-dimensional object, although the object suggested by this three-dimensional interpretation of the drawing cannot be constructed in three-space, and the impossibility of this interpretation is immediately seen by the observer.

A *likely figure* is the figure whose interpretation, selected by an observer, is in fact impossible, but it is not noticed by the observer to be impossible.

An *unlikely figure* is the figure whose interpretation, selected by an observer, is in fact possible, but is considered by the observer as impossible.

It is argued that all these figures, being illusions of spatial interpretation of pictures (situated somewhere between low-level "optical" illusions and higher-level "semantical" illusions, e.g., like those occurring in the Rorschach test), are more relevant to psychology of vision, and related computer image interpretation research, than to geometry or mathematics in general.

In fact, only the spatial interpretation of the drawing of the figure, not the figure itself, can be reasonably called impossible. Eventually, all impossible figures do have possible spatial interpretations, so that the actual question with impossible figures is: why our interpretation mechanisms do not find those possible interpretations for some, otherwise rather simple drawings. This question is also closely related to the problem of modelling the notion of "naturalness" of an interpretation. Therefore, on the one hand impossibility effects can be fully explained only when we have learned the rules under-

The full text of this paper appears in *Signal Processing* [25]

lying our spatial interpretation mechanisms, and on the other hand they provide a rich source of information about the inner workings of these interpretation mechanisms.

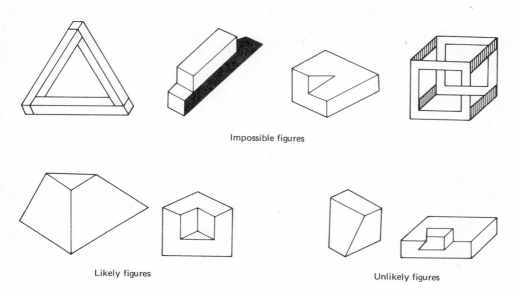

Impossible figures

Likely figures   Unlikely figures

Figure 1

The analysis of the impossible figures illusion allows us actually to formulate several properties of our spatial interpretation procedures, namely:

(1) An essentially two-stage structure of these procedures:
  (a) *Analytical interpretation* based mainly on local depth cues of various sorts.
  (b) *Global synthesis*, including verification, adjustment, and correction of local evidence, and based at least in part on fitting to models (memorized general patterns).
(2) Basic "interpretation assumptions" filtering out the whole bulk of geometrically feasible, although "unnatural" interpretations:
  (a) *Simplicity* assumption – the interpretation should be as simple as possible, preferably not more complex than the drawing.
  (b) *Minimal change* assumption – geometrical features and relations present in the drawing (e.g., straightness, parallelism, intersection relations) are preserved also in the three-dimensional interpretation (provided it does not violate the simplicity assumption too much).
  (c) *General position* assumption – the object is depicted in the drawing from a nonsingular point of view, i.e., such that a slight change of the assumed point of view (centre of projection) does not change significantly either the structure (e.g., topological) of the drawing, or important features of its elements (e.g., straightness, parallelism, etc.).

(3) A set of basic "impossibility causes", i.e., those important local features whose inconformities detected at the verification stage produce the majority of impossibility effects:

(a) figure/background distinction,
(b) depth relations estimation,
(c) planarity/nonplanarity impression.

Some relevant references are listed below.

REFERENCES

1. L. S. Penrose, R. Penrose, Impossible objects: a special type of visual illusion, *Brit. J. Psychol.* 49, 1958, pp. 31–33.
2. D. A. Huffman, Impossible objects as nonsense sentences, In: B. Meltzer, D. Michie, eds., *Machine Intelligence 6*, Edinburgh Univ. Press, Edinburgh, 1971, pp. 295–323.
3. M. B. Clowes, On seeing things, *Artificial Intelligence*, 2, 1971, pp. 79–116.
4. R. L. Gregory, *The Intelligent Eye*, Weidenfeld & Nicholson, London, 1970.
5. A. Thiéry, Ueber geometrisch-optische Täuschungen, *Philosophische Studien*, 11, 1895, pp. 307–370.
6. T. M. Cowan, The theory of braids and the analysis of impossible figures, *J. Math. Psychol.*, 11, 1974, pp. 190–212.
7. T. M. Cowan, Organizing properties of impossible figures, *Perception*, 6, 1977, pp. 41–56.
8. A. W. Young, J. B. Deregowski, Learning to see impossible, *Perception*, 10, 1981, pp. 91–105.
9. D. Waltz, Understanding line drawings of scenes with shadows, In: P. H. Winston, ed., *The Psychology of Computer Vision*, McGraw-Hill, New York, 1975, pp. 19–91.
10. A. K. Mackworth, How to see a simple world: an exegesis of some computer programs for scene analysis, In: E. W. Elcock, D. Michie, eds., *Machine Intelligence 8*, Ellis Horwood Ltd., Chichester, 1977, pp. 510–537.
11. T. Kanade, A theory of Origami world, *Artificial Intelligence*, 13, 1980, pp. 279–311.
12. T. Kanade, Recovery of the three-dimensional shape of an object from a single view, *Artificial Intelligence*, 17, 1981, pp. 409–460.
13. S. W. Draper, The use of gradient and dual space in line drawing interpretation, *Artificial Intelligence*, 17, 1981, pp. 461–508.
14. Z. Kulpa, Obiekty nieistniejące (Nonexisting objects, in Polish), *Problemy*, No. 9, 1976, pp. 20–22.
15. J. Hochberg, V. Brooks, The psychophysics of form: reversible-perspective drawings of spatial objects, *Amer. J. Psychol.*, 73, 1960, pp. 227–354.
16. D. H. Schuster, A new ambiguous figure: a three-stick clevis, *Amer. J. Psychol.*, 77, 1964, p. 673.
17. J. O. Robinson, J. A. Wilson, The impossible colonnade and other variations of a well-known figure, *Brit. J. Psychol.*, 64, 1979, pp. 363–365.
18. V. F. Koleichuk, Nevozmozhnye figury? (Impossible figures?, in Russian), *Tekhnicheskaya Estetika*, No. 9, 1974, pp. 14–15.
19. B. Ernst, *The Magic Mirror of M. C. Escher*, Ballantine Books, New York, 1976.
20. R. Arnheim, *Art and Visual Perception*, Univ. of California Press, San Diego, 1974.

21. B. V. Raushenbakh, *Prostranstvennye Postroyenya v Zhivopisi* (*Spatial Constructions in Visual Art*, in Russian), Nauka, Moscow, 1980.
22. F. Attneave, Multistability in perception, *Scientific American*, 225, 1971, pp. 62–71.
23. M. Gardner, Mathematical games — A Möbius band has a finite thickness, and so is actually a twisted prism, *Scientific American*, 239, No. 2, 1978, pp. 12–16.
24. Z. Kulpa, Oscar Reutersvärd's exploration of impossible lands, In: *150 Omöjliga Figurer – Oscar Reutersvärd*, exhibition catalogue, Malmö Museum, Malmö, 1981.
25. Z. Kulpa, Are impossible figures possible?, *Signal Processing*, 5, 1983 (in press).

Z. Kulpa
Institute of Biocybernetics and Biomedical Engineering
00-818 Warsaw
Poland

10 A system for the analysis of the boundaries of a shape

V Di Gesù

1. INTRODUCTION

In this paper an algorithm to find the contours of a bidimensional image coded on a squared grid is shown. The extraction of the image contours gives, as is well known, much information about the shape and topological connection of an image, which helps in understanding a scene through drawing the boundary relationship [1] as a basis for syntactic pattern recognition [2–4].

The second section introduces some general definitions. The third section explains in detail the algorithm proposed to extract the boundary, which is based on the local properties of a 2 × 2 window. The use of a 2 × 2 window speeds up the boundary-following algorithm. The fourth section shows some of the structural information that can be inferred on the image by using only the information contained in the boundary described as a succession of points. Experiments have been performed on data of biological [5], biomedical [6], and astronomical [7] interest.

2. GENERAL BACKGROUND

In this section some definitions useful in describing the boundary-following algorithm, are given.

Def. 1. A square grid R is the set of ordered pairs (i, j) for i=1, 2, ..., n and j=1, 2, ..., m.

Def. 2. An image I on R is a mapping I: R→S, where S⊂N⁺ (the set of positive integer numbers). Whenever S={0, 1} the image is said to be a binary image.

In the following only binary images are considered; although this assumption is restrictive for some applications, there are many others in which the structural information related to the silhouette prevails. In the case of an image defined on many grey-levels, thresholding techniques can be used to lead the image analysis back to the study of a set of silhouettes.

Def. 3. A scanning window (SW) is a 2 × 2 subgrid of R, M_{ij}, where the pair (i, j) clash with the first element of the subgrid (Figure 1). To each SW, M_{ij}, the following evaluation function is related:

$$\text{Val}(M_{ij}) = \sum_{k,1}^{4} a_k 2^{(4-k)}$$

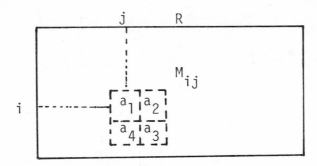

Figure 1. The scanning-window M_{ij}
centred on (i, j).

From the definition of Val it follows:

$Val(M_{ij}) = 0 \Rightarrow$ empty-zone of I,
$Val(M_{ij}) = 15 \Rightarrow$ full-zone of I,
$Val(M_{ij}) \leqslant 0$ and $Val(M_{ij}) < 15 \Rightarrow$ boundary zone of I.

3. ALGORITHM FOR BOUNDARY DETECTION

The algorithm is based on the following properties:

Prop. 1. If the scanning route of SW is top to bottom and left to right, the first value of SW in an external border, if it exists, is $Val(M_{ij})=2$ or $Val(M_{ij})=10$. The first value of SW in an internal border, if it exists, is $Val(M_{ij})=13$.

Prop. 2. The SW value $Val(M_{ij})=5$ and $Val(M_{ij})=10$ are related to one folded component of I (Figure 2(a)) or to a pair of contiguous components of I (Figure 2(b)). Therefore, the starting point is chosen in accordance with the route of Prop. 1 and the following route is determined from the value of SW as indicated in Table 1. In the table the values $Val(M_{ij})=5$ and $Val(M_{ij})=10$ are omitted, since for Prop. 2 they correspond to two different topological configurations. Therefore, the route followed, in this case, does not

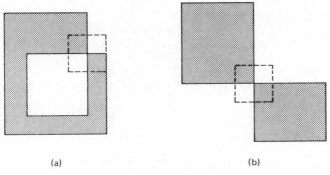

(a)   (b)

Figure 2. A folded component and
contiguous components of I.

Table 1

Val($M_{i_0 j_0}$)	i	j
1	i_0+1	j_0
2	i_0	j_0+1
3	i_0	j_0+1
4	i_0-1	j_0
6	i_0-1	j_0
7	i_0	j_0-1
8	i_0	j_0-1
9	i_0+1	j_0
11	i_0	j_0+1
12	i_0	j_0-1
13	i_0+1	j_0
14	i_0	j_0-1

depend only on the current value of SW but also on the previous value, Pval, as is shown in the following:

$$\text{if Val}(M_{ij})=5 \text{ or } 10 \begin{cases} \text{Pval=9 or 13 or 1 then } j \leftarrow j-1, \\ \text{Pval=6 or 4 or 7 then } j \leftarrow j+1, \\ \text{Pval=12 or 8 or 14 then } i \leftarrow i-1, \\ \text{Pval=2 or 11 or 3 then } i \leftarrow i+1. \end{cases}$$

The boundary is represented as a linked list in which the records have as an informative part the coordinate pair (i,j) and a direction d coded from $\emptyset$ to 7 as in a Freeman notation 3. Each border of an image is labelled "ext" if it is an external boundary and "int" if it is an interior border. The next section shows how to describe an image, starting from this information.

At the end of this section some considerations on the computational complexity of the algorithm are made. In the case of a serial processor the computational time is of the order of $0(mn+\tfrac{1}{3}(\sqrt{mn}))$. The representation of an image by its borders is also, on average, memory saving. In fact the memory space to code a binary image is $0(mn)$ bits and the number of bits required to code the pair (i,j) and d is $0(\ln_2 m + \ln_2 n + 3)$ bits. It follows that the boundary representation is space saving if the following inequality is satisfied:

$$K(\ln_2 m + \ln_2 n + 3) < mn,$$

where K is the total number of boundary-points in the image. For example, if m=n=1024 the inequality is satisfied for K < 45590, this is widely verified in real situations.

4. BOUNDARY DESCRIPTION

In this section a summary is given of the algorithms that have been implemented to analyze the boundary of an image to deliver the relationships between them that are useful in the description of the image.

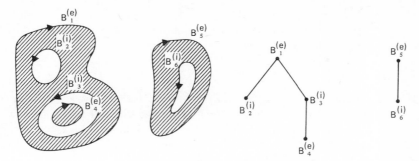

Figure 3. Example of a DTF.

4.1. The descriptive tree forest

The boundaries of an image I are labelled, as has been stated previously, by "ext" and "int" and are denoted by $B^{(e)}$ and $B^{(i)}$. It is easy to see that each $B^{(e)}$ corresponds to a component of I. It is useful to establish the degree of connection of each component and the relation "contained" between them. For this purpose it is necessary to compute a descriptive tree forest (DTF) in which the root of each tree belonging to the DTF corresponds to a maximal component of I (i.e., a component that is not included in other components). The direct arcs, connecting nodes of type $B^{(e)}$ ($B^{(i)}$) to nodes of type $B^{(i)}$ ($B^{(e)}$), represent the relation "contained" between boundaries of different types. Therefore, given a node labelled "ext" in the tree, the number of its direct descenders ($B^{(i)}$) is the connection number of the components pointed by it, while its descenders labelled "ext" give the number of components contained in it. The algorithm necessary to build the DTF of I is based on the following property:

$B^{(i)} \; \{ \; B^{(e)} \; (B^{(e)} \; \{ \; B^{(i)}) \Leftrightarrow$ the rectangle that circumscribes $B^{(e)}$ ($B^{(i)}$) contains $B^{(i)}$ ($B^{(e)}$) and the number of intersections, n_r, between a straight line, parallel to the Y axis, and $B^{(e)}$ ($B^{(i)}$) below $B^{(i)}$ ($B^{(e)}$) is odd,

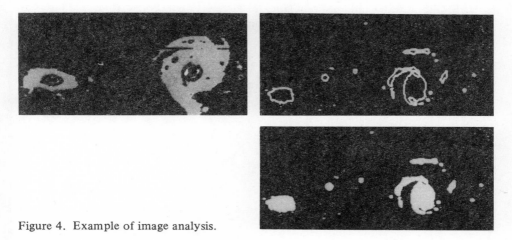

Figure 4. Example of image analysis.

where the symbol "ƺ" means "contained in". A suitable use of the DTF could be as an instrument for syntactical classification and structural description in image processing. An example of a DTF is given in Figure 3.

4.2. Shape descriptors

For a DTF it is useful to have a global description of I; nevertheless, a set of component descriptors can be joined to give better knowledge of the shape of each component of I, some of which, such as the number of concavities, the component area, the component perimeter, the barycentre, and the moments, have been computed using the data structure described previously.

5. IMPLEMENTATION AND EXPERIMENTS

The boundary-following algorithm and all the image analysis procedures described in this paper have been implemented in an interactive image description system (IDS). IDS is written in FORTRAN 77 and is supported on DEC-VAX and IBM3033 systems. The DEC-VAX version supports graphic utility of the VS11 display and printer plotter devices. The version implemented on IBM3033 supports CALCOMP graphic utility.

Some applications of the IDS system have been performed on simulated and real data, the last concerning astronomical and biomedical images. Figure 4 shows the input, thresholding and result of an analysis of an astronomical image in which an elliptical and a spiral galaxy are present.

REFERENCES

1. T. Pavlidis: Springer Verlag, New York, 1977.
2. E. L. Brill: WESCON Conference, Session 25, 1968.
3. H. Freeman: IEEE trans. EC-10, 260–268, 1961.
4. J. Raviv, D. N. Streeter: IBM Res. Rep., RC-1577, 1965.
5. N. J. Pressman: Univ. of Pennsylvania, Tech. Rep. UCRL-52135, 1979.
6. C. S. Guptill: Int. Conf. on Techniques for Pictorial Applications, Firenze, June 1979.
7. V. Di Gesù, B. Sacco, G. Tobia: Conf. on Trattamento dell'informazione della SAIT, Roma, 1980.

V. Di Gesù
Istituto di matematica dell'Università di Palermo

11 A track-following algorithm for contour lines of digital binary maps

L Caponetti, M T Chiaradia, A Distante and M Veneziani

1. INTRODUCTION

Object tracking is a standard tool for image segmentation and encoding which employs sequential algorithms based on raster-tracking and border-following methods [1]. The first method extracts objects by tracking them row by row in TV raster mode, the second executes the tracking in all directions and processes the pixel neighbourhoods. Raster-tracking algorithms have, however, the disadvantage of depending on raster orientation and direction. Thus in some applications they must be combined with border-following methods [2, 3].

This paper analyzes problems of contour-line map trackings and presents a track-following algorithm which uses raster-scanning and border-following methods [4, 5]. This algorithm allows simultaneous following and thinning of the contour lines of a digital binary map. It is a one-pass algorithm in the sense that the pixels are scanned once with the only exception being the adjacent line elements.

2. CONTOUR-LINE MAPS AND DATA ACQUISITION

A contour-line map, such as an isolevel map, conveys information by means of thin lines on a contrasting background. The contour lines have the following properties (Figure 1):

  (i) they can never overlap;
  (ii) they can be close or
  (iii) their extremes can touch the borders of the plane in which they lie;
  (iv) they can be broken.

Figure 1. Contour-line map.

A map can be digitized either using the raster or vector method. In the first case the map is scanned in TV line-by-line fashion and converted into an array of numbers or grey-levels. However, even though the raster representation preserves the spatial form of a map and is a useful mode for displaying, it is not efficient in terms of computer memory. In the second case the map is digitized by an automatic line-following device and described in terms of a series of x, y pairs in a coordinate system.

It must be noted that most of the digitizers work in the raster-scan mode and thus it is necessary to transform the digital map from raster into vector format. Our algorithm works on maps digitized by a raster-scanning device. The following problems have been solved:

2.1. Transform of a digital map into a binary map

The digitization process causes the digital line maps to have more than two grey-levels, therefore maps have to be converted into a binary map by threshold techniques during the acquisition process. Through these line thickness has been reduced without introducing line interruptions.

2.2. Line thinning and following

The contour lines normally occupy only a small fraction of the total area and their thickness can be greater than a pixel. In order to obtain an efficient data representation the track-following algorithm performs line following and line thinning at the same time. The algorithm works correctly on variable-width contour lines of the map. The maximum value of the width must be known.

3. THE TRACK-FOLLOWING ALGORITHM

The track-following algorithm performs sequentially the tracking of the contour lines. For each contour line the algorithm consists of two steps:

(1) In the first step it starts in the raster-scanning mode and continues until it comes across the first pixel of the contour line.

(2) In the second step it changes into line-tracking mode using either vertical or horizontal direction only. The thickness of the line is thinned down to only a pixel by detecting the line centre.

The contour line is considered as a sequence of horizontal or vertical segments; each segment can have a width less or equal to T (where T is the maximum value of the line thickness). A line segment is identified by means of the extremes coordinates (left-right or upper-lower borders). The algorithm recognizes the current direction of the contour line on the basis of the width of the current horizontal vertical segment.

To initialize the tracking direction, three different kinds of contour-line data have been examined:

(1) The first concerns the contour line whose first line element lies on the upper plane border (Figure 2, line a). In this case the contour line is initially followed in the horizontal mode.

(2) The second concerns the contour line whose first line element is inside the plane. The contour line can be initially followed in horizontal or vertical mode in counter-clock sense. Moreover, if the contour line terminates on one of the plane borders (Figure 2, line b) the track following starts again in clockwise mode for the second line part. Particular configurations of this case can happen when the contour line is closed (Figure 2, line c) or the contour line is broken.

(3) In the third case the first line element lies on the lateral plane borders. The contour line is initially followed in vertical mode.

The following variables are introduced: NCOLS is the pixel number in each raster line, NLINES is the line number, CURLIN is the current line, CURCOL is the current column, MAP is the input value of the current pixel, LINE-VALUE is the contour line grey-level, WIDTH is the current segment width, CENTER is the current segment centre. Using these notational conventions the track-following algorithm is shown in Figure 3.

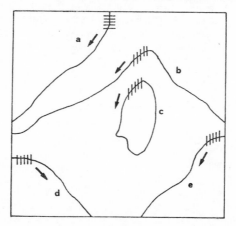

Figure 2. Four different kinds of contour-line data.

In the first step the algorithm analyzes the map in raster mode until the first line element is detected, then it recognizes one of the above configurations to initialize the tracking mode.

In the second step it changes into line-tracking mode by detecting the coordinates of the second line segment in the initial direction. The segment width is compared with the T value: if the segment width is less or equal to T then the segment centre becomes the current line pixel otherwise the direction is changed into the other and the procedure is repeated, as in the previous case, by testing the value of the width segment. Now the algorithm continues to track in the current direction with transactions from one direction into the other (Figure 4). When two successive segments with different directions are recognized the segment centres are interpolated.

If the recognized pixel is on the plane border or equal to the first pixel, or if the contour line is broken, the algorithm stops. There is a limited number of legitimate configurations which are recognized as adjacent contour lines (Figure 5).

```
for I=1 to NLINES;  for J=1 to NCOLS;
  begin
  if      MAP(I,J) = LINE_ VALUE then do
     begin
     set   CURLIN = I; set CURCOL = J; set ENDTRC = false
     Initialize MODE with horizontal or vertical direction
     while ENDTRC = false do
        begin
        while MODE = 'HORIZONTAL' and ENDTRC = false do
           begin
           compute LEFT, RIGHT, CENTER, WIDTH
           if WIDTH>T then do
                   begin   MODE = 'VERTICAL'; FLAGOR = 1
                   end
           else do
                   begin
                    Save center coordinates (CURLIN, CENTER)
                    Interpolate current center segment with previous one
                    Examine next segment in order to set ENDTRC
                    Evaluate CURLIN and CURCOL
                   end
           end
        while MODE = 'VERTICAL' and ENDTRC = false do
        begin
         Compute UPPER, LOWER, CENTER, WITH
         if WIDTH>T then do
                 begin
                 MODE = 'HORIZONTAL'; FLAGVER = 1
                 end
            else do
                    begin
                     Save center coordinates (CENTER, CURCOL)
                     Interpolate current segment center with previous
                     Examine next segment in order to set ENDTRC
                     Set FLAGVER=0
                    end
        end
        if FLAGOR = 1 and FLACVER = 1 then
         Evaluated adjacent configuration
        end
     end
  end; end
```

Figure 3. The track-following algorithm.

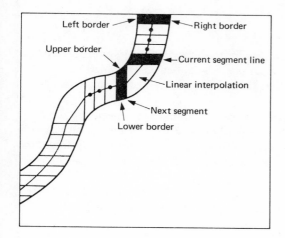

Figure 4. Example of track following.

Figure 5. Examples of adjacent contour lines.

(a)

(b)

Figure 6. (a) Original digital map.
(b) Track following of contour lines after the algorithm.

4. RESULTS AND CONCLUSIONS

Our algorithm has been implemented on a VDC-501 display connected to a PDP11/45 computer running under an RSX11M operating system. The VDC-501 system consists of 512x512x10 bits refresh memory. The refresh memory is used as working storage and it is addressable in eight directions to minimize the I/O time as required during the program running.

The algorithm is a one-pass algorithm and the execution time depends on the map complexity. For example, a map of 512x512 pixels with a contour line density of 10% requires about 20 s. The algorithm uses only buffer space to store the map in vector format. No extra memory is needed to store labelling or reference pointers of the contour lines.

The algorithm has been applied to the picture shown in Figure 6(a); the results are shown in Figure 6(b). The digital map has been obtained by using a flying spot scanner with 20 μm resolution.

REFERENCES

1. A. Rosenfeld and A. C. Kak, Digital Picture Processing, Academic Press, New York, 1976.
2. R. Cederberg, Chain coding and segmentation for raster scan devices, Computer Graphics and Image Processing 10, 1979, 224–234.
3. B. Kruse, A Fast Algorithm for Segmentation of Connected Components Binary Images, Proceedings of the First Scandinavian Conference on Image Analysis, 1980.
4. P. E. Danielsson, An Improvement of Kruse's Segmentation algorithm, Computer Graphics and Image Processing 17, 1981, 394–396.
5. H. Freeman, Map and Line-drawing Processing in J. C. Simon and R. M. Haralick, Digital Image Processing, D. Reidel, Holland, 1981, 363–382.

L. Caponetti, M. T. Chiaradia, A. Distante and N. Veneziani
GNCB-CNR
Università di Bari
Italy

12 An approach to the decomposition of complex figures

L P Cordella and G Sanniti di Baja

1. INTRODUCTION

The problem of describing the shape of a silhouette, i.e., of a uniform grey-level figure on a background, has been faced with many different approaches [1–11], mainly of structural type. The most common one considers the contour of the figure, finds on it a number of "critical points" (generally neck points, concavities, etc.) and, according to suitable rules, joins pairs of such points in order to get a partition of the figure into a set of non-overlapping components. The shape of the figure is then described in terms of the shapes of its components and of their spatial relationships. It is assumed that the shape of the components is more elementary than that of the figure and then easier to be described. Within this approach, methods like the graph-theoretic clustering one [6, 7] could be included, according to which, pairs of suitable contour points are joined by segments entirely internal to the figure, in order to detect regions of high compactness.

A different approach to shape decomposition is essentially based on the evaluation of local thickness [11]. The medial line of the figure, labelled according to the distance function from the background, is first appropriately partitioned into sets of pixels whose labels satisfy some given conditions; then, the regions obtained by applying a reverse distance transformation to such sets are considered and a decision is taken about the attribution of overlapping parts of the regions so as to achieve figure components.

We recognize that both curvature of the contour and local thickness are useful features in order to achieve an adequate description of a complex figure and that the labelled medial line (LML) of a figure holds the information about such features. In fact, thickness information is held by the labels associated to the pixels of the LML, while contour inflections are reflected by both label and direction changes along the LML (see later). It appears, then, reasonable to partition the LML into subsets along which no significant label and/or direction changes are detectable. In this way a correspondence is established between LML parts and simply shaped figure components. However, since the digital medial-line transform is not invariant with respect to noise and to isometric operations (rotation, translation, scaling), the previous partition would, consequently, not be invariant. Some appropriate merging among subsets seems suitable to overcome this drawback. After the partition process has been performed, a description of the figure can be given in terms of the LML components and of their interrelationships.

In the following the main steps of our partitioning procedure will be illustrated and some examples given.

2. LML PARTITIONING

Let us consider a binary picture and let us call a figure any connected set of 1-elements and a background the set of 0-elements. Moreover, let us label every pixel of the figure by its 8-distance from the background [12]. By the LML of the figure we mean the union of those pixels whose label is locally maximal and of those necessary to ensure the LML connectedness.

The presence of all the local maxima into the so-defined LML implies that this does not necessarily have unit width everywhere, but guarantees that all the information about the figure is preserved. In fact, the entire figure could be reconstructed by applying a reverse distance transformation to the LML. As will be clear in the following, a unit width medial line is more convenient for our purposes. The reduction of the LML to unit width by further thinning means some information about the original figure is lost. However, it can be reasonably assumed that such information is not significant for description purposes. We will then use a unit width LML and, for simplicity, will still refer to it as the LML of the figure.

In Figure 1 the pixels of a figure are labelled according to their 8-distance from the background. Pixels belonging to the LML are surrounded by a frame: among them, those characterized by a locally maximal label result to be equidistant from at least two points of the background, while the remaining ones are those necessary to ensure LML connectedness.

A preliminary segmentation of the medial line into simple curves can be immediately obtained by dividing it into branches, that is into connected sets of pixels joining two particular kinds of points: the end points and the branch points. The former have in their 8-neighbourhoods only one component of pixels belonging to the background and are connected to only 1 pixel of the medial line. Branch points, on the contrary, have more than two components of the background in their 8-neighbourhoods and are connected to more than 2 pixels of the medial line.

A further split of the branches thus found is achieved in the second phase of our procedure. This phase involves two steps. During the first step, the previously found

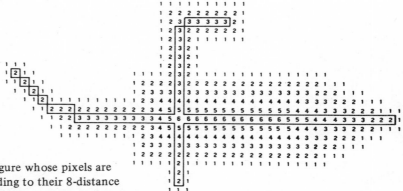

Figure 1. A figure whose pixels are labelled according to their 8-distance from the background. LML pixels are framed.

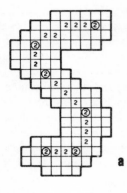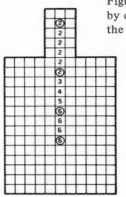

Figure 2. Contour inflections reflected by direction (a) and label (b) changes of the LML.

a

b

branches are divided into arcs by using curvature information, while during the second step the obtained arcs are split into segments by using width information.

The first step is accomplished as follows. Starting from every branch point (or in the case of a lack of branch points from a suitably chosen pixel) the branches which have as one extreme that branch point are successively traversed while a polygonal approximation of merging type is performed. In practice, while tracing the branch we consider as candidate vertices those pixels in correspondence to which a direction change is detected. As soon as a candidate vertex is found such that, with respect to the straight line joining it to the branch point, some intermediate candidate vertex has a distance which exceeds a given tolerance, the previous candidate vertex is taken as the second extreme of the arc started from the branch point. We keep track of its position and also consider it as the first extreme of the next arc. Starting from this pixel, the polygonal approximation is continued until a new vertex, an end point, or a branch point is met. At the end of this first step, the branches of the resulting LML are divided into arcs which, according to the chosen tolerance, are straight lines. Each arc corresponds to a planar part of the figure whose medial axis is oriented, in the (x, y) reference system, so the straight line joins the two extremes of the arc. On the other hand, strong curvature inflections on the contour of the figure may exist which are not detectable along a branch of the medial line from a geometrical point of view. Such inflections are reflected by width change rather than by orientation change (see for instance Figure 2). Hence, we perform the second step in order to achieve a further split of the previously found arcs.

A simple hypothesis, to start with, would suggest splitting the arcs into a number of segments each made by only one connected set of equilabelled pixels, excluding those sets made by a pixel whose label is not locally maximal. If we divide the arcs in this way, the planar parts corresponding to the obtained segments should be characterized by a constant width. However, it does not seem convenient to follow this hypothesis since it would produce partition into too many parts. Hence, we devise grouping together adjacent connected sets of equilabelled pixels provided that the segment so obtained allows a simple description of the region corresponding to it.

In order to perform the second step of our procedure, every previously found arc is represented on a reference system where the abscissae and the ordinates are, respectively, the order number in which every pixel is encountered when traversing the arc and its attached label. In this way a digital curve in the (pixel, label) plane is obtained. The curve is traced and a polygonal approximation guided by label change is performed, according to the same criteria used in the first step.

At the end of this last partitioning, the LML which results is divided into segments, each of which is characterized by a constant slope in both the (x, y) and the (p, l) planes. However, it is easy to see that segments characterized by a unitary slope in the (p, l) plane are unnecesssary for descriptive purposes (see, for instance, the central segment in Figure 2(b)). In fact, their pixels are all nonlocal maxima except one extreme. In practice, in order to minimize effects due to nonsignificant contour inflections, segments having a slope between 3/4 and 1 are also considered unnecessary: all such segments vanish or

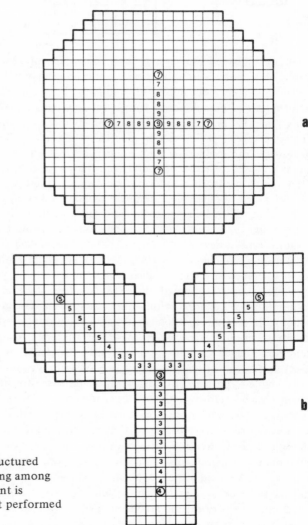

Figure 3. Two differently structured figures and their LMLs. Merging among segments sharing a branch point is performed in (a), while it is not performed in (b).

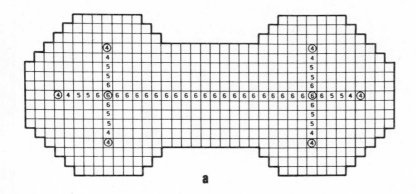

a

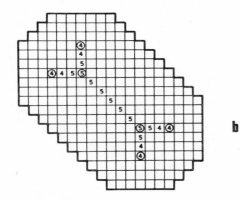

b

Figure 4. Merging between groups of segments is performed in (b), while it is not performed in (a).

collapse into the extreme maximal label according to whether such an extreme is preserved, or not, as part of another segment.

Although each LML segment allows a simple description of the region corresponding to it, the partition obtained does not seem to be completely satisfactory in order to achieve a reliable description of the whole figure. In fact, the medial line transformation may be biased by digitization noise. Moreover, due to the non-Euclidean metric adopted when working on a square grid, different medial lines are obtained when the figure is rotated or when a scale change is performed. As a consequence, the description or the recognition tasks become rather complicated. A reduction in the number of medial line components is likely to facilitate the next step of describing the figure. It now seems convenient to perform an appropriate grouping among the previously found segments. Not all the segments of the partitioned LML have to be taken into account, only the ones sharing a branch point. In fact, the reason for considering branch points as candidate vertices was of an operative nature: we needed simple curves on which to perform polygonal approximations. It seems reasonable to merge the segments corresponding to compact parts of the figure whose shape is essentially characterized by the label of the branch point. For illustration, let us consider Figure 3, where two differently structured figures are shown, whose LMLs both contain a branch point. In these and similar cases,

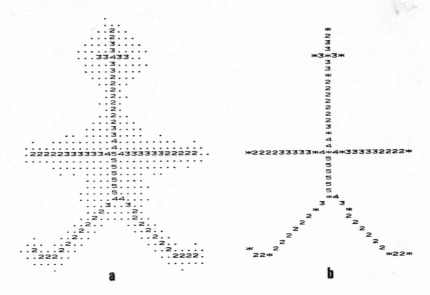

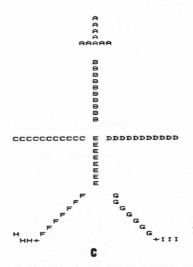

Figure 5. An example of LML partition:
(a) the input figure (dotted) and its LML
(superimposed labels); (b) LML
decomposition into segments (whose
extremes are starred); (c) the final
partition (every LML component is
assigned a different label).

a simple general rule seems applicable: segments sharing a branch point are merged provided that their lengths are sufficiently short with respect to the label of the branch point, and that label values increase along a segment in the direction leading to the branch point. From a perceptual point of view, this means that the branch point is, in such case, the nucleus of a compact region whose rough description is not influenced by the merged segments which only contribute to its detailed description. Analogously, merging among two or more groups of segments linked by single segments is also performed. This further process allows, for example, the description of Figure 4(a) as being composed of two lobes (corresponding to the two groups of segments) and a neck (corresponding to the segment joining the two branch points). On the contrary, Figure 4(b) is not decomposed.

The three main steps of the outlined procedure are illustrated in Figure 5, where the LML superimposed on the input (a), its partition into segments (b), and the final LML components after grouping among segments (c) are shown. Every component in Figure 5(c) is assigned a different label. Adjacent segments share a common vertex (crossed) playing the role of a hinge. Although the segments unnecessary for descriptive purposes are not represented in this printout, we keep track of their existence so as to establish the interrelationships among medial line components.

3. CONCLUSIONS

The medial line partitioning so far outlined could be accepted as a starting point for achieving a structural description of the whole figure. For instance, a labelled graph can be devised whose nodes are the LML components. Adjacency among the LML components is mapped into links between nodes. Every node is characterized by the features of the represented LML component (such as slope of the segment in the (p, l) plane, maximum label value, etc.), while links are characterized by the type of relation between LML components (for instance, angle between adjacent segments).

Nevertheless, for both reliability and effectiveness it seems convenient to perform further merging and/or splitting which could not have been simply implemented until now. Our present research is directed to improving the outlined method for describing a figure and to performing experimental work to verify if, according to our approach, satisfactory shape matching or similarity determination can be achieved.

ACKNOWLEDGEMENTS

The help of Mr Umberto Cascini and Mr Salvatore Piantedosi in preparing the illustrations is gratefully acknowledged.

REFERENCES

1. H. Blum, A transformation for extracting new descriptors of shape, in: W. Wathen-Dunn (ed.), *Models for the perception of speech and visual form* (Cambridge, MA, MIT Press, 1967) 362–380.
2. K. Maruyama, A study of visual shape perception, R-72-533, Dept. of Comp. Sci., Univ. of Illinois, Urbana (1972).

3. H. Y. Feng and T. Pavlidis, Decomposition of polygons into simpler components: feature generation for syntactic pattern recognition, *IEEE Trans. on Computer* C-24 (1975) 636–650.
4. T. Pavlidis, A review of algorithms for shape analysis, *CGIP* 7 (1978) 243–258.
5. H. Blum and R. N. Nagel, Shape description using weighted symmetric axes features, *Pattern Recognition* 10 (1978) 167–180.
6. L. G. Shapiro and R. M. Haralick, Decomposition of two-dimensional shapes by graph-theoretic clustering, *IEEE Trans. on PAMI* PAMI-1 (1979) 10–20.
7. L. G. Shapiro, A structural model of shape, *IEEE Trans. on PAMI* PAMI-2 (1980) 111–126.
8. C. M. Bjorklund and T. Pavlidis, Global shape analysis by k-syntactic similarity, *IEEE Trans. on PAMI* PAMI-3 (1981) 144–155.
9. W. S. Rutkowski, Shape segmentation using arc-chord properties, *CGIP* 17 (1981) 114–129.
10. A. J. Nevins, Region extraction from complex shapes, *IEEE Trans. on PAMI* PAMI-4 (1982) 500–511.
11. C. Arcelli and G. Sanniti di Baja, Shape splitting using maximal neighborhoods, *Proc. 6th Int. Conf. on Pattern Recognition*, Munich (1982) 1106–1108.
12. A. Rosenfeld and J. L. Pfaltz, Sequential operations in digital picture processing. *J. ACM* 13 (1966) 471–494.

L. P. Cordella and G. Sanniti di Baja
Istituto di Cibernetica del CNR
Via Toiano 6
80072 Arco Felice
Napoli
Italy

13 Image segmentation by discrete relaxation

M T Capria, S Di Zenzo and M Poscolieri

1. INTRODUCTION

Consider a case of context-independent classification, and assume the feature vector to be of length n. We then have a n-dimensional feature space and will partition it into a certain number of decision regions, one for each class of interest.

If we disregard one of the components, say the nth, all of the feature space becomes projected over the subspace spanned by the first $n-1$ components. What we get is a $(n-1)$-dimensional feature space where decision regions are no longer pairwise disjoint: they will in general be pairwise overlapping.

Disregarding the context (for image: the spatial domain) under some aspects resembles disregarding a component of the feature vector. So one way of handling the context seems to be to allow for mutually overlapping decision regions in feature space and use contextual information to resolve the ambiguities that arise as a consequence of that.

An efficient way of using contextual information is provided by relaxation techniques. In what follows, the use of discrete relaxation with higher-level constraints will be examined.

2. GENERALIZED COMPATIBILITIES

Basic notions on relaxation can be found in [1-4]. In this section we shall discuss a generalization of the compatibility coefficients $c(i, j; h, k)$ used in discrete relaxation. It is known that these coefficients express constraints on the simultaneous classification of pairs of pixels; namely, $c(i, j; h, k)$ is taken to be either 0 or 1 according to whether the labelling of pixel i as belonging to class j is incompatible or compatible with that of pixel h as belonging to class k. Here i and h denote pixel positions, $i=(i_1, i_2)$, $h=(h_1, h_2)$.

Various generalizations of this concept to express higher-order constraints, i.e., constraints involving more than two pels at the same time, have been proposed. We have adopted the following scheme. Let p_{ij} denote the probability that pixel i will belong to class j. Let $p_i=(p_{i1}, ..., p_{ic})$ be the vector of the probabilities of pixel i (c number of classes). In discrete relaxation these probabilities are taken to be 0 or 1.

Write $p_{ij}^{(m)}$ for p_{ij} as computed at mth iteration. Set:

$$p_i^{(m)} = (p_{i1}^{(m)}, ..., p_{ic}^{(m)}), \tag{1}$$

$$p_i^{(m)} = \{p_{i+k,j}^{(m)}: \; -s \leq k \leq s, \; 1 \leq j \leq c\}$$

$$= \{p_{i+k}^{(m)}: \; -s \leq k \leq s\}. \tag{2}$$

In eq. (2) i+k stands for (i_1+k_1, i_2+k_2), and $-s \leq k \leq s$ stands for $-s \leq k_1, k_2 \leq s$. Notice that $p_i^{(m)}$ is a three-dimensional matrix of type $(2s+1) \times (2s+1) \times c$ made of the probability vectors of the pels that lie within a $2s+1 \times 2s+1$ window centred at pel position i. In its standard form, discrete relaxation uses the following iterative scheme:

$$p_{ij}^{(m+1)} = \min_h (\max_k c(i, j; h, k) p_{hk}^{(m)}). \tag{3}$$

If in this equation the pixels relevant to i are all in a $2s+1 \times 2s+1$ window centred at i, then the equation itself is a special case of

$$p_i^{(m+1)} = f(p^{(m)}), \tag{4}$$

where f is a function on the set of all the matrices of type $(2s+1) \times (2s+1) \times c$ over $\{0, 1\}$ into the set $\{0, 1\}^c$. We shall add the condition:

$$p_i^{(m+1)} \leq p_i^{(m)}. \tag{5}$$

This condition is intended to hold term by term. It ensures convergence in a finite number of steps of each of the probability vectors. Its meaning is the following: if at any iteration pixel i ceases to be a candidate to class j, then it shall never be resumed as a candidate to that class.

The scheme (4), (5) stands on a level of broad generality and can be made specific in a number of different ways. The explicit form of the function f is dependent on the expected shape properties of the regions belonging to the various classes. If, for example, the regions of class j are expected to be convex, then f might be such that $p_{ij}^{(m+1)} = 1$ provided pixel i has at least three consecutive 8-adjacent pixels u, v, w such that $p_{uj}^{(m)} = p_{vj}^{(m)} = p_{wj}^{(m)} = 1$.

3. IMAGE SEGMENTATION BY DISCRETE RELAXATION

Overlapping decision regions have already been suggested by Rosenfeld and relaxation has been proposed by Rosenfeld *et al.*, so our claim of novelty is very limited. We tried to find a proper mix of the things along with some refinements that can improve the quality of the results. Our basic algorithm is described in this section, the refinements in the next.

Suppose that training sets are available for c classes. It is then possible to determine c decision regions in the feature space, for example, c ellipsoids. We do that without the constraint of pairwise disjointness (which incidentally provides the advantage that the size and shape of each region is freely determined by the corresponding observations only).

In general there will be pels which plot into points of the feature space that belong to only one decision region. In situations of reasonably good separation it is expected

·that at least half the pels enjoy this property: these pels are first classified based on their feature vectors only. (Except for very special images, these 'prototype pels' are expected to be 'dense' in the picture. Unless otherwise said, the majority of the pels will have some prototype pels in their neighbours.)

Set $p_{ij}^{(0)}=1$ if and only if pixel i plots into decision region j. Then relaxation is started. At each iterate the probability vectors of all the picture elements are updated. Only changes from 1 to 0 are allowed, so the uncertainty in the classification of every pixel cannót increase. Those pels for which only one component of the probability vector is 1 at the end of the relaxation process are considered as classified. The process ends when it becomes stationary.

In order to have the algorithm fully specified, we give a set of contextual rules which, taken altogether, provide a possible realization of the function f in eq. (4).

Let N(i) be the set of the pel positions 8-adjacent to i. Let C(i) be the set of those j such that $p_{ij}^{(m)}=1$. For every pixel position i:

(i) If $\sum_{h\epsilon N(i)} p_{hj}^{(m)} \leq 2$, set $p_{ij}^{(m+1)}=0$.

(ii) If $p_{ij}^{(m)}=1$ and $p_{hj}^{(m)}=|p_h^{(m)}| = 1$ for at least six of the $h\epsilon N(i)$, set $p_{ik}^{(m+1)}=\delta_{kj}$.

(iii) If there exists j in C(i) such that $p_{hk}^{(m)}=\delta_{kj}$ for at least six of the $h\epsilon N(i)$,

set $p_{ik}^{(m+1)}=\delta_{kj}$.

(iv) Suppose $p_{ij}^{(m)}=1$. If

   (a) N(i) contains three consecutive pels u, v, w such that $p_{uk}^{(m)}=p_{vk}^{(m)}=p_{wk}^{(m)}=\delta_{kj}$,

   (b) for every $h\epsilon N(i)$, $p_{hk}^{(m)}=1$ implies k=j or k not in C(i),

     set $p_{ik}^{(m+1)}=\delta_{kj}$.

These criteria still stand on a ground of large generality. They could be supported by very weak requirements on the local geometry of regions together with probability consider-ations. More specific and stronger criteria can be formulated if some knowledge of the expected shape properties of the regions is available. The above contextual criteria have been adopted in the experimentation (described below).

There is experimental evidence that, under certain circumstances, relaxation is able to fire a kind of cascade process: the deletions of certain labellings from those possible for certain pels may drag along other such deletions for the neighbouring pels, and so on.

4. REMARKS

Our first remark is about the estimation of decision regions. This point is crucial for the quality of the results. Decision regions can be estimated either parametrically or nonpara-metrically. Nonparametric raises the problem of storage occupation and, besides, requires more care to be handled correctly. However, there are cases where it is strongly advisable to leave the parametric environment, e.g., when the class probability density functions may have more than one mode. We implemented two methods, one parametric (ellip-soidal decision regions), the other nonparametric (decision regions of any shape estimated by the Parzen method).

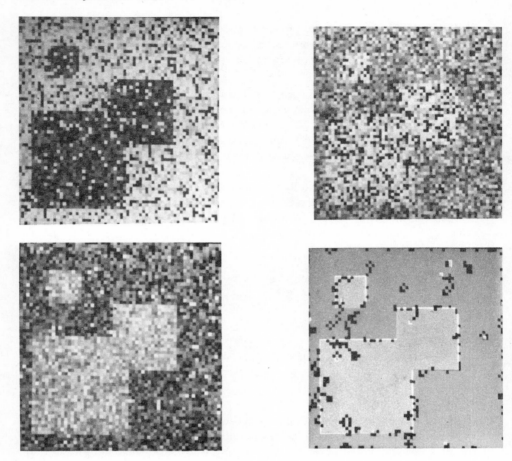

Figure 1

(a) Ellipsoids as decision regions. The training observations are used to train a multi-variate normal probability density function $p(x|j)$ for each class j. This in turn provides a family of ellipsoids in the feature space as surfaces of equal probability. As decision ellipsoid for class j one may take the smallest member of this family which contains all the training observations of class j. A more effective approach is described below.

(b) The Parzen method [5] with Gaussian kernels provides a class probability density function $p(x|j)$ for each class j as superposition of normal densities. This again provides a family of surfaces of equal probability (not constrained to some analytical form). Again, one may take the smallest member of this family that contains the training observation, or proceed as suggested below.

Once a class probability density $p(x|j)$ has been estimated, a decision region for class j is correctly obtained as the subset of the feature space, where $p(j|x) = p(x|j)p(j)/p(x)$ is above some threshold. So, the equation of the decision region for class j is

$$p(x|j)\, p(j)/\Sigma_{k=1}^{c}\, p(x|k)p(k) \geqslant \epsilon. \tag{6}$$

$p(j|x)$ is an estimate of the probability that the property vector x will belong to class j. This is the proper function of x to be thresholded. It behaves as a degree of membership (though its nature is very different): it is 1 when x almost surely belongs to j and 0 when almost surely it does not, and takes intermediate values when there is uncertainty.

However, if $p(j|x)$ is to be used, then there is the need to estimate $p(j)$, j=1, ..., c. $p(j)$ represents the fraction of the individuals of class j within the collection being investigated (a collection of picture elements in the case of an image).

That brings in one of the peculiar aspects of pattern recognition of collections, namely the one connected with *a priori* probabilities. Our second remark is indeed about this point. From (6) it can be seen that the size of each decision region depends on the fractions $p(j)$, j=1, ..., c. These can be readjusted after each iteration of the relaxation process. This amounts to expanding/shrinking the decision regions according to the rate of growth of each of them. Then, after each iteration, those p_{ij} for which pixel i no longer plots within decision region j are set to 0, thus contributing further entropy decrease.

5. EXPERIMENTAL RESULTS

We have tested the above algorithm on a two-class problem, applying it to artificial colour images representing light squares on a dark background. Actually, two images have been generated under different condition of separation between the probability densities of the object and the background along the three bands.

In the first image, the Bhattacharya distances between the two densities along the three bands were 0.64, 1.26, 1.38, reflecting a relatively good separation between the two classes. The analogous distances for the second image were 0.29, 0.26, 0.18, thus reflecting a situation of very bad separation. Both images were 64x64 in size. Border pels have been disregarded, so only 3844 have been considered for classification.

The results obtained are summarized in the following table:

	Image 1	Image 2
Total number of pels	3844	3844
Correctly classified	3587	3289
Misclassified	10	15
Rejected	27	56
Ambiguous	220	485

By 'ambiguous pels' we mean those pels i for which $p_{ij} = 1$ for more than one class j at the end of the relaxation process (i.e., when the iterative process has become stationary).

To understand how much uncertainty was removed by relaxation we can compare the results with the situation before entering relaxation:

	Image 1	Image 2
Total number of pels	3844	3844
Correctly classified	2885	2597
Misclassified	21	18
Rejected	97	132
Ambiguous	841	1097

It can be seen that in a case of good separation, the number of rejected/ambiguous pels reduces by a factor of about 4, while in a case of bad separation it reduces by a factor greater than 2. Figures 1(a)–(c) show the three bands of image 2 (the one with very bad conditions of separation). Figure 1(d) shows the final result of the classification.

REFERENCES

1. A. Rosenfeld, R. A. Hummel, and S. W. Zucker, Scene labeling by relaxation operations, IEEE Tr. Sys., Man, Cyber. 6, 1976, 420–433.
2. J. O. Eklundh, H. Yamamoto, and A. Rosenfeld, A Relaxation Method for Multi-spectral Pixel Classification, IEEE Tr. on PAMI 1, 1980, 72–75.
3. G. Fekete, J. O. Eklundh, and A. Rosenfeld, Relaxation: Evaluation and Applications, IEEE Tr. on PAMI 3, 1981, 459–469.
4. A. Rosenfeld and A. C. Kak, Digital Picture Processing (second edition), Academic Press, New York, 1982.
5. E. Parzen, On Estimation of a Probability Density Function and Mode, Ann. Math. Statist. 33, 1962, 1065–1076.

M. T. Capria and M. Poscolieri
Istituto di Astrofisica Spaziale del CNR
Via E. Fermi
Frascati
Roma
Italy

S. Di Zenzo
IBM Scientific Center
Via del Giorgione 129
Roma
Italy

14 A mathematical method for feature extraction from images

L Preuss

1. INTRODUCTION

This paper presents possible applications of an extended form of the Legendre transformation to feature extraction. In its simplest form, i.e., for a functional relationship involving only two variables, the Legendre transform replaces a curve representing that relation by the ensemble of all its tangents in the original space, and then maps each of these tangents into one point of the transformed, or Legendre space. It follows, intuitively, that any nearly straight segment of the curve generates a bundle of near-identical tangents which will be imaged in approximately the same point of the Legendre space. The number of tangents reflects the length of the segment, while their mutual similarity reflects the straightness of that segment.

If, now, it was possible to attribute to the points in Legendre space a weight proportional to both these magnitudes, then this would provide a useful tool for image processing in general, and for feature extraction in particular. An additional reason for the use of a Legendre transform in this context derives from the fact that (in contradistinction to the direct use of a derivative, for instance) it conserves the total amount of information [1].

However, in its classical form, the Legendre transformation applies only to single-valued, continuous, and derivable functions, as already implied by the curves and tangents mentioned above. Such functions are not available in image analysis, where the observed connections between an independent variable and measured values of the dependent variable are necessarily discrete and will generally be distributed in elongated clusters, rather than on an ideal line.

Therefore, I have introduced an extension of the Legendre transform applicable to discrete, multivalued distributions. Its rationale is the same as was set forth for the Legendre transform by Callen [1]. Although the extension applies to any number of dimensions, it will be sketched here for a relation between two variables only, since a full mathematical treatment is outside the scope of this note.

2. A HEURISTIC APPROACH

One can visualize an empirical relationship between an independent variable x and a dependent variable y as a more or less fuzzy, ribbon-shaped domain in which the observed connections between both variables are particularly dense. In terms of information, this means that any tangent to either side of the ribbon embodies a locally most effective division of the plane with respect to the connections represented therein.

In the above context, the term "tangent" is already used with a generalized meaning which connotes a quantitative definition which avoids the strict dichotomy requiring that a straight line either is a tangent to a given curve, or not. Thereby, it becomes possible to attribute to any straight line in the plane a continuously variable amount of "tangentiality", which will be called its weight F. The weight is proportional both to the number of observations within a small band centred on the line, and to the efficiency with which the band locally divides the plane. A measure of this efficiency is provided by the lengthwise stagger [2] within the band, where the latter may have a vertical extension of some two to five units of the y variable, for instance.

To obtain the dual of a distribution, the plane of origin is divided into parallel bands with a given slope S, and the weight F of each band is then attributed to that cell of the transformed space which has a value of the independent variable equal to S and a value of the dependent variable Y equal to the intercept along the y axis of the midline of the band in the original space. The operation is then repeated for all values of the slope S, yielding a distribution F(Y, S) in the transformed space. Numerical results will now be shown on the basis of a simple example.

3. AN ILLUSTRATIVE EXAMPLE

Figure 1(a) shows the contour of a cipher 5 written by hand with a felt pen. This was selected for analysis because its left-hand margin has a few characteristics which are of interest here. It contains two strokes which at first sight would appear straight. On closer inspection, however, both deviate noticeably from a straight line. Also, the curve has two regions (identified in Figure 1(a) by the letters M and N) with curvatures that are not very different, although an observer would classify region M as an acute corner separating two elements, and region N as integral part of an elliptical segment.

The aim now is to decompose such a curve by a mathematical process which introduces no empirical threshold value, neither an arbitrarily predetermined one, nor one obtained through a learning process.

Obviously, one main parameter of interest here is the curvature of the analyzed line, which is defined as its change of orientation per unit length. In order to introduce this

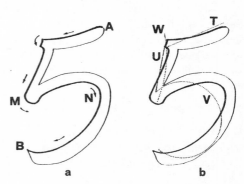

Figure 1   a   b

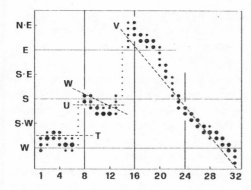

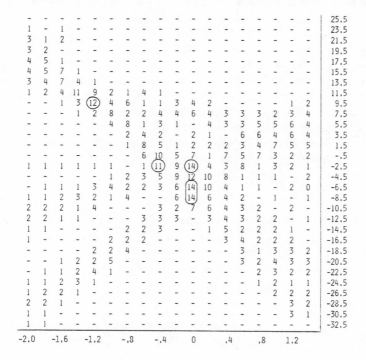

Figure 2

parameter as a derivative, the orientation of the curve was measured starting from point A and moving along the curve towards point B, in 32 discrete steps. Within each step the orientation was measured in units of $11.25°$.

These measurements result in the dots shown in the diagram of Figure 2, where the distance, as measured from A along the curve (and following all its bends) is entered along the x axis, whilst the orientations of the curve in Figure 1(a) are entered along the y axis of Figure 2. It should be kept in mind that this latter diagram has only a preparatory function in that it provides the proper variables for a subsequent Legendre transformation.

The irregularities in Figure 2 are typical for results obtained when scanning an image with a reasonably high but not extreme resolution. In spite of these irregularities, one

-	-	-	1	-	-	-	-	-	-	-	-	-	-	-	-	-	-	25.5
1	-	1	-	-	-	-	-	-	-	-	-	-	-	-	-	-	-	23.5
3	1	2	-	-	-	-	-	-	-	-	-	-	-	-	-	-	-	21.5
3	2	-	-	-	-	-	-	-	-	-	-	-	-	-	-	-	-	19.5
4	5	1	-	-	-	-	-	-	-	-	-	-	-	-	-	-	-	17.5
4	5	7	1	-	-	-	-	-	-	-	-	-	-	-	-	-	-	15.5
3	4	7	4	1	-	-	-	-	-	-	-	-	-	-	-	-	-	13.5
1	2	4	11	9	2	1	4	1	-	-	-	-	-	-	-	-	-	11.5
-	-	1	3	(12)	4	6	1	1	3	4	2	-	-	-	-	1	2	9.5
-	-	-	1	2	8	2	2	4	4	6	4	3	3	3	2	3	4	7.5
-	-	-	-	4	8	1	3	1	-	4	3	3	5	5	6	4		5.5
-	-	-	-	-	2	4	2	-	2	1	-	6	6	4	6	4		3.5
-	-	-	-	-	1	8	5	1	2	2	2	3	4	7	5	5		1.5
-	-	-	-	-	-	6	10	5	7	1	7	5	7	3	2	2		-.5
1	1	1	1	1	1	-	1	(11)	9	(14)	4	3	8	1	3	2	1	-2.5
-	-	-	-	-	1	2	3	5	9	(12)	10	8	1	1	1	-	2	-4.5
-	1	1	1	3	4	2	2	3	6	(14)	10	4	1	1	-	2	0	-6.5
1	1	2	3	2	1	4	-	-	6	(14)	6	4	2	-	1	-	1	-8.5
2	2	2	1	4	-	-	-	3	2	7	6	4	3	2	-	2	-	-10.5
2	2	1	1	-	-	-	3	3	3	-	3	4	3	2	2	-	1	-12.5
1	1	-	-	-	-	2	2	3	-	-	1	5	2	2	2	1	-	-14.5
1	-	-	-	-	2	2	2	-	-	-	-	3	4	2	2	2	-	-16.5
-	-	-	-	2	2	4	-	-	-	-	-	3	1	3	3	2		-18.5
-	-	1	2	2	5	-	-	-	-	-	-	3	2	4	3	3		-20.5
-	1	1	2	4	1	-	-	-	-	-	-	-	2	3	2	2		-22.5
1	1	2	3	1	-	-	-	-	-	-	-	-	1	2	1	1		-24.5
1	2	2	1	-	-	-	-	-	-	-	-	-	-	2	2	2		-26.5
2	2	1	-	-	-	-	-	-	-	-	-	-	-	-	3	2		-28.5
1	1	-	-	-	-	-	-	-	-	-	-	-	-	-	3	1		-30.5
1	1	-	-	-	-	-	-	-	-	-	-	-	-	-	-	-		-32.5

| -2.0 | -1.6 | -1.2 | -.8 | -.4 | 0 | .4 | .8 | 1.2 |

Table 1

recognizes the mapping of the upper (approximately westerly) stroke in the interval from x=0 to 6, that of the subvertical stroke in the interval from x=8 to 12, and that of the main loop to the right of x=16. If this loop was exactly circular, the cluster sloping downward from x=16 to 32 would be exactly straight.

Figure 2 represents the function, or more accurately the distribution, which is subjected to an extended Legendre transformation, resulting in the density distribution shown in Table 1. Here the x-variable is a measure of the slope in the preceding diagram, that is of the curvature in the original image. The values on the ordinate correspond to the intercept along x=14 in the diagram of Figure 2. For clarity the density values have been heavily rounded, since only their relative values are relevant. One finds two large maxima in the column for zero curvature, another one for x= −1.2, and finally one for x= −0.4. We shall not discuss here a few minor ripples (mostly due to an insufficient resolution) which appear in regions of low densities, at the lower right of the table for instance.

If the maxima are ordered by decreasing values then they correspond − in the space from which the extended Legendre transformation originates − to the straight segments T, U, V, and W entered in Figure 2, taken in that order. In the analyzed image itself, these segments identify the elements shown in dotted lines in Figure 1(b) (and labelled by the same letters) as possible features. These are: an upper and a lower straight segment (T and U), a circular loop (V), and, finally, a second, slightly curved variant (W) of the subvertical stroke.

As already stated, no threshold value was introduced in the calculations. In particular, no maximal value of the curvature was determined beyond which a line segment of the image would be identified as a corner separating two elements.

The fit of the elements found with the original curve is admittedly only just acceptable. Obviously, the given curve cannot be decomposed into straight and circular parts unless a high number of them is used. If, for instance, the loop was approximated by several arcs, a very good fit could be obtained. But then the elliptical loop would be decomposed into an aggregate of at least three different arcs and this would destroy its quality as a homogeneous feature, which in the present context is more fundamental than the precision of the fit. In fact, the comparative looseness of the latter is also borne out by the fact that, in the transform, the maximum which corresponds to the loop is smaller than those attributed to the shorter but better matched strokes.

4. CONCLUSION

The simplicity of the decomposition obtained is due to the fact that the method introduced here characterizes features as local maxima of efficiency in a representation of information, in the sense of Shannon. The method thus not only provides a procedure for feature extraction, but also an implicit definition of the very term "feature".

ACKNOWLEDGEMENT

This work was supported by the Swiss National Science Foundation (Grant Nr. 2,854-0,80), which is gratefully acknowledged.

REFERENCES

1. H. B. Callen, Thermodynamics (Wiley, New York, 1960).
2. L. G. Preuss, A Class of Statistics based on the Information Concept; Comm. in Statistics, Vol A9 Nr. 15 (1980).

L. Preuss
Feldeggstrasse 74
8008 Zurich
Switzerland

15 Some experiments in parameter encoding of video signals

L Capo, A Giglio and G Zarone

1. INTRODUCTION

The aim pursued by image coding is "to minimize the number of bits required to transmit pictures for a given level of distortion". The hitherto deeply investigated methods fall into "waveform encoding", i.e., they try to replicate the original waveforms by mainly removing the statistical redundancy present in a high degree in a video signal.

The rate distortion theory gives a solution to this problem. However, there are some difficulties in using the results of this theory. Above all it does not tell us how to synthesize a practical coder. Moreover, it is difficult to compute the necessary "rate distortion function" for many realistic models of the picture source and distortion criteria, due to the lack of exhaustive statistical models for video signals and to the lack of a widely accepted distortion criterion [1].

A completely different approach is introduced by parameter encoding, whose aim is not the faithful reconstruction of the original waveforms, but the extraction and transmission of only the basic features necessary for a specific application. Since the ultimate recipient of the information is a normal observer, great importance must be given to perceptual and psychophysical aspects of human vision in determining the basic features to be taken into account.

It is well known that a lot of information is given to the viewer by the contours of the image objects [2, 3]. Only the contour set, in fact, allows the recognition of a picture immediately and unambiguously, although it does not provide the quality required by TV standards. The contour may be segmented in prefixed geometric parts (i.e., circle arcs); these parts may be represented by fairly efficient parameter codes. Rendition of details, shading, etc., within object boundaries requires additional data to be transmitted. Therefore, a dramatic bit rate reduction is expected to be achieved by characterizing an image with only a few parameters and some additional information.

The present paper deals with some experiments in contour extracting and coding. Because of the extremely high temporal redundancy of typical videotelephonic signals, not only intraframe but also interframe coding techniques will be investigated. Then, the coding of regions within boundaries (implemented by interpolative methods) will be examined. Finally, the results, obtained by applying these techniques to a real videotelephonic sequence, will be given. The simulated coding process is represented by the block scheme of Figure 1.

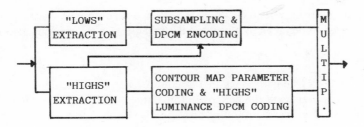

Figure 1. Block scheme of the investigated coder.

2. CONTOUR EXTRACTION

Contour extraction is made up of three steps: detection, thinning and tracking.

2.1. Detection

Detection consists in recognizing the pels which are near sharp luminance transitions and are possible contour elements. It takes place with application of a local detection operator. The application to any pixel of a detection operator provides a "gradient" matrix; the highest values of this matrix are at contour lines. The sizes of pel neighbourhood, whose elements concur to determine a gradient value, depend on the required accuracy, on the number of quantization levels, and on the S/N ratio [4]. In our experiments we used some known operators, such as Laplacian, Sobel, and Robert Cross operators [5]. Moreover, we tested the following operator:

$$G = \max(|GX|, |GY|),$$

where $GX = L(0)-L(N)$, $GY = L(6)-L(N)$, $L(N)$ is the luminance value of the pel at which G is to be computed, and $L(0)$ and $L(6)$ are the luminance values of adjacent pels on the same row and on the same column (see Figure 2).

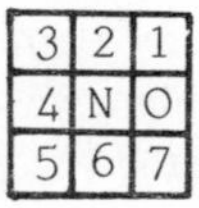

3	2	1
4	N	O
5	6	7

Figure 2. Elementary cell.

Note that the neighbourhood size of the last operator is less than the neighbourhood size of the other above-mentioned operators. It allows us to achieve higher accuracy in contour detection. In general, as the neighbourhood sizes decrease, the vulnerability of the contour extraction process to noise increases. However, this drawback can be neutralized by an appropriate choice of tracking algorithm, as we will see later.

We emphasize that:

(1) High gradient values cluster near contour lines.

(2) Because of special illumination conditions or because of noise, either contour pels may be characterized by low gradient values or flat area pels may be characterized by high gradient values.

(3) Moving-object blurring, due to camera integration, does not make sharp contours.
These problems need to be taken into account in implementing the tracking algorithm.

2.3. Thinning

The clustering of high gradient values in the contour neighbourhood would force the
tracking algorithm to duplicate the actual contours and would introduce false contours.
In order to avoid this drawback, a thinning operation is necessary.

It is likely that a pel of the actual contour is located at the highest value in a cluster.
The thinning operation was, therefore, implemented as follows. A gradient value is reset
to zero if its value is not the maximum once it is compared with the gradient values
pertinent to the adjacent elements on the same row and on the same column, or (see
Figure 2)

$$\text{IF}((G(N).NE.M1).AND.(G(N).NE.M2)) \text{ THEN } G(N)=0,$$

where $M1 = \max(G(N), G(0), G(4))$ and $M2 = \max(G(N), G(2), G(6))$.

2.3. Tracking

Contour tracking consists of recognizing the pels which belong to the same contour. In
accomplishing this aim we exploited the following properties:
(1) The contour must be continuous.
(2) A contour element must have a high detection operator value.
(3) The correlation between following pel luminances of the same contour is high
(congruence property).
(4) The contour lines have regularity requirements (structural property).
The tracking procedure operates according to two different modes: a "start-point" mode
and a "contour-tracing" mode.

In the "start-point" mode, the gradient matrix is scanned. In the scanning every
gradient value is compared with a threshold. The threshold getting through locates a
start point. As soon as a start point is located, the "contour-tracing" mode is performed.
This consists of searching the next contour element. The region where this element is
searched is the elementary cell, that is the eight pels neighbouring the element just traced.
For each of these eight elements we calculated the function:

$$f = G + F + W1 + W2$$

where G is the above detection operator; F is a non-increasing function of luminance
difference between present and previous contour elements, therefore, it takes into account
the high correlation existing between luminance values of adjacent pels of the same
contour; $W1 \neq 0$ for the only element in line with previous ones of the same contour; and
$W2 \neq 0$ for the only element which makes contour closing easier.

The element that maximizes this function is taken as the next element. The search
stops when function f falls below a given threshold. This locates the contour "end point".
Once a contour is completely traced, the "start-point" mode is performed again.

The presence of F, W1, and W2 weights allows us to favour the direction for which
contour prosecution is more likely and to neutralize noise effects and contour singularities

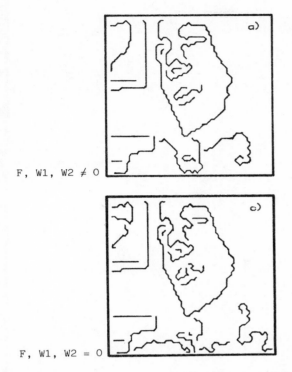

Figure 3. The effect of terms F, W1, and W2.

F, W1, W2 ≠ 0

F, W1, W2 = 0

(gaps and ramification points). The structure of function f is additive to keep computation time low. However, no significant improvement was obtained by giving f a multiplicative structure. The optimum W1 and W2 values were found by heuristic search. They
are W1=2 and W2=1. Figure 3 shows the improvement obtained in contour extracting by
using the terms F, W1, and W2.

3. CONTOUR CODING

3.1. Intraframe contour coding

The well-known Freeman coding requires 3 bits per contour element. This figure can be
lowered to 2 bits if one adopts a differential technique. A further improvement can be
accomplished by an entropy coding. However, Freeman coding may represent an intermediate step even in a more sophisticated coding, since it presents some very interesting
properties [6].

3.2. Interframe contour coding

More efficient contour coding is obtained if a prediction of present contour, based on the
previous frame contour, is available. In this the circumstance that displacements are very
small (in general < 4 pels/frame) can be exploited. Besides, the translations are nearly
rigid. Prediction is achieved by computing the displacement of the contour as a whole.

Figure 4. An example of overlap between
the predicted (L1) and present (L2)
contours.

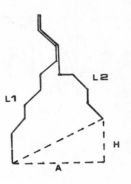

Figure 5. Actual overlap between the
predicted (L1) and present (L2) contour.
1, L1 − (L1∩L2); 2, L2 − (L1∩L2);
3, L1∩L2.

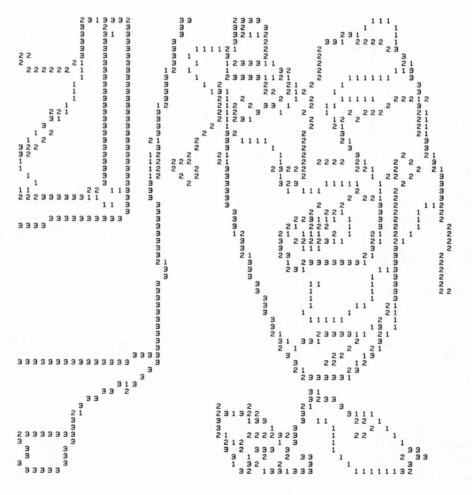

The present contour will be different from the predicted one owing to deformations due to movement (since contours are two-dimensional projections of 'three-dimensional objects). Only significant deformations have to be transmitted.

Displacement as a whole is computed by a correlation technique [7]. Let $C1(x, y)$ and $C2(x, y)$ be the functions that describe the previous and present contours.

$$C1(x, y), C2(x, y) = \begin{array}{l} 1 \text{ if the } (x, y) \text{ pel belongs to contour,} \\ 0 \text{ else.} \end{array}$$

It is decided that a displacement $D^* = (\Delta x^*, \Delta y^*)$ occurs if the $(\Delta x^*, \Delta y^*)$ pair makes a correlation function between the previous and present contour maximum.

$$CORR(\Delta x, \Delta y) = \sum_{x, y} C1(x+\Delta x, y+\Delta)C2(x, y),$$

where $x=1, ..., x_{max}, y=1, ..., y_{max}$, and $\Delta x, \Delta y = -4, ..., 4$.

$$\Delta x^*, \Delta y^*: CORR(\Delta x^*, \Delta y^*) = \max_{\Gamma} CORR(\Delta x, \Delta y),$$

where $\Gamma = |-4, ..., 4| \times |-4, ..., 4|$. Once the displacement is computed, overlap between the predicted and present contours appears, generally, as in Figure 4.

The segment that joins end points of two contours is defined "closure". In order to keep computation time low, we assumed

$$CH = \max(A, H)$$

as an estimate of the closure length.

A deformation is regarded as significant, or not, by implementing the test:

IF(CH.GT.T1) THEN the deformation is significant,
ELSE IF(AR.GT.T2) THEN the deformation is significant,
ELSE the deformation is not significant,

where T1 and T2 are two appropriate thresholds and AR is the area delimited by closure, present and predicted contours. An actual overlap of the present and predicted contours is depicted in Figure 5.

Table 1. Contour coding average cost

	Whole sequence	Frame 2
Intraframe (bits/pel)	0.28	0.24
Interframe (bits/pel)	0.22	0.12

By using this technique an average 20% bit rate saving has been obtained (see Table 1). A refinement of the displacement estimate is expected to improve the overall performance.

```
                              contour
                                 ↓
        A  A  A  A  Z  Y  X│X  Y  A  Z  A  A  A  A  A  Z  A  A
        A  Z  A  A  A  Y  X│X  Y  A  A  A  A  Z  A  A  A  A  A
        A  A  A  A  Z  Y  X│X  Y  A  Z  A  A  A  A  A  Z  A  A
        A  Z  A  A  Y  X│X  Y  A  A  A  A  A  Z  A  A  A  A  A
        A  A  A  Y  X│X  Y  A  A  A  Z  A  A  A  A  A  Z  A  A
        A  Z  Y  X│X  Y  A  Z  A  A  A  A  A  Z  A  A  A  A  A
        A  A  Y  X│X  Y  A  A  A  A  Z  A  A  A  A  A  Z  A  A
        A  Z  Y  X│X  Y  A  Z  A  A  A  A  A  Z  A  A  A  A  A
```

Figure 6. "Highs" (X and Y pels) and "Lows" (A and Z pels) characterization.

4. "HIGHS" SYNTHESIS AND "LOWS" RECONSTRUCTION

4.1. "Highs" synthesis

Not only the contour position, but the luminance of elements X and Y (which are near the contour, see Figure 6) need to be transmitted. Because of the spatial masking phenomenon, it is much more important to replicate the exact contour position than the exact luminance value at the contour. Therefore, an alternative coding scheme, where the X values were not transmitted but reconstructed at the receiver by interpolation on the ground of Y values, was investigated. The X and Y luminances were coded by a spatial DPCM.

Table 2. Main results

		Intraframe			Interframe
Frame					
HF/SF		1/5	1/7	1/7	1/7
VF/SF		1/2	1/2	1/2	1/2
Transmitted highs pels		X, Y	X, Y	Y	Y
1–8					
Lows luminance	(bits/pel)	1.06	0.74	0.74	0.27
Highs					
Position	(bits/pel)	0.31	0.31	0.31	0.12
Luminance	(bits/pel)	1.38	1.45	0.44	0.40
Total	(bits/pel)	2.75	2.50	1.49	0.79
MSSTE		1.22	1.80	2.85	3.41
9–16					
Lows luminance	(bits/pel)	1.03	0.72	0.72	0.33
Highs					
Position	(bits/pel)	0.26	0.26	0.26	0.10
Luminance	(bits/pel)	1.21	1.18	0.50	0.38
Total	(bits/pel)	2.50	2.16	1.48	0.81
MSSTE		1.03	1.69	2.42	3.80

4.2. "Lows" reconstruction

In order to code the regions inside the detected contours, the Z luminances, obtained by subsampling (see Figure 6) were transmitted. Several horizontal (HF) and vertical (VF) spatial sampling frequencies were utilized. The A pels were constructed at the receiver by interpolation on the ground of Y and Z grey levels. The Z values were coded by a spatial DPCM, following a given path (intraframe coding) or by a temporal DPCM (interframe coding).

5. RESULTS

The test signal is a typical videotelephonic sequence made up of 16 non-interlaced pictures. Every frame consists of 64 pels × 64 pels. Every pel luminance was digitized by a uniform quantizer with 256 output levels (8 bits/pel). Some statistics on this signal are given by Arena and Zarone [8].

To characterize the performance of the investigated coding techniques tested by the above signal, we used bit rate and MSSTE (Mean Square Supra Threshold Error). The last one is a parameter closely related to the subjective opinions of the observers on picture quality [9].

Table 2 shows the results obtained by intraframe and interframe coding. We kept the first eight and the last eight frames of the sequence apart. SF is the original spatial sampling frequency.

6. CONCLUSIONS

Some contour extraction algorithms, intraframe and interframe contour-coding techniques and lows reconstruction (by interpolative methods) have been investigated in the framework of the parameter picture coding. The available results are partial. The experimented methods need to be optimized. This study, in fact, is just a preliminary report on a deeper analysis to be carried out. The parameter encoding deserves particular consideration because it will undoubtedly allow us to achieve extreme bit rate reduction [10].

REFERENCES

1. A. N. Netravali, J. O. Limb, Picture coding: a review, *PIEEE*, vol. 68, March 1980, 366–406.
2. D. Gabor, P. C. J. Hill, Television band compression by contour interpolation, *PIEEE*, May 1961, 303–315.
3. D. H. Hubel, T. N. Wisel, Receptive fields, binocular interaction and functional architecture in the cat's visual cortex, *J. Physiol.*, 160 (1962) 106–154.
4. J. K. Aggarwal, R. O. Duda, A. Rosenfeld, Segmentation: boundaries, in: J. K. Aggarwal, R. O. Duda, A. Rosenfeld (eds), *Computer methods in image analysis*, IEEE Press, New York (1977) 181–182.
5. R. C. Gonzales, P. Wintz, Digital image processing, (Addison-Wesley Pub. Co., 1977).
6. H. Freeman, On the encoding of arbitrary geometric configurations, *IRE Trans. on Electr. Comput.*, vol. EC-10 (1961) 260–268.

7. D. I. Barnea, H. F. Silverman, A class of algorithms for fast digital image registration, *IEEE Trans. on Computer*, vol. C-21 (1972) 179–186.
8. L. Arena, G. Zarone, 3-D filtering of tv signals, *Alta Frequenza*, vol. 46 (1977) 108–116.
9. D. Sharma, A. N. Netravali, Design of quantizers for DPCM coding of picture signals, *IEEE Trans. on Comm.*, vol. COM-25 (1977) 1267–1274.
10. B. G. Haskell, R. Steele, Audio and video bit rate reduction, *PIEEE*, vol. 69 (1981) 252–262.

L. Capo, A. Giglio, and G. Zarone
Istituto Elettrotecnico
Facoltà d'Ingegneria
21 via Claudio
80125 Napoli
Italy

16 Scale-invariant image filtering with point and line symmetry

C Braccini and A Grattarola

1. INTRODUCTION

In the field of digital signal processing, and of image processing in particular, there is an increasing interest for application problems that cannot be dealt with by means of the classical linear shift-invariant processing methods, for which well-known implementation techniques have been developed, along with efficient algorithms and architectures. A particular class of such application problems is considered in this paper, namely the one characterized by the constraint that no distortion of the processed image must be induced by size changes of the input image. It is easy to verify that no shift-invariant processing, within the class of linear systems, has such a property, which we call scale-invariance (or form-invariance).

Scale-invariant processing is of fundamental interest in image filtering whenever an image of unknown size (e.g., a scene "seen" from an unknown distance) has to be first preprocessed, e.g., for noise reduction or edge enhancement, and then recognized by means of the available techniques for scale-independent detection [1, 2]. This is the case, to give one of the most representative examples, in the area of artificial vision, where a basic problem is the automatic recognition of objects arbitrarily located (with respect to the acquisition system) in space, and therefore of two-dimensional patterns independently from their position, rotation and scaling. Considering only the scale independence, it is clear that any preprocessing operation performed on the acquired pattern (of unknown scale) must be such that scale changes of the input are transformed into scale changes of the output, in order for the scale-invariant recognition methods to be applicable.

It has been shown elsewhere [3] that, within the class of linear shift-variant (LSV) systems, a whole family of scale-invariant filters actually exists, as described in the next section. It is worth while pointing out that the kernels characterizing some subclasses of such a processing describe a variety of physical problems of great interest in the field of image processing, like the modelling of certain optical degradation-restoration systems [4], the image reconstruction from line integrals [5], and the modelling of the peripheral visual system in man [6].

After the most general class of scale-invariant filters is introduced, in Section 2, and some related implementation aspects are discussed, the class of filters of interest in the applications mentioned above is analyzed in Section 3, where the possibility of exploiting their symmetry properties in order to reduce the computational complexity is discussed. In particular, it is shown that a suitable coordinate mapping can be used to transform

LSV processing based on kernels that are shift-invariant with respect to the angular coordinate into a shift-invariant filtering. Such kernels include filters with point and (radial) line symmetry. Finally, some processing examples are given in Section 4, illustrating the effects of two scale-invariant filters, with different line symmetries, on the input image.

2. SCALE-INVARIANT IMAGE PROCESSING

A two-dimensional linear system is said to be scale-invariant (or form-invariant) if, for any couple of input and output signals $f(x, y)$ and $g(x, y)$, the output produced by the scaled version $f(a_1x, a_2y)$ of the input is $Ag(b_1x, b_2y)$, with A, b_1, and b_2 real functions of the real numbers a_1 and a_2.

Letting $w(u, v; x, y)$ be the shift-variant weighting function characterizing the linear system, the scale-invariance constraint can be associated with the input-output relationship

$$g(x, y) = \int\limits_{-\infty}^{+\infty}\!\!\int f(u, v)w(u, v; x, y)\, du\, dv \tag{1}$$

to obtain an equation in the unknown function $w(\ )$. It can be shown, on the basis of a result derived for the one-dimensional case [3], that the most general class of solutions to such an equation is

$$w(u, v; x, y)=p(u, v, x, y)w_0(u/x^{s_1}, v/y^{s_2}, x/y^z), \tag{2}$$

where $p(\ )$ represents the product of arbitrary powers of its arguments, s_1 and s_2 are arbitrary real numbers, and w_0 is any regular function of its variables. The values of s_1 and s_2 determine the output scale factors, which are $b_1 = a_1^{1/s_1}$ and $b_2 = a_2^{1/s_2}$, and the parameter z through the constraint $b_1 = b_2^z$. The values of the exponents appearing in $p(\ )$ simply affect the amplitude factor A.

It is apparent that the implementation of the shift-variant filters defined in eq. (1) with the general kernels of eq. (2) can only be performed in the spatial domain by means of a discrete version of eq. (1): this technique implies, in general, a computational complexity of the order of N^4 for an $N{\times}N$ input image. However, several subclasses of scale-invariant filters can be defined, that characterize important application areas and for which more efficient implementation techniques can be devised.

Let us consider first the subclass of weighting functions

$$w(u, v; x, y)=u^{-c_1}v^{-c_2}x^{-d_1}y^{-d_2}w_0(x/u^{t_1}, y/v^{t_2}). \tag{3}$$

By letting $\widetilde{F}(s_1, s_2)$ be the Mellin transform [7] of $f(x, y)$, defined as

$$\widetilde{F}(s_1, s_2) = \int\limits_{0}^{\infty}\!\!\int\limits_{0}^{\infty} f(x, y)\, x^{s_1-1}y^{s_2-1}\, dx\, dy \tag{4}$$

and by taking the Mellin transform of both sides of eq. (1), one has in this case

$$\widetilde{G}(s_1, s_2)=\widetilde{F}(q_1+t_1s_1, q_2+t_2s_2)\widetilde{W}_0(s_1-d_1, s_2-d_2) \tag{5}$$

with $q_i=1-c_i-d_it_i$, $i=1, 2$. Equation (5) shows that the Mellin transform can be used to obtain an algebraic relationship from the integral equation (1), with kernels of class (3), as happens with the Fourier transform in the shift-invariant case. The basic differences are that eq. (5) holds for causal inputs (which is not a restriction in practice) and the variables s_1 and s_2 are scaled and shifted according to the parameters of each specific window defining the system within class (3). Such a dependence disappears (i.e., $q_i=0$, $t_i=1$, $d_i=0$) for the particular subclass

$$w(u, v; x, y)=u^{-1}v^{-1}w_0(x/u, y/v) \tag{6}$$

for which

$$\tilde{G}(s_1, s_2) = \tilde{F}(s_1, s_2)\tilde{W}_0(s_1, s_2). \tag{7}$$

Equations (5) and (7) suggest an implementation technique for the systems defined in eqs. (3) and (6), based on the computation of the Mellin transform. Such a computation can be performed by exploiting the relationship between the Mellin and the Fourier transforms. In fact, by means of the change of variables

$$x = Xe^{\mu}, \qquad y = Ye^{\nu} \tag{8}$$

eq. (4) becomes

$$\tilde{F}(s_1, s_2)=X^{s_1}Y^{s_2} \int\limits_{-\infty}^{+\infty}\!\!\!\int f_1(\mu, \nu)e^{\mu s_1}e^{\nu s_2}\, d\mu\, d\nu \tag{9}$$

which is recognized as essentially the Fourier transform of $f_1(\mu, \nu)=f(Xe^{\mu}, Ye^{\nu})$ for $s_1=-j\omega_1$ and $s_2=-j\omega_2$. The shifts and scalings appearing in eq. (5) are dealt with by suitably resampling f_1 and by weighting it by an exponential sequence before the Fourier transform is computed. Therefore, the classical FFT algorithm can be used, provided the mapping of eq. (8), leading from $f( )$ to $f_1( )$, is performed with no degradation in the processed signals. Considering only the resolution aspect, it can be checked that an increase by a factor of approximately $\ln N$ is needed in the number of samples of $f_1( )$ computed from the samples of $f( )$, if a comparable resolution is desired. Therefore, the computational complexity associated with this procedure becomes of the order of $(N \ln N)^2 \log N$ (for N a power of 2), plus the operations needed to perform the mapping of eq. (8), which are in the order of $(N \ln N)^2$.

To conclude the analysis of the implementation of classes (3) and (6), it is worth noticing that the procedure presented above, consisting of the log mapping followed by the product of the Fourier transforms of the input and of the window, can be interpreted as the implementation of a shift-invariant fast convolution technique in the mapped (μ, ν) domain. In fact, the particular dependence of w_0 in eqs. (3) and (6) on the ratio of the output and input coordinates, is such that the log-mapping operation transforms the shift-variant kernel into a shift-invariant response, i.e., depending on the difference of the new (mapped) variables. For example, in the simpler case of eq. (6), letting $\mu = \ln x$, $\nu = \ln y$, $\mu' = \ln u$, $\nu' = \ln v$, eq. (1) becomes

$$g_1(\mu, \nu) = \int\limits_{-\infty}^{+\infty}\!\!\!\int f_1(\mu', \nu')h(\mu-\mu', \nu-\nu')\, d\mu'\, d\nu' \tag{10}$$

with $g_1(\mu, \nu)=g(e^{\mu}, e^{\nu})$ and $h(\mu-\mu', \nu-\nu')=w_0[\exp(\mu-\mu'), \exp(\nu-\nu')]$. Although no explicit saving in computation is involved with respect to the previous approach, the implementation of certain form-invariant kernels by means of shift-invariant convolutions like eq. (10) might be particularly suitable for image-processing systems operating in the (mapped) spatial domain.

3. SCALE-INVARIANT PROCESSING WITH POINT AND LINE SYMMETRY

The motivation to investigate particular subclasses of the weighting functions defined by eq. (2) is twofold, namely analyzing classes of processing oriented to relevant applications and seeking more efficient implementation schemes than in the general case, by exploiting the properties of each specific kernel. In this section we consider the form-invariant weighting functions that characterize some important image-processing applications, like the ones mentioned in the introduction.

In order to introduce these subclasses of kernels, whose common property is the rotation-invariance, let us start from the most general form-invariant circularly symmetrical functions derived from eq. (2).

First, we consider those weighting functions belonging to such a class that are shift-variant simply in the sense that they "expand" with the distance $r = (x^2+y^2)^{1/2}$ of the current point from the origin (0, 0). They are obtained from eq. (2) by setting z=1, so

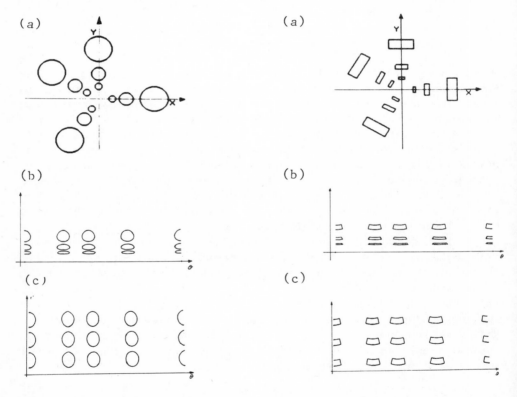

(a)

(b)

(c)

(a)

(b)

(c)

Figure 1. Effects of the mappings from Cartesian (a) to polar (b) to ln r−θ (c) coordinates of a point-symmetrical window linearly expanding with r. A set of constant-value contours is sketched in each plane.

Figure 2. Same as Figure 1, for a radially symmetrical window (that is rotationally invariant). The effects of the mappings are the same as in the case of Figure 1, and in particular, processing is shift-invariant in (c).

that the constraint holds $b_1=b_2$, or $s_1/s_2=(\ln a_1)/(\ln a_2)$. In the general case of $a_1 \neq a_2$ the expression one gets is

$$w(u, v; x, y) = p(u, v, r)w_r(u/r^{s_1}, v/r^{s_2}) \qquad (11)$$

and the corresponding "expansion" of w_r follows two different exponential laws along the two axes u and v. In the particular case of $s_1=s_2=1$ $(=-1)$ the expansion (contraction) is linear with r. This means that the processing of the input image is performed with a resolution that decreases (increases) from the origin to the periphery. If the property of circular symmetry around the current point is introduced, the new class of weighting functions is defined

$$w(u, v; x, y) = r^{-2}w_c \{[(u-x)^2+(v-y)^2]/r^2\} \qquad (12)$$

$$= r^{-2}w_c [1+\rho^2/r^2-2\rho \cos(\theta-\phi)/r],$$

having set $x=r \cos \theta$, $y=r \sin \theta$, $u=\rho \cos \phi$, and $v=\rho \sin \phi$ and having chosen $p(\ )=r^{-2}$ in order to obtain the amplitude factor A equal to 1. The contours of constant value in eq. (12) are, for any choice of the current point (x, y), i.e., of r, circles centred in (x, y) and with diameter proportional to r. In other words, the kernels of class (12) have a point symmetry around (x, y) and are identical all over the plane except for their size, as sketched in Figure 1(a).

The point-symmetrical windows of eq. (12) belong to a more general class, characterized by the dependence on the ratio of the input and output radial coordinates and on the difference of the angular variables, i.e., to the class

$$r^{-2}w_1(\rho/r, \theta-\phi) \qquad (13)$$

for which the input-output relationship (1) can be rewritten, in polar coordinates, as

$$g_1(r, \theta) = \int_0^\infty \int_0^{2\pi} f_1(\rho, \phi)r^{-2}w_1(\rho/r, \theta-\phi)\rho \, d\rho \, d\phi \qquad (14)$$

being $g_1(r, \theta)=g(r \cos \theta, r \sin \theta)$ and $f_1(\rho, \phi)=f(\rho \cos \phi, \rho \sin \phi)$.

The weighting functions defined in eq. (13) are the kernels of the integral equations modelling the image reconstruction from line integrals and the restoration of some shift-variant optical distortions [4, 5], while the windows of eq. (12), i.e., those having circular symmetry, model the behaviour of the peripheral visual system [6]. In this case, the shape of the windows is assumed to be the difference of two Gaussian (DOG) functions with different standard deviations (that depend on r). It should be noticed that class (13) includes the windows

$$r^{-2}w_2(\rho/r, |\theta-\phi|) \qquad (15)$$

that have a particular line symmetry, i.e., with respect to lines joining the current point with the origin. A sketch of such radially symmetrical windows is shown in Figure 2(a). The general case of line symmetry, with respect to an arbitrary direction (like the vertical one sketched in Figure 3(a)) is not included in the class of eq. (13) and can be represented by eq. (11). The interest for line-symmetrical image processing stems from the fact that it is sometimes desirable to perform a direction-selective filtering. A typical example is the detection of edges having a particular orientation: if a noise smoothing has to be performed first, it is advisable to low-pass the image with filter masks extended along the selected direction (in order not to smooth out the edges of interest), which becomes a symmetry axis of the masks.

Coming now to the implementation techniques of eq. (14), it should be noticed that the transform method discussed in the previous section still applies, suitably modified to take into account that eq. (14) represents a so-called "Mellin-type" convolution with respect to the radial coordinates, while it is a shift-variant (or "Fourier-type") convolution with respect to the angular coordinates. Therefore, taking the mixed Fourier-Mellin

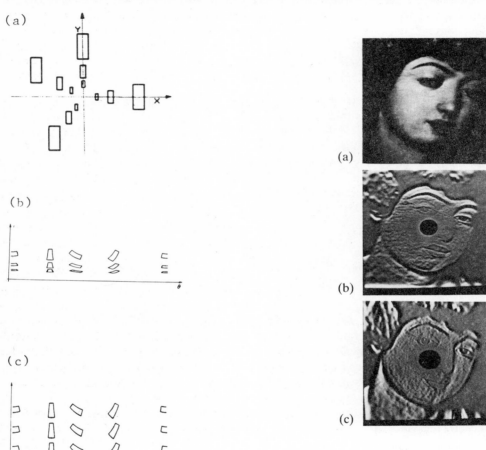

Figure 3. Effects of the polar mappings on a window having line symmetry along directions which are not radial. The sketches of the constant-value contours show that only in (c), i.e., in the ln r−θ plane, is a partial shift-invariance obtained for constant θ.

Figure 4. Scale-invariant processing with line symmetry: (a) 512×512 original image, 8 bit/pel; (b) output of the filters, corresponding to the sketch of Figure 3(a), low-pass along the horizontal axis and band-pass along the vertical axis; (c) same as (b) with filters rotated 90°. The black hole covers the area where no processing takes place (see text).

transforms of both sides of eq. (14) one gets

$$\hat{G}(s, \omega) = \hat{F}_1(s, \omega)\hat{W}_1(2-s, \omega), \tag{16}$$

s being the Mellin variable corresponding to r and ω the Fourier variable corresponding to θ.

Equation (16) defines an implementation method of eq. (14) based on the computation of the Fourier-Mellin transform and thus requiring an order of $N^2 \ln N \log N$ operations, plus the computation of half of the mapping (8) (i.e., with respect to r only). To this saving with respect to the case of eq. (5) — due to the fact that the Mellin transform has to be computed only along one direction — the additional computation of the mapping from the (x, y) to the (r, θ) plane is associated, requiring a number of operations proportional to N^2. Moreover, such a mapping implies a nonstructured access to the data, which represents a serious drawback for the most common image-processing systems.

It is worth mentioning that the advantages of dealing with circularly symmetrical functions like the ones of eq. (12), i.e., the fact that they are completely represented by any one-dimensional section, are lost as soon as the mapping to the polar plane is performed.

As already noticed in the previous section, further insight into the procedure based on eq. (16) is gained if it is interpreted as a fast convolution in the mapped domain. To this end, Figures 1(b) and 2(b) sketch the result of the first mapping, from (x, y) to (r, θ), in the case of point-symmetrical and radially symmetrical windows. It can be noticed that the mapped windows are, in both cases, shift-invariant for any fixed r. The result of the second mapping, from r to ln r, is shown in Figures 1(c) and 2(c): the processing is now shift-invariant in both directions. For the case of nonradial line symmetry (Figure 3(a)), the results of the two mappings are shown in Figures 3(b) and (c), from which it is apparent that a partial shift-invariance (for constant θ) is achieved in the (ln r, θ) plane. These considerations suggest that eq. (14) can be implemented by a "classical" shift-invariant processing system (e.g., a parallel architecture) provided an efficient technique (with particular reference to the data access) is devised to perform the mapping. In the case of the windows of Figure 3(a), the same conclusion holds for each column of the (ln r, θ) plane. In order to exploit the partial shift-invariance of the windows, one might operate in the (ω', θ) plane, ω' being the Fourier variable corresponding to ln r: the resulting computational complexity is $N^3 \log N$.

As a final comment, it must be pointed out that many of the conclusions drawn for the implementation of the kernels of eq. (13) are valid for windows where the ratio of the radial coordinates is inverted, i.e., for the class

$$w_2(r/\rho, \theta-\phi) \tag{17}$$

which, however, exhibits different symmetries with respect to the case of eq. (13). For instance, the class, dual of class (12),

$$\rho^{-2}w_c[1+r^2/\rho^2-2r\cos(\theta-\phi)/\rho] \tag{18}$$

has what might be called a "pseudocircular" symmetry around the current point. In fact, for any fixed r, the contours of constant value are still a family of circles, but they are

not concentric around (x, y) as in the case of eq. (12). All the centres lie on the line
joining the origin to (x, y), at a distance from the origin proportional to r, and each radius
is proportional to r: therefore, the size of the windows is again linearly increasing with r.
The main interest for the class of windows of eq. (18) is due to their relationship with the
scaled transforms [8].

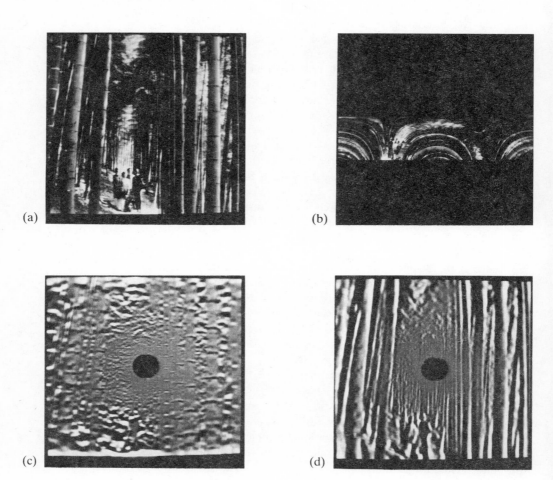

(a) (b) (c) (d)

Figure 5. Results of the same processing
as Figure 4 applied to the new original
image (a). The polar representation of (a)
is shown in (b), θ being the horizontal
axis. (c) Output of the line-symmetrical
filters which are low-pass in the horizontal
direction and band-pass in the vertical
direction. (d) Same as (c) with filters
rotated 90°.

4. PROCESSING EXAMPLES

In this section some processing examples are presented, to illustrate the effects of the form-invariant filtering when directional filters are used, i.e., filters with line symmetry. We have chosen a very simple kind of line symmetry, namely the symmetry with respect to the coordinate axes x and y, as schematically represented in Figure 3(a), so that the results of the processing (in particular of the edge enhancement) will be particularly noticeable. The chosen weighting functions are not shift-invariant with respect to the angular variable and their size increases linearly with the distance r of the current point from the origin. Therefore, they belong to the class

$$w(u, v; x, y) = r^{-2}w_0(|u-x|/r, |v-y|/r) \tag{19}$$

and have been chosen, for the sake of simplicity, to be separable, i.e., of the form

$$w_0(|u-x|/r, |v-y|/r)=w_{01}(|u-x|/r)w_{02}(|v-y|/r), \tag{20}$$

so that the space-domain implementation of eq. (1) is easier than in the general case as far as both the memory requirement of the mask and the computational complexity (of the order of N^3) are concerned.

   A first set of examples is shown in Figure 4 and refers to the original image of Figure 4(a). This photograph, as well as all the others, is taken from the monitor connected to a VDS-701 image-processing system and to a PDP-11/34 host computer. The image resolution is 512×512 pixels, 8 bit per pixel. Figure 4(b) is the result of a processing characterized by a smoothing behaviour along the x axis and an edge crispening behaviour along the y axis. More specifically, the w_{01} of eq. (20) is a constant, while w_{02} is the difference of two Gaussian functions having standard deviations in the ratio 1 : 3 and a band-pass characteristic in the frequency domain. It can be noticed in Figure 4(b) that processing has enhanced the contours that are horizontal or close to horizontal, while smoothing out the vertical details. A dual result is obtained if the above-described masks are rotated by 90°, as can be checked in Figure 4(c). In both cases, the resolution of the enhanced details decreases linearly from the centre to the periphery. The black hole in the central region corresponds to the area where, due to the small values of r, the size of the masks is too small with respect to the pixel distance, so that no processing actually takes place.

   While the "edge content" is roughly omnidirectional in the original image of Figure 4(a), it is concentrated along the directions close to the vertical in Figure 5(a), that is the second original image which undergoes the same processing previously described. Figure 5(b) shows the same original image (actually its portion lying within the maximum inscribed circle) in polar coordinates, r being the vertical axis and θ the horizontal one; if the log-mapping were applied to the r axis, the resulting image could be processed by a set of filters (one for each angle) that are shift-invariant with respect to the ln r coordinate, according to the sketch of Figure 3(c). Figure 5(c) shows the result of processing aimed to perform a low-pass along the x axis and to enhance the horizontal edges. A more meaningful result is obtained by means of a 90° rotation of the weighting functions: a lot of the information content is preserved, at a variable resolution degree, decreasing from the centre toward the periphery (Figure 5(d)).

5. CONCLUSIONS

In implementing form-invariant filters, the point and line symmetries can be exploited to some extent to gain efficiency. Several cases concerning applications of great interest in image processing have been analyzed, and two alternative implementation schemes (in the transform domain and in the space domain) have been discussed. It has been pointed out that the coordinate transformation is the critical step in reducing LSV filtering to a shift-invariant convolution, therefore efficient schemes for such a mapping are currently being sought, with particular reference to structured access to the data.

ACKNOWLEDGEMENTS

This work has been supported by the National Research Council (C.N.R.) of Italy.

REFERENCES

1. C. Braccini, G. Gambardella, A. Grattarola, Digital image processing by means of generalized scale-invariant filters, Proc. NATO Advanced Research Workshop on Issues in Acoustic Signal/Image Processing and Recognition, S. Miniato, Aug. 3–6, 1982.
2. D. Casasent, D. Psaltis, Deformation invariant, space-variant optical pattern recognition, in: E. Wolf (ed.) *Progress in Optics*, vol. XVI (North-Holland, 1978) 291–356.
3. C. Braccini, G. Gambardella, Linear shift-variant filtering for form-invariant processing of linearly scaled signals, *Signal Processing* 4 (1982) 209–214.
4. G. Robbins, T. Huang, Inverse filtering for linear shift-variant imaging systems, *Proc. IEEE* 60 (1972) 862–872.
5. E. Hansen, Image reconstruction from projections using circular harmonic expansion, Ph.D. Thesis, El. Eng. Dept. Stanford Univ. (1979).
6. C. Braccini, G. Gambardella, G. Sandini, A signal theory approach to the space and frequency variant filtering performed by the human visual system, *Signal Processing* 3 (1981) 231–240.
7. I. H. Sneddon, The use of integral transforms, (McGraw Hill, New York, 1972).
8. C. R. Carlson, R. W. Klopfenstein, C. H. Anderson, Spatially inhomogeneous scaled transforms for vision and pattern recognition, *Opt. Lett.* 6 (1981) 386–389.

C. Braccini and A. Grattarola
Istituto di Elettrotecnica
Università di Genova
Viale F. Causa 13
16145 Genova
Italy

17 Implementation of a digital filter having transfer function sinc u

A Andronico, G Bastianini, P L Casalini and A Tonazzini

1. INTRODUCTION

In the last years the techniques of digital filtering of one- and two-dimensional signals have replaced corresponding analogic methods. This is due to:

(1) The flexibility of digital computers.
(2) The advantages of stability, precision and simplicity given by the digital techniques.
(3) The possibility of carrying out dedicated software and hardware to be used as stand-alone systems.

Most works on digital filtering are based on the representation of one- and two-dimensional signals by means of discrete sequences of numbers; a mathematical theory of numerical transforms, discrete systems, and FIR and IIR filters is well developed [1, 2]. Nevertheless, in our first approach to digital filtering, we prefer to start again from the theoretical study of analog filtering and its mathematical background. Therefore the study and the project of the filters we present are based on the classical definitions of Fourier transforms; in fact their use can be justified by the Theory of Distribution [3]. We describe the use of a digital computer in order to simulate analog processing procedure.

2. THE METHODOLOGY USED

Filtering an analog image means to modify its spectrum in such a way as to eliminate, or attenuate, some undesired frequency components, and to transmit other unaltered ones. This operation is entirely characterized by the equation:

$$G'(u, v) = G(u, v)H(u, v), \tag{1}$$

where $G'(u, v)$ and $G(u, v)$ are the Fourier transforms of the filtered and the original image, and $H(u, v)$ is the filter transfer function.

Equation (1) may be expressed in this equivalent form:

$$g'(x, y) = \int\!\!\int_{-\infty}^{+\infty} g(a, b)h(x-a, y-b)\, da\, db, \tag{2}$$

where $g'(x, y)$ and $g(x, y)$ are, respectively, the filtered and the original image, and $h(x, y)$ is the filter impulse response [4]. According to the shape of $H(u, v)$ the filters are classified as low-pass, band-pass, high-pass and stop-band. In order to implement these filters on a digital computer, and to carry out to digitized images (matrix $g(n, m)$ obtained by a two-dimensional sampling), we have used the definition given by eq. (2).

In fact, eq. (1) allows a large choice for H(u, v) but two discrete Fourier transforms are to be computed and large amounts of main memory and computing time are needed, even if FFT algorithms are used [5]. Moreover, we should have worked with complex value functions.

In order to implement a filter on a digital computer, the following steps have been executed:

(1) A transfer function H(u, v) has been chosen such that its inverse Fourier transform h(x, y) is analytically known.

(2) A convolution matrix has been obtained by means of two-dimensional sampling of h(x, y).

(3) Linear convolution has been executed between the convolution matrix and the digitized image, according to the discrete convolution equation [6]:

$$g'(m, n) = \sum_{k=1}^{m} \sum_{l=1}^{n} g(k, l)h(m-k+1, n-l+1). \tag{3}$$

This process minimizes utilization of the central processor memory, because the two-dimensional convolution may be easily reduced to a succession of one-dimensional convolutions, and so we may store image files as sequential-row-access files on magnetic disk. This method, therefore, is suitable for a mini-computer such as the one used for this work (HP21MX).

The limits of this procedure are essentially in the choice of the transfer function, which is subject to some bounds. Nevertheless, the CPU time is reduced compared to that given by eq. (1). As will be seen, we have obtained valuable results with small convolution matrices.

3. THE CHOICE OF THE TRANSFER FUNCTION

We have synthesized three fundamental classes of filters (low-pass, band-pass, high-pass) by using transfer functions which are separable in the rectangular coordinates (u, v). Moreover, we have chosen both components of H(u, v) of the same shape so that the filter has the same behaviour on the frequency components in the x and y directions.

In fact, *a priori*, it is not possible to make a sharp distinction between these components, moreover, such a transfer function represents a first, evident, extension from the one-dimensional case to the two-dimensional one. We have chosen both components of H(u, v) between the even function, so that we have a real inverse Fourier transform.

In a first analysis, we have considered linear combinations of sinc u functions (sinc u = (sin πu)/πu. These transfer functions define stable FIR filters which do not produce phase-distortion in the processed image. Nevertheless, they do not separate very near frequencies.

4. RESULTS

We have experimented with these filters on test images expressed by:

$$g(x, y) = \sum_{i=1}^{k} A_i \cos(2\pi F_i x) \cos(2\pi F_i y), \tag{4}$$

(a)   (b)

Figure 1. (a) Original image (frequency
components: 0.01 and 0.4); (b) filtered
image (passband: 0–0.08).

for which the behaviour of the filter in the frequency domain is analytically foreseeable.
The tests we have executed on these images have permitted evaluation both of the errors
caused by the analog-to-digital conversion and of those caused by various approximations.
An important part of these errors is the error of aliasing [7] due to sampling of the filter
impulse response which is not band-limited. This error causes a loss of energy, first of the
signal impulse response itself, then of the processed image.

  Since we are interested in the visual aspect of filtered images, we obtain an estimation of the total error by comparing the filtered image to the ideal result created *ad
hoc* by using eq. (4). The aliasing error has been, furthermore, evaluated in a quantitative

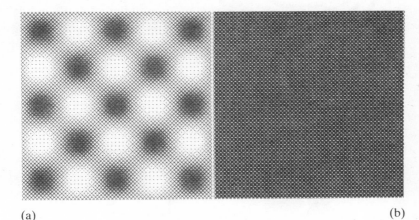

(a)   (b)

Figure 2. (a) Original image (frequency
components: 0.01, 0.2, and 0.4);
(b) filtered image (passband: 0.137–
0.382).

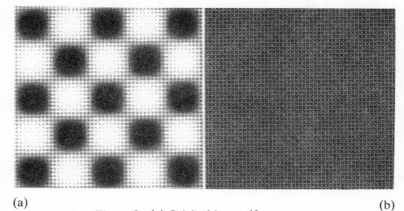

(a)

(b)

Figure 3. (a) Original image (frequency components: 0.01 and 0.4); (b) filtered image (passband: 0.4−0.5).

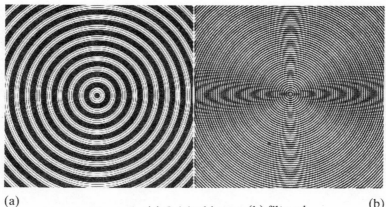

(a)

(b)

Figure 4. (a) Original image; (b) filtered image (high-pass).

(a)

(b)

Figure 5. (a) Original image; (b) filtered image (high-pass).

manner. Both these errors are within acceptable limits. As an example we give the equation of the transfer function used to synthesize a low-pass filter:

$$H(u, v) = \text{sinc}\,\frac{u}{F}\ \text{sinc}\,\frac{v}{F}, \tag{5}$$

where the passband in both directions is intended to be the interval 0, F_c with

$$\text{sinc}\,\frac{F_c}{F} = 0.707,$$

similar to the simplest of the electric filters [8].

We executed this elaboration using the SADAF 2 structure [9] of I.E.I.-C.N.R. of Pisa for digital image processing. We implemented these programs on the IASY system connected with the PAS system of video-representation [10, 11], by which we could observe and compare the original and the filtered images. Using this structure we obtained Figures 1-3. Each of them illustrates one of the implemented filters. The frequencies are relative to sampling frequency.

We note that these results have been obtained with square convolution matrices whose maximum dimension is five and, consequently, whose computing time was very short. We tested these filters on generic images, which were not determined by eq. (4), and we obtained satisfactory results. Some of these results are shown in Figures 4 and 5.

REFERENCES

1. Oppenheim A. V., Schaefer R. W.: "Digital Signal Processing"; Prentice Hall, 1975.
2. Seminario fiorentino sui filtri numerici: "Corso introduttivo sui filtri numerici"; 20 Settembre 1972, Teorema ed.
3. Schwartz L.: "Méthodes Mathematiques pour les Sciences Phisiques"; Hermann, 1965.
4. Goodman J. W.: "Introduction to Fourier Optics"; McGraw Hill Book Company, 1968.
5. Bini, Capovani, Lotti, Romani: "Complessità numerica"; Boringhieri, 1981.
6. Pratt W. K.: "Digital Image Processing"; J. Wiley & Sons, 1978.
7. Cariolaro G.: "Teoria dei segnali determinati"; Ed. Universitaria Patron, 1972.
8. Carassa F.: "Communicazioni elettriche"; Boringhieri, 1977.
9. L. Azzarelli, M. Chimenti, F. Fabbrini: "Risorse e organizzazione del servizio per l'elaborazione di immagini"; Rapporto Interno n. 22, 1980 I.E.I. Pisa.
10. L. Azzarelli, M. Chimenti, O. Salvetti: "Un sistema di elaborazione automatica di immagini del territorio"; Annual Conference A.I.C.A., Pavia 1981 vol. 1 pp. 337-345.
11. S. Cerri: "Il sistema di immagini policromatico PAS: architettura e programmi di utilità"; Nota Interna (in corso di pubblicazione), 1982 I.E.I. Pisa.

A. Andronico, G. Bastianini, P. L. Casalini, and A. Tonazzini
IEI–CNR
Via S. Maria 46
56100 Pisa
Italy

18 Virtues of optimizing edge line appearance models
E E Triendl

1. INTRODUCTION

The virtues of this approach are:

1. It gives a complete edge description (angle, position, quality, levels A and B).
2. Infinite resolution for ideal edges (weak dependence on sampling frequency).
3. Independent from sampling geometry (it can handle random "rasters").
4. Feeds on results of simple edge detectors (no need to process every pixel).
5. Suitable for various types of edges and lines: after finding a reasonable edge probability, the optimizer can search for the best edge or line model to fit the data.
6. Suitable for colour and multispectral data.
7. Reconstruction of the picture at arbitrary resolution, as far as edges and constant grey levels are concerned, is possible due to the complete description of local neighbourhoods in the scene domain.
8. It holds for preprocessed pictures, e.g., for constant aperture texture parameters.
9. Results lend themselves to implementation in the form of KD-trees.
10. Suitable for pyramid structures.

A step edge arises in nature when two homogeneous regions are juxtaposed. A region is uniform because of the uniformity of some feature across the region, e.g., intensity, colour, texture, etc. For an ideal step edge each region has a constant feature value, the regions have different feature values and the boundary between the regions is a straight line, as shown in Figure 1.

Sampling and digitization produce a transition ramp between the two uniform regions: that is pixels have been produced which represent the sum of radiation intensities around a pixel centre weighted by the point-spread function of the scanning device

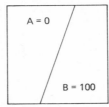

0	0	70	100
0	20	90	100
0	40	100	100
0	70	100	100

Figure 1. Ideal step edge.   Figure 2. Sampled image.

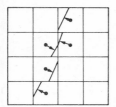

Figure 3. Visualization of edge response.

(e.g., for Landsat images this is a circular spot with a diameter of 75 m). The value of a pixel near an edge is a function of the distance of the edge from the pixel and can be computed from the point-spread function (Figure 2).

Using the point-spread function, appearances of ideal step edges at various positions and angles may be synthesized and correlated with the factual data. If data like those shown in Figure 2 have actually been produced by an ideal step edge, the maximum of correlation with the synthetic edge appearance will assume the maximum value of 1 for a certain angle and distance from the pixel centre.

The optimizing edge-appearance model maximizes the square of the normalized cross correlation between synthesized edge and factual data in a subimage and thus computes the edge triple PHI, RAD, COR, which are angle, position, and quality or correlation, respectively. Initial values for the optimizer and the decision whether or not to look for an edge may be derived from the vector gradient. The optimization procedure is well behaved and finds the maximum in about six steps of correlation. The mean grey levels of the two regions forming the edge may be computed subsequentially to get the edge quintuple (PHI, RAD, COR, MEAN, DIF). Figure 3 shows a visualization of the edge triples (PHI, RAD, COR).

Since the edge quintuples (angle, distance, correlation, grey-level mean, grey-level difference) give a full description of the edges, that includes location of the edges with subpixel accuracy, it is possible to reconstruct the image at a finer resolution [1]. The same idea can be applied to produce texture edges once the texture areas have been characterized by some texture parameter [2]. Quadratic raster is not required; any random, but reasonably homogeneous arrangement of sampling points will do [3].

2. FORMULAS

Let (x, y) be continuous coordinates and (i, j) sampling points, then the picture $P(i, j)$ is derived by folding the scene $S(x, y)$ with the point-spread function $SPREAD(x', y')$ of the scanning device:

$$P(i, j) = \int\int_{x, y = -\infty}^{\infty} S(x, y)SPREAD(x-i, y-j) \, dx \, dy. \qquad (1)$$

Assuming an ideal edge

$$S(x, y) = \begin{cases} 1 & \text{if } x > RAD \\ 0 & \text{if } x \leqslant RAD \end{cases}$$

$$P(i, j) = \int_{x=RAD}^{\infty} \int_{y=-\infty}^{\infty} SPREAD(x-i, y-j) \, dx \, dy. \qquad (2)$$

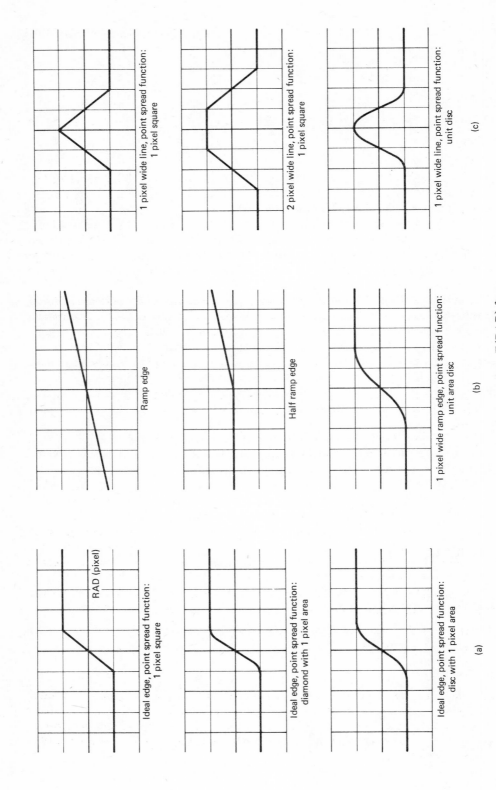

RAD (pixel)

Ideal edge, point spread function:
1 pixel square

Ideal edge, point spread function:
diamond with 1 pixel area

Ideal edge, point spread function:
disc with 1 pixel area

(a)

Ramp edge

Half ramp edge

1 pixel wide ramp edge, point spread function:
unit area disc

(b)

1 pixel wide line, point spread function:
1 pixel square

2 pixel wide line, point spread function:
1 pixel square

1 pixel wide line, point spread function:
unit disc

(c)

Figure 4. Cross sections T(RAD) for
(a) ideal edges, (b) ramp edges, (c) lines.

The response at location RAD to an edge at the origin is the edge cross section T(RAD):

$$T(RAD) = \int_{x=-RAD}^{\infty} \int_{y=-\infty}^{\infty} SPREAD(x, y) \, dx \, dy. \tag{3}$$

Translation and rotation by an angle PHI gives the appearance of an ideal (0, 1) edge at location (i, j):

$$F(i, j) = T(i \cos(PHI) - j \sin(PHI) - RAD). \tag{4}$$

Finally, maximization of the square of the normalized synthetic edge F and picture P

$$P(i, j) = P(i, j) - MEAN(P(i, j)),$$
$$F(i, j) = F(i, j) - MEAN(F(i, j)),$$

$$q = \frac{\left(\sum_{i,j} (F(i, j) P(i, j)) \right)^2}{\sum_{i,j} F(i, j)^2 \sum_{i,j} P(i, j)^2} \tag{5}$$

with respect to PHI and RAD yields the edge triple (PHI, RAD, COR). The hill-climber uses Lagrange interpolation (parabolic fit) at the maxima of PHI and RAD (see below). The interpolation formula for values FA, FB, FC of arguments A, B, C is:

$$LAG = \frac{(B^{**}2 - C^{**}2)FA + (C^{**}2 - A^{**}2)FB + (A^{**}2 - B^{**}2)FC}{(B-C)FA + (C-A)FB + (A-B)FC} \tag{6}$$

Subsequently the grey-level A+ on one side of the edge is obtained by the weighted sum for F>0 and F<0, respectively, for the other side.

$$A+ = \frac{\sum (P(i, j) F(i, j); \text{ if } F(i, j) > 0)}{\sum_{i,j} (F(i, j); \text{ if } F(i, j) > 0)}. \tag{7}$$

3. EDGE CROSS SECTIONS

The digital edge cross sections depend on the nature of the scene domain edge (ideal, ramp, line) and the point-spread function. Point-spread functions investigated were (0, 1) step functions ("holes through which the sensor looks") with circular, square, and diamond shapes and an area of 1 square pixel.

Figures 4(a)–(c) show cross sections T(RAD) for ideal edges (square, diamond, and circular shaped point-spread functions), ramp edges and lines with various widths.

4. CORRELATION DATA

4.1. Ideal edges

To decide on the optimization strategy, something has to be known about the function to be optimized. The printout of the squared 5 × 5 pixel window, shows that this function is well behaved for COR ≥ 0.3 below 40 degrees. Figure 5(a) shows the "autocorrelation" (in fact F has to be computed for every PHI and RAD) of ideal edges for a circular

Figure 5. Correlation of an ideal edge
appearance at location RAD=0 and angle
PHI=80 with ideal edge appearances
(PHI, RAD) and various point spread
functions. Operator size = 5×5 pixel.
(a) Unit circle point-spread function,
(b) square point spread, (c) cross-
correlation circle *vs.* square point-spread
function.

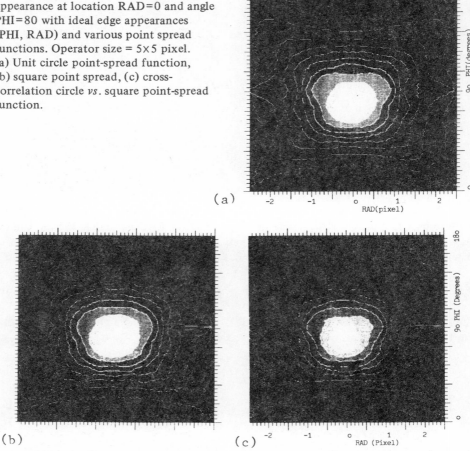

(a)

(b)   (c)

point-spread function (area: 1 square pixel). As might be guessed from the similarity of
the edge cross sections, it is very similar to the "autocorrelation" for a square point-
spread function (Figure 5(b)) and to the "cross correlation" between an ideal edge
sampled by a square shaped point-spread function and by a diamond shaped one (maximum of COR = 0.999). The shape (not the area) of the point-spread function can therefore be disregarded and the circle employed where a square would be appropriate.

4.2. Ramp edges

The "autocorrelation" of a 1 pixel ramp edge is rather similar to the ideal edge, but has
a broader maximum (Figure 6(a)), while the "autocorrelation" of a half ramp edge
(Figure 6(b)) shows a very broad maximum with respect to location. A 1 pixel ramp edge
will be detected by the ideal edge model with COR = 0.996, while cross-correlation with
a half ramp yields a maximum COR = 0.896 (0.891 for a full ramp) but no maximum
(Figures 7(a) and (b)). Thus the ideal edge model will not respond to half ramps and ramps.

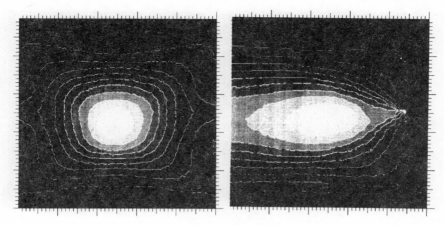

Figure 6. "Autocorrelation" ot ramp
edge appearances. (a) 1-pixel ramp edge,
(b) half-ramp edge.

4.3. Influence of noise, response to a dot

Contrary to gradient type edge detectors the edge-appearance model will not respond to
a 1 pixel dot, since its maximum of COR = 0.030 Gaussian noise (Figure 8(a)) yields a
harmless COR = 0.156, an ideal edge appearance superimposed with 100% random noise
(Figure 8(b)) still yields COR = 0.550, while COR = 0.986 for a 1:10 signal-to-noise ratio.

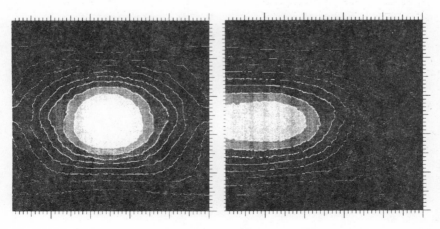

Figure 7. "Cross-correlation" of ramp
edges appearances with the appearance
of ideal edges. (a) 1-pixel ramp edge,
(b) half-ramp edge.

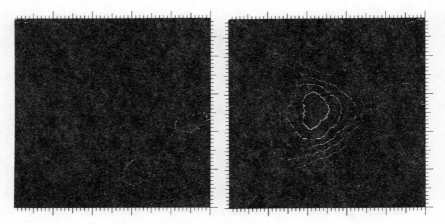

Figure 8. "Cross-correlation" of noise
and noisy edge with ideal edge
appearance. (a) Random noise, (b) ideal
edge + 100% noise.

Table 1

		Line width (pixel)	
	Ideal edge	1	2
Ideal point spread			
Circle	1.000		
Square	0.999		
Diamond	1.000		
Ramp edge			
1 pixel	0.996		
Half	0.891		
Full	0.866		
Line			
Thin	0.793	0.995	
1 pixel	0.793	1.000	0.742
2 pixel		0.742	1.000
1-pixel dot	0.030		
Edge + noise			
Noise	0.156		
100%	0.550		
50%	0.801		
10%	0.986	0.699	

4.4. Lines

Since any type of scene domain edge cross section may be used, lines are also possible. The "autocorrelation" of a 1 pixel wide line (Figure 9(a)) shows a steeper hill than for edges and secondary maxima with COR = 0.150 at RAD = ± 1.5, therefore, the hill-climber will have to start closer to the maximum: less than 0.5 pixel away. Also the response to line width is more critical: the appearance model for 1 pixel lines will respond with COR = 0.995 to infinitely thin lines and with COR = 0.742 to 2 pixel wide lines. An important question is how the edge model responds to lines. The "cross-correlation" of ideal edge with a 1 pixel line (Figure 9(b)) shows two maxima with COR = 0.793 at a distance of 1.5 pixel. This is quite obvious, since a line at the rim of the 5 × 5 aperture cannot be distinguished from an edge (the other side is cut off) at 0 and 90 degree orientation. At a distance of 1 pixel COR is below 0.450, at 0.5 pixel it is below 0.200.

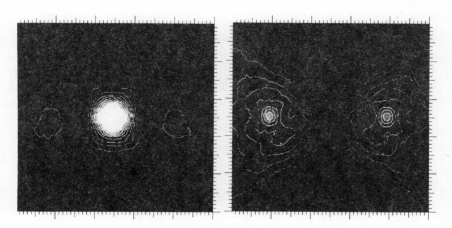

Figure 9. Correlations of a 1 pixel wide
line. (a) Line *vs*. line, (b) line *vs*. edge.

4.5. Dependence on the size of the operator

Up to now all examples were for 5 × 5 pixel operators. For larger apertures the maximum of correlation is wider for position (26% for 9 × 9) and more narrow for angular differences (−13% for 9 × 9) as seen from the autocorrelation of the appearances of ideal edges in a 9 × 9 field (Figure 10).

4.6. Summary of correlation maxima

Correlation maxima between various types of edges and lines are summarized in Table 1. No maxima are encountered for ramp and half-ramp edges. The maxima for lines are about 1.5 pixel off.

Figure 10. "Autocorrelation" of 9 × 9 ideal edge appearance.

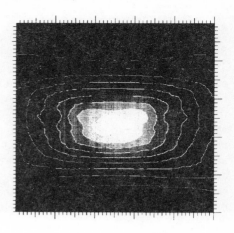

5. HILL CLIMBING

If — for a noiseless edge — an analytic solution would be found it would require three values of COR(PHI, RAD) to compute the edge triple (PHI, RAD, COR) at the maximum of COR. For the types of hills COR(PHI, RAD) expected a coordinate strategy with Lagrange interpolation (parabola fit) at the maxima is used. Since PHI is known with good accuracy from the vector gradient the hill-climber searches first for the maximum with respect to position RAD and then, if RAD is within limits of say ± 0.7 pixels, for the maximum with respect to angle PHI. Such an edge is detected with three steps per coordinate (six steps per edge) in almost all cases.

6. COMPUTATIONAL COMPLEXITY

The software implementation of the edge appearance operator is three loops deep. We first perform a general count of arithmetic operations, followed by a count when all shortcuts are used and will also consider less general implementations. The loops of the operator are:

(a) Global initialization, set tables for edges cross section(s) T, sine, cosine, arctangent.
(b) Operator initialization (per pixel): read P, mean of P, normalize P to zero mean, compute SUM P**2 and initial PHI, decide to go on or return "no edge".
(c) Optimize: step RAD and PHI, apply Lagrange interpolation at the maximum. It is assumed that an average of six steps with two Lagrange interpolations are required.
(d) Compute synthetic edge and correlation formula.

Under these assumptions, we count the following numbers of arithmetic operation per detected edge in loops (b), (c), and (d). The operator size is m = sxs (first column) and m = 5×5 (second column):

(b) Operator initialization

mul	m	25
div	2	2
add/sub	3m+8	23

(c) Optimization

mul	18	18
div	2	2
add/sub	26	26

(d) Correlation and synthetic edge (Formulas (4) and (5) per step (to be multiplied by 6)

mul	4m+2	102
div	2	2
add/sub	6m−3	147

Sum total for a detected edge

mul	25m+20	645
div	16	16
add/sub	39m+16	991

For fast implementation proper scaling is required to perform integer calculations with short word lengths. A reasonable scaling for the general case with 1/16 of pixel resolution and 128 intervals of PHI corresponding to 180 degrees, uses the following conventions and tables:

RAD   = 16 RAD
SIN(PHI) = 16 sin(PHI)   Table size = 128 × 5 bit
COS(PHI) = 16 cos(PHI)   Table size = 128 × 5 bit
T(RAD)   = Edge cross section table, 5 bit, size = 160 for 5 × 5 operator.

To speed up the generation of the synthetic edge, we waive normalization of F to zero mean, normalizing the cross section T instead, thus generating F with a near zero mean. This leads to a negligible broadening of the correlation maximum. Further, we replace the multiplications i cos(PHI) and j sin(PHI) in the edge cross section formula (4) by a sequence of additions and subtractions. To avoid index calculations this sequence is compiled into a loopless series of instructions together with the correlation formula (5).

(d) Fast correlation and synthetic edge

mul	2m+2	52
div	1	1
add/sub	3m+1	76

Sum total for a detected edge

mul	13m+30	355
div	10	10
add/sub	21m+35	560

Dropping generality and restricting ourselves to ideal edges with an aperture slightly less than 1 square pixel we find that the synthetic edge assumes only three values, which may be scaled to (−1, 0, 1) for certain positions and angles. For example, for RAD=0 and

Table 2. Synthetic ideal edges at RAD=0

PHI=0°					PHI=26.4°					PHI=45°				
1	1	1	1	1	1	1	1	1	1	1	1	1	1	0
1	1	1	1	1	1	1	1	1	0	1	1	1	0	-1
0	0	0	0	0	1	1	0	-1	-1	1	1	0	-1	-1
-1	-1	-1	-1	-1	0	-1	-1	-1	-1	1	0	-1	-1	-1
-1	-1	-1	-1	-1	-1	-1	-1	-1	-1	0	-1	-1	-1	-1

angle PHI = 0, 26.6, 45, 63.4, 90, 116.6, 135, 153.4 degrees (Table 2). Restricting the hill-climber to these angles and to the corresponding values of RAD (approximately 0.5 pixel intervals) allows precalculation of the synthetic edges with their sum squares and replacement of multiplications by additions and subtractions. The corresponding step sizes of the hill-climber, 26.6 degrees for PHI and approximately 0.5 pixel for RAD, are in tune with the 18 degrees and 0.6 pixel steps used in the present implementation. In this case we count three multiplications, one division and m subtractions for the inner loop. Using recursion for the mean values of P and P**2 in loop (b) the total numbers of operations per recognized edge pixel are:

mul	37	37
div	10	10
add/sub	7m+42	217

7. RESULTS FOR A SYNTHETIC SCENE

To demonstrate the operator a scene has been synthesized which contains edges and lines (width = point-spread function) in all directions, positive and negative, crossing

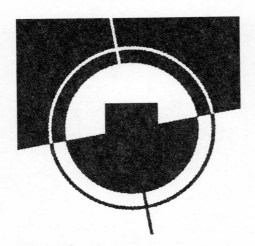

Figure 11. Synthetic scene.

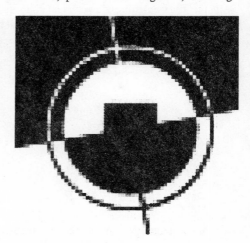

Figure 12. Sampled picture 63 × 63 pixel.

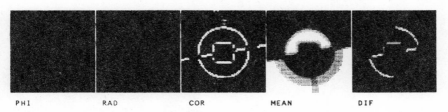

PHI   RAD   COR   MEAN   DIF

Figure 13. Edge quintuple for 5 × 5 ideal
edge operator.

edges and lines and corners (Figure 11). Sampling with a square-shaped point-spread function results in the 63 × 63 pixel digital picture shown in Figure 12. Application of the 5 × 5 ideal edge operator results in edge quintuples (PHI, RAD, COR, MEAN, DIF) rendered in Figure 13 as if they were five channels of a multispectral picture.

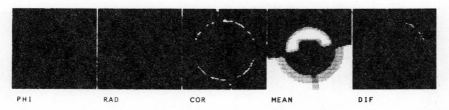

PHI   RAD   COR   MEAN   DIF

Figure 14. Edge quintuples for ideal line
operator.

Figure 14 shows the corresponding result for ideal 1 pixel wide lines. Negative values of RAD and DIF are rendered black (= 0) in Figures 13 and 14.

8. VISUALIZATION OF RESULTS

There are many ways to display the data contained in the edge quintuple. The simplest is to render pixels with an edge passing through them, i.e. ABS(RAD) < 0.5, with a grey value corresponding to COR (Figure 15).

To exploit the full capability of the edge model both the size of the operator and its super-resolution have to be taken into account. Correlation COR=1 means that we have an exact description of the scene within the size of the operator. The edge may be represented by a line with angle PHI drawn at a distance RAD from the centre of the operator through its whole field of view, e.g., the domain corresponding to a 5 × 5 pixel window in the sampled image. For lesser correlations we may draw a shorter line with less contrast. Results of this line overstrike method are shown at eightfold resolution in Figure 16(a) for ideal edges and in Figure 16(b) for the 1 pixel line operator.

Similar to the line overstrike method the original scene is reconstructed by rendering the original edges using MEAN and DIF (mean value alone, where no edge is near). Here we may not simply overstrike, but have to use weighted sums and roll-off functions to avoid false edges at the rim of the operator. Edge overlap is used in the remote-sensing example (Section 9).

Figure 15. COR for pixels with an edge
passing through.

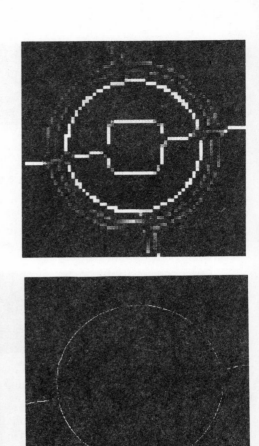

(a)

(b)

Figure 16. Line overstrike method at
eightfold resolution. (a) Edge, (b) 1 pixel
wide line.

Figure 17. Line grouping method at
eightfold resolution for edges with
COR > 0.5.

The line overstrike method will, of course, produce widened and jagged lines at bends and corners. Grouping neighbouring line elements in the super-resolution picture by summing up locally the linear functions that produce them and rendering the zero crossings produces the best results so far, but has some trouble with crossing edges. Figure 17 shows all edges found with COR > 0.5.

9. REMOTE-SENSING AND RECONSTRUCTION EXAMPLE

A 64 pixel subimage of a Landsat scene (river Rhine flowing through the town of Mannheim, Figure 18) is processed by a 5x5 ideal edge-model operator. The edges rendered by the line overstrike method are shown in Figure 19, while Figure 20 shows a reconstruction by edge overlap at eightfold resolution.

Figure 18. 64 pixel subimage of a Landsat frame.

Figure 19. Edges rendered by line overstrike.

Figure 20. Reconstruction by weighted edge overlap.

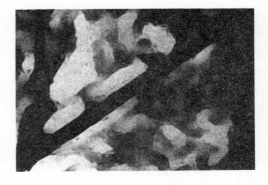

Figure 21. Sampling at random points. (a) Original painting, (b) random sampling.

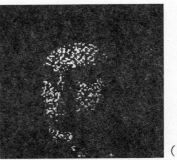

(a) (b)

10. RANDOM-RASTER EXAMPLE

Indices (i, j) in F(i, j) and P(i, j) of the correlation formula are replaced by the coordinates of a random sampling (with constant point-spread function). Figure 21(b) shows a random sampling of the painting "Bildnis eines jungen Mannes" by A. Duerer (Figure 21(a)). The size of the edge operator was set to cover in the order of 30 random samples. Figure 22(a) shows the edges, Figure 22(b) the reconstruction.

(a) (b)

Figure 22. Results for random sampling. (a) Edges, (b) reconstruction.

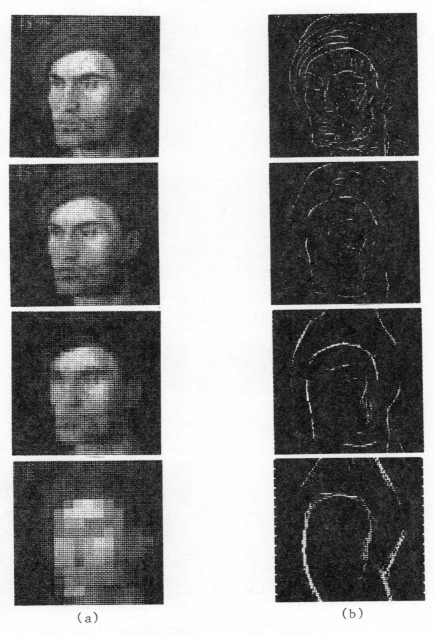

Figure 23. Edges in a pyramid structure.
(a) Pyramid structure, (b) edges.

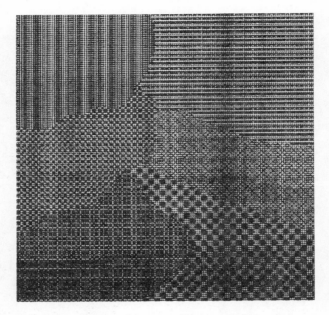

Figure 24. Synthetic textures with irregular boundaries.

Figure 25. Texture parameter image.

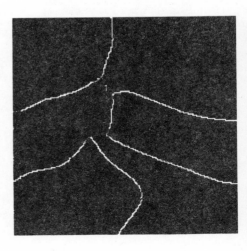

Figure 26. Result of edge operator applied to first parameter.

11. PYRAMID-STRUCTURE EXAMPLE

A pyramid-structure image, where each level represents the mean of 2×2 pixels in the next lower level is particularly suited for the edge model since the edge operator at the higher level has enough positional accuracy for the next two or three lower levels (Figure 23).

12. TEXTURE EDGES

Preprocessing with a constant size texture operator has an effect similar to that of a point-spread function with the same size. A sample of six texture parameters applied to a synthetic texture picture (Figure 24) yields the reduced multichannel image (Figure 25).

13. SYNTHETIC APERTURE RADAR (SAR) EXAMPLE

SAR images are characterized by heavy speckle noise (Figure 27). Preprocessing may consist either of low-pass filtering and reduction or texture-parameter extraction. Low-pass filtering and rendition of the edges at the resolution of the original picture are used for the example shown in Figure 28.

Figure 27. Synthetic aperture radar image.

Figure 28. Edges of SAR image at original resolution.

14. DATA STRUCTURES

Usually the edge quintuples are stored in an array registered with the original picture. Using 16 bits per pixel to allow for negative values results in a tenfold increase in storage requirement. Since edges occur sparsely a two-dimensional tree representation is more effective. It is also faster for simple grouping operations [4].

In more general cases, combinations of edge and line models with various sizes or in pyramid structures, array storage becomes increasingly difficult. If the edge model is seen as an attempt to explain pictorial data within a neighbourhood then — even for a single size of neighbourhood — we may have an edge, a line, a ramp, several edges, or unexplained pixels, that we may want to characterize by texture parameters.

REFERENCES

1. Triendl E. How to Get the Edge into the Map, in Proc. of the 4. ICPR, pp. 946–950, 1978.
2. Triendl E. and T. Henderson, A Model for Texture Edges, in Proc. 5. ICPR, 1980.
3. Triendl E. Modellierung von Kanten bei unregelmaessiger Rasterung, Proc. 1. DAGM Symposium, pp. 260–264, Springer, 1978.
4. Henderson T. and E. Triendl, The K-D Tree Representation of Edge Descriptions, Proc. 6. ICPR, 1982.
5. Davis L. S., A Survey of Edge Detection Techniques, Computer Graphics and Image Processing 4, 1975.
6. Hueckel M., An Operator which locates Edges, in Digital Pictures, J. Assoc. Comput. Mach. 18, 1971.

E. E. Triendl
DFVLR
Institute for Optoelectronics
8031 Wessling
Federal Republic of Germany

19 Nonlinear image restoration with phase constraints

G Garibotto

1. INTRODUCTION

The superior information content of the phase over the amplitude is a well-known property which has been used in image coding [1] and two-dimensional filtering [2]. In fact, the phase of the Fourier transform of a signal describes the displacement of the frequency components and, irrespective of their amplitude, it allows a satisfactory reconstruction of edges and details of the signal.

Theorems have been proved in order to establish uniqueness conditions for the phase-only reconstruction of one-dimensional sequences as well as two-dimensional images [3]. Essentially, it is necessary to verify that the Z-transform of this function is not factorable into symmetric factors. This condition is satisfied by most two-dimensional sequences because of the general impossibility to factor a two-variable polynomial into lower-order terms, and this property has been experimentally proved for pictorial images [4].

Algorithms have been proposed [4] to recover a two-dimensional sequence in closed form by finding the solution to a set of linear equations. Unfortunately, the practical application of this technique is strongly limited by computational problems, and an iterative solution has proved to be much more efficient. In this second case it is necessary to compute two-dimensional FFTs of size $(M1, M2)$ so that $Mi \geqslant 2Ni-1$ $(N1, N2)$ being the image support [4]. The amplitude estimate at step p is combined with the ideal phase function; the inverse transform is truncated to the actual image size $(N1, N2)$ and the process is repeated until convergence to the original function is obtained. Moreover, an adaptive iterative procedure has been proposed in [4], to improve the convergence rate of this technique. Potential applications have been suggested for blind deconvolution, image coding, etc. [5].

Unfortunately, even in the best conditions, a large number of iterations (≈ 30) is requested for the reconstruction of small-size images (128x128 support and 256x256 frequency samples).

In this paper the two-dimensional image support has been sectioned into overlapping windows (64x64 pixels) in order to perform an approximate phase reconstruction by a microcomputer (LSI 11/23) with minimum memory capabilities. The phase information is used as a constraint in the restoration of images, blurred by a zero-phase function.

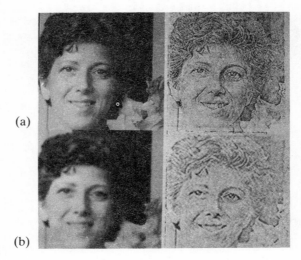

(a)

(b)

Figure 1. Sectioned image reconstruction from phase-only information (three steps of iteration). (a) Original image (256×256 samples) and reconstruction, (b) blurred image (average of 9×9 samples) and reconstruction.

(a) 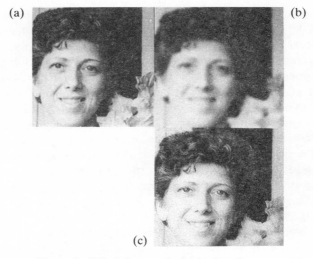 (b)

(c)

Figure 2. Blind deconvolution using the ideal phase function. (a) Original image, (b) severely blurred image (aver of 13×13 samples), (c) reconstruction at the first iteration.

2. ITERATIVE PHASE RECONSTRUCTION FOR SECTIONED IMAGES

As previously stated, the uniqueness theorem for phase-only reconstruction [4] requires the computation of FFTs on a frequency support $Mi \geqslant 2Ni-1$ which is more than twice image size. This technique can also be used on a sectioned image, by processing each window individually.

On the other hand, the phase constraints can also be profitably used if the FFT size is limited to $Mi \geqslant Ni$, but in this case no convergence is guaranteed to the optimal unknown solution. In fact, a phase function on the support (N1, N2) corresponds to a family of two-dimensional sequences which differ by convolution with a zero-phase function. This ambiguity is partially reduced by overlapping subimages and setting additional linear and nonlinear constraints on the grey-level distribution.

The previously discussed iterative phase-only reconstruction is modified as follows. For each subimage the ideal phase is computed on the effective window support (not the enlarged one). Starting from an initial guess of the amplitude function (in the worst case a constant value), the correct phase is used to recover an estimate of the signal. A threshold to the admissible range (min, max) is performed, and the average grey level is constrained to a selected value for each window. One iteration is terminated when all the overlapping windows have been processed in this way. Then a certain amount of white noise is added to the image and a further iteration is started.

Figure 1(a) shows a result of this procedure after three steps of iteration. Of course, convergence to the original image is not achieved any more, but the most relevant information about edges and texture has been correctly recovered, and satisfactory brightness continuity between adjacent windows has been achieved.

3. PHASE CONSTRAINTS IN IMAGE RESTORATION

The previously discussed procedure can be used to improve the quality of an image distorted by a zero-phase blurring function, in which case the phase of the signal should be almost unaffected. This hypothesis is quite acceptable in image recording by defocused lenses and blind deconvolution can be attempted using the undistorted phase function.

Unfortunately, the phase function obtained by the blurred image is always severely distorted by at least two reasons: truncation to the image support and quantization noise. In Figure 1(b) the blurred image is quantized to 8 bits and the approximate sectioned procedure, described in the previous section, does not allow a satisfactory reconstruction of edges and details. Much more satisfactory results are obtained using the original undistorted phase function as shown in Figure 2. The blurred image of Figure 2(b) is combined with the ideal phase and the result of Figure 2(d) is achieved at the first iteration of the proposed approximate procedure.

The blurred phase information can be profitably used as a constraint in nonlinear edge enhancement. At first the distorted image is processed in order to enhance the high frequency components which have been smoothed out by the blurring function. This result represents the initial amplitude estimate for the iterative phase reconstruction.

Two different techniques have been considered to achieve nonlinear edge enhancement, and both are based on the computation of the grey-level histogram h (1) within a local window (LxL), centred on the current sample.

(a) (b) (c) (d)

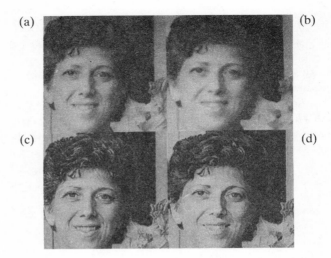

Figure 3. Nonlinear image enhancement
using a window size of 5×5 samples.
(a) Original image, (b) result of median
filtering, (c) unsharp median filtering,
(d) extremum filtering.

(a) (b) (c) (d)

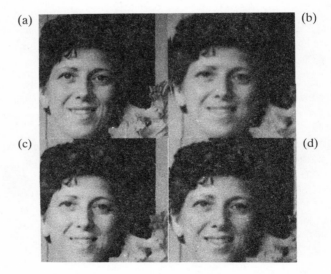

Figure 4. Image restoration with phase
constraints. (a) Original image, (b) blurred
image, (c) nonlinear enhancement of the
distorted image by the extremum filter,
(d) phase reconstruction after two
iterations.

3.1. Unsharp median filtering

The median $M(n_1, n_2)$ of the signal $f(n_1, n_2)$ is computed in a fast way according to the technique described in [6] (see Figure 3(b)). The output $g(n_1, n_2)$ is obtained as follows:

$$g(n_1, n_2) = \alpha M(n_1, n_2) + \beta(f(n_1, n_2) - M(n_1, n_2)).$$

Figure 3(c) shows an example of this processing, which is quite useful in edge enhancement of noise-free images.

3.2. Extremum filtering [7]

This filter produces an output which is the extremum value of the local histogram, closer to the grey level of the input sample. The processed image is an almost piece-wise constant signal with emphasis of edges and sharp discontinuities. An example of this processing is shown in Figure 3(d).

In both cases the high-frequency content which is added to the distorted image is no longer a white noise but it is signal-dependent and the modified amplitude function is quite close to the ideal image. Unfortunately these nonlinear operations may introduce artifacts in the processed image, as any enhancement procedure generally does, and the proposed phase reconstruction is used to avoid inconsistent variations of the output signal. An example of this processing technique is shown in Figure 4. The blurred image in Figure 4(b) has been filtered by the "extremum" operator (Figure 4(c)). After two iterations of the proposed phase reconstruction the result of Figure 4(d) is obtained. The high-frequency components introduced by the enhancement filter are controlled by the blurred phase function and still an appreciable improvement in the quality of the processed image has been achieved. Similar results have been obtained using the unsharp median filter.

4. CONCLUSIONS

In this paper an ideal situation of noise-free images has been considered. When the distorted image is also affected by additive noise, the computed phase function is no longer a reliable estimate of the original one, even in the hypothesis of a zero-phase blurring process. An additional phase distortion is determined by image quantization, as previously pointed out. The most relevant source of errors is related to the lack of information about the original image support. In the previous examples the blurred support should have been wider than that of the original image. Unfortunately this constraint cannot be used in the proposed sectioned approach, and the blurred phase function is used just as a control to avoid inconsistent results in nonlinear image enhancement.

REFERENCES

1. A. G. Tescher, The role of phase in adaptive image coding, USCIPI report no. 510, I.P.I., Univ. of Southern California, Dec. 1973.
2. G. Garibotto, 2-D recursive phase filter for the solution of two-dimensional wave equations, *IEEE Trans. on Acoust. Speech and Signal Process.* vol. ASSP-27 (1979) 367–373.

3. M. H. Hayes, J. S. Lim, A. V. Oppenheim, Signal Reconstruction from phase or magnitude, *IEEE Trans. on Acoust. Speech and Signal Process.* vol. ASSP-28 (1980) 672–680.
4. M. H. Hayes, The reconstruction of a multidimensional sequence from the phase or magnitude of its Fourier transform, *IEEE Trans. on Acoust. Speech and Signal Process.* vol. ASSP-30 (1982) 140–154.
5. A. V. Oppenheim, J. S. Lim, Importance of phase in signals, presented at 1980 IEEE l'Aquila Workshop on Digital Signal Processing.
6. G. Garibotto, L. Lambarelli, Fast on-line implementation of two-dimensional median filtering, *Electronics Letters* 13 (1979) 24–25.
7. J. M. Lester, J. F. Brenner, W. D. Selles, Local Transform for biomedical image analysis, *Computer Graphics and Image Processing* 13 (1980) 17–30.

G. Garibotto
3M Italia Ricerche SpA
17016 Ferrania (SV)
Italy

Panel: "Which Computer Architecture for Image Processing?"

Since we cannot present the actual turns of discussion and exchange of arguments by the different participants, the panelists give here a brief description of their main position statements. Overlapping among the independent contributions of the panelists is justified by common arguments and completeness of the single contribution.

INTRODUCTION AND PRELIMINARY STATEMENTS

V. Cantoni

Computational problems that arise in image processing may have special properties that are difficult to take advantage of using a general purpose system. Many algorithms involve the execution of the same operations on all pixels of a large image and often need a very high computation speed. For example, in the industrial environment, computer vision and part inspection may involve real time video data of 256x256 pixels with 6 bits of information for each pixel; since even the simplest picture-processing operation usually involves at least one access to the data in the 3x3 pixel window surrounding each pixel in the input picture, processing rates of 100 million operations per second are required, and these are beyond the reach of conventional computers. Moreover, the Landsat earth observation satellite generates about 200 images per day using a multispectral scanner, each scene contains about 8 million pixels. Assuming 1000 operations per pixel and desiring a response time of less than a second, a computational power of 8 billion instructions per second is required to process these images. The only practical solution, so far, has been to make use of parallelism in the system architecture.

Image-processing problems may be broadly divided into two characteristic classes: low-level image processing and high-level vision (or image understanding, image interpretation, image analysis, etc.). Low-level image processing deals with restoration problems, geometric correction, edge detection, object segmentation, feature extraction, noise removal, etc. Such tasks usually possess the following features: (i) the output has the same matrix size as the input; (ii) identical operations are applied to all pixels; (iii) computation for a pixel involves only the pixels in the local neighbourhood of that pixel. High-level vision involves classifying segments or features of the image into known classes. For these cases, the image is often no longer considered as a large matrix of pixels. More convenient data structures to represent segments of the image can be a set of parametric measures or a relation graph. In these cases a set of sequential processes which may be conducted in parallel, independently, are often required.

Many special purpose systems based on multiprocessor architectures have been developed for low-level image processing and also some for high-level image processing. These architectures may be divided into four types:

Processor arrays in which, in the limit, there is one processor assigned to each pixel in the image, so that computation time only depends on the computational depth and not on the image size. Processors, built using single bit arithmetic-logic units, communicate between near neighbours (CLIP IV, MPP, ILLIAC IV, BASE 8, etc.).

Pipeline architectures in which data streams in raster scan format from the buffer memory into a pipeline of processing stages. A sequence of delay units provides access to defined windows of data (usually 3x3 square). Ignoring an initial delay in which the first pixel has traversed all the processors, as data streams in, processed data streams out: the total computation time will depend linearly on the number of pixels and not on the computational depth. (Cytocomputer, FLIP, PICAP II, etc.)

MIMD systems in which several processors and also block memory communicate through a suitable interconnection scheme. There are many kinds of interconnection networks: high-speed bus, under the control of one master processor; reconfigurable network; crossbar. MIMD processors are not very efficiently organized for low-level image processing since much hardware is devoted to individual control units and reliable asynchronous data communication between processor units. There is also a problem with sharing near-neighbour data between the processor units. Some feature extraction algorithms, those serial in nature, such as contour following, are especially well suited to MIMD architecture as well as classification schemes, especially those involving syntactic methods (ZMOB, PASM, PUMPS, C.mmp, etc.).

Functional unit systems which consist of a host system and a set of special function processing units. A special function processing unit is a special purpose hardware system for implementing a single function (or a set of related functions). The general goal is to match the speed of the function unit with the capabilities of the host system (TOSPICS).

Many image-processing functions, such as convolution and discrete Fourier transform, may be defined by an inner product operation. The inner product computer is a special function unit for rapidly computing the inner product operation (DSR). Systolic design is an architectural concept proposed for VLSI. Systolic architectures are particularly suited to chip implementation of operation in signal and image processing such as filtering, correlation and discrete Fourier transform (ESL). Image sensor can be made with charge-coupled devices that have the possibility of including some low-level processing on the same substrate of the sensor.

Among the future systems to be carried out are two new ideas:

Pyramid computers having a multilevel data structure with reduced spatial resolution at the higher levels. Characteristics of these architectures are the efficiency and elegance with which some image-processing problems can be solved, and the fact that data from different resolutions can be instantaneously supplied with a slight increase of hardware (max. 30%) (PCLIP, Array/Net).

Architectures of computers based on the dataflow concept that can provide efficient support for the image structure.

REFERENCES

1. V. Cantoni, S. Levialdi: "Matching the task to an IP architecture", Proc. VI Int Conf. on Pattern Recognition, Munich, 1982, p. 254–257.
2. M. J. B. Duff: "Special Hardware for Pattern Processing", Proc. VI Int. Conf. on Pattern Recognition, Munich, 1982, p. 368–379.
3. A. P. Reeves: "Computer architecture for IP in U.S.A." Proc. I. Conf. on Image Analysis and Processing, Pavia, 1980, p. 117–126.

V. Cantoni
Pavia University

POSITION STATEMENTS

A. VLSI components and fault-tolerance methodologies for image processors

Renato Stefanelli

1. INTRODUCTION

In addition to the classification of the computing architectures presented by Cantoni, I should like to mention the possibility, for the image-processing application as well as for the other computer applications, of a distinction between:

(1) dedicated, not programmable, machines (application-dedicated machines), and
(2) general purpose, programmable, image processors (research and software development-dedicated machines).

The first ones must have a high performance in order to match the requirements of the application. An application-dedicated machine must have a very high speed in order to match the real time requirements, and must be highly reliable in order to ensure the correctness of the results.

In the second ones efficiency may be of lower importance; the system must be equipped with efficient and standard languages (operating on standard data representation), software tools etc.

In both cases, but mainly in the first one, the high speed and high throughput requirements yield to complex, parallel, non-Von Neumann structures.

Then two problems arise:

(1) the use of very large VLSI components, and
(2) the use of fault-tolerance methodologies.

Both of them must be examined when the structure of an image processor is to be chosen.

2. VLSI

The main problems arising in VLSI circuit design are concerned with the external data communications (limited number of I/O pins), the internal data communications (data transfer between each elementary circuit of the chip) and faults management.

The data to be processed flow through the input pins and, after the elaboration, they are extracted from the output pins. In modern VLSI circuits the ratio between the number of internal gates and the number of I/O pins is growing; in order to obtain good utilisation of all the elementary circuits of the chip, a large computation for each input data must be performed; this implies also an accurate choice of the algorithms. Moreover, a high ratio computation versus I/O time can be achieved by the use of serial type arithmetic units.

In order to obtain a high computing capability, a high ratio between computing chip area and internal communication area must be achieved; then a regular structure of not very little processing elements, each one connected to the adjacent ones, may be preferred; serial communications (between serial type arithmetic units) can also reduce the communication area.

The large chip area of a VLSI circuit can produce both a low production yield (faults during production) and low reliability (faults in run time); then:

(1) "Design for testability" methodologies must be used in order to facilitate the test of the integrated circuit; in the tested integrated circuit faulty elements are identified and their functionality substituted by some spare elements built in the chip by cutting, e.g., some internal connections by a laser beam.
(2) "Fault-tolerant" methodologies must be adopted.

3. FAULT TOLERANCE

Faults, in run time, must first be detected and then corrected; each functional element in the chip must then be provided with additional circuits in order to test the correctness of each result; the error information is provided to other additional circuits that correct the result.

When image-processing systems are concerned, the two following main methodologies can be adopted:

(1) If the circuit contains a low number of large arithmetic units or large memories, the faulty result can be corrected by adding a computed correction. As an example, Figure 1 presents a multiplier; residue codes Ra, Rb, and Rr are computed from the two factors and from the (eventually faulty) result; a syndrome S is then computed $S = Rr - r(Ra \times Rb)$; the correction to be added to the faulty results, in order to perform the correct product, is the output of a ROM, where syndrome S is the input.
(2) If the circuit is composed of a regular structure of equal elements, the whole structure can be reconfigured, as soon as a fault is recognized; the structure must be equipped with spare elements and with automatic reconfiguration circuits. As an example, Figure 2 presents a rectangular regular structure (the last column on the right is composed of spare elements); dashed elements are supposed to be faulty; Figure 3 presents the same structure after reconfiguration; Figure 4 presents the elementary cell with the two multiplexers MUX1 and MUX2 performing the reconfiguration rules and the added circuits that propagate the reconfiguration and fatal failure signals.

When large image processing systems are to be designed, the following two main considerations must be kept in mind:

(1) As the dimension of each VLSI circuit is growing (up to one million transistors), in order to obtain a good performance for a long time period, much more than a fault

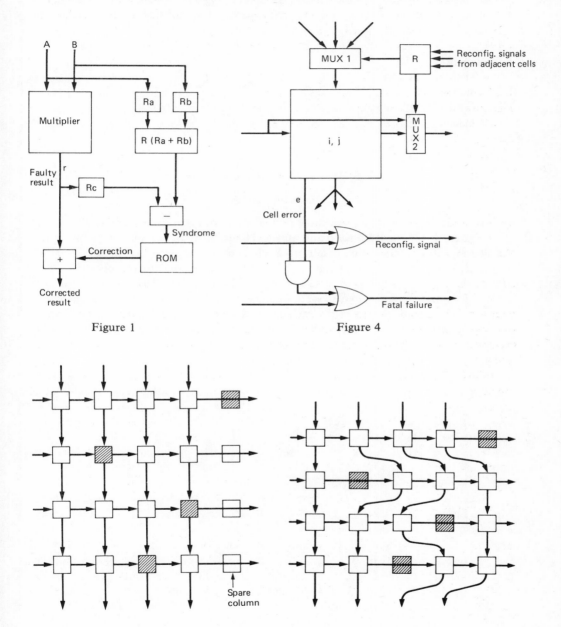

Figure 1

Figure 4

Figure 2

Figure 3

must be detected and corrected.

(2) Fault-tolerance methodologies must be carefully applied because the chip area can be largely increased, as shown in the examples in Figures 1 and 4, thus increasing the probability of faults. Then, the real performance of a fault method must always be carefully evaluated.

R. Stefanelli
Dipartimento di Elettronica
Politecnico di Milano
Piazza Leonardo da Vinci 32
20133 Milano
Italy

B. Data-flow architectures for IP

Concettina Guerra

Data-flow machines represent a possible alternative to Von Neumann machines. The data-flow concept is old; in 1966 Karp and Miller [1] introduced a computational model for parallelism, called computation graph, from which data-flow machines can be considered to directly descend. In a computation graph nodes correspond to the operations and directed arcs to the pathways for data and control. Data are viewed as flowing on the arcs from one node to another in a stream. As soon as a node has data on all the incoming arcs, the corresponding operation is performed and the resulting data are output on the outcoming arcs. A major feature of the computation graph is its determinacy, that is the output of a computation for given input is well defined and independent of the order in which the operations, when enabled, are executed.

In a data-flow machine an instruction is enabled whenever all the necessary data are available to it; when enabled an instruction consumes its input values and computes new values. There is no single control but the control is ensured by the flow of data. A high degree of parallelism can be obtained with this machine by assigning all the ready instructions to different processors. Among the many projects at present active in the development of architectural schemes embodying the data-driven computation concept, let us mention two groups at MIT, the Dennis group, which is the oldest one in this area, and the Arwind's group, which is building a machine using VLSI technology.

Recently, many projects have begun both in Europe and in Japan. A project of the Nippon Electric Co., called the Template-Controller Image Processor (TIP), is especially designed for image-processing applications, where the dramatically high amount of data demands high parallelism. TIP [2] combines the data-driven computation concept with the pipeline technique. In this way TIP overcomes the disadvantages of the pipeline processor, that is (i) the lack of flexibility to adapt to different processing requirements and (ii) the need for all processing elements to be synchronized by the same clock to avoid improper propagation through the pipe. TIP is essentially composed of three rings (see Figure 1): (i) a ring of processing units, each implementing a specific function; (ii) a ring, called the addressor unit, containing generator, reader, writer, bit-operator, and distributor; and (iii) a main ring which interconnects the other two. The addressor

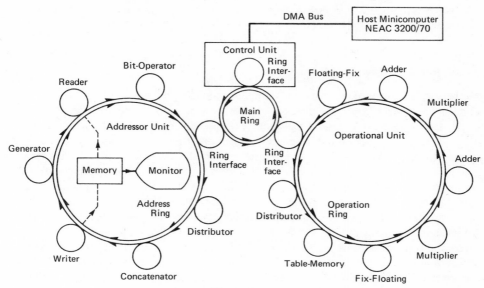

Figure 1

unit reads data from memory and provides them, via the main ring, to the processing units; it also writes back into the memory the output data from the processing units. During the initialization phase, the addressor unit receives from the host computer a set of templates which are defined externally. After the interpretation, these templates are sent to the processing units to control data flow and image definition. Data flows through the rings, each data item bearing an identifier, one or more destination flags. When the data reaches the right processor, i.e., the processor indicated by its destination flag, an operation is performed according to the template the data identifier matches; the original data disappears unless it contains other destination flags. The performance of the TIP system on a typical image-processing task, the Fourier transform, is evaluated in terms of the numbers of data transfers through the ring bus between processing modules.

REFERENCES

1. R. M. Karp, R. E. Miller, "Properties of a Model for Parallel Computation: Determinacy, Termination Queueing", SIAM Journal on Applied Mathematics, vol. 14, 1966, pp. 1390–1411.
2. S. Hanaki, T. Temma, "Template — controlled Image Processor (TIP project)", in Multicomputers and Image Processing, K. Preston, L. Uhr (eds.), Academic Press, 1982.

C. Guerra
Istituto Automatica
Università di Roma
Roma
Italy

C. **Computer architectures specialized for iterative neighbourhood transforms**

Stanley R. Sternberg

1. INTRODUCTION

Mathematical morphology [3] is a unique approach to image analysis which permits conceptualization of image-processing transforms in terms of complete pictorial concepts rather than fragmented parts. The structure of mathematical morphology is derived from a set of basic principles which ultimately determine the global characteristics of a unique way of thinking about images. These notions include the concept of a structuring element, a predefined shape which is employed as a probe to test out the spatial nature of the image undergoing analysis. Applications of different structuring elements can establish size, perimeters, holes, concavities or connectedness, etc. Structuring elements can be grey-scale for operation on grey-scale images, detecting edges, filtering noise, etc. Image transformations employing structuring elements are implemented as sequences of logical neighbourhood transformations in computer architectures which emulate cellular automata.

A cellular automaton consists of an unbounded two-dimensional space together with a neighbourhood relation which gives, for each cell in the space, a list of cells which are its neighbours. A cellular system is specified by assigning to each cell a cell state and a rule which gives the state of a cell at discrete time t+1 as a function of the state of its neighbourhood at time t, called the transition function. Neighbourhood processors are a class of devices that operate upon digital images, or more generally data matrices by sequences of neighbourhood operations. The term tessellation automaton can be applied to this class of computer architectures, where the neighbourhood relation and transition functions are the same for all cells in a bounded space but the transition rules may be modified each time step.

Parallel-array processors of the type typified by CLIP (Cellular Logic Image Processor [1]) represent direct realizations in hardware of tessellation automata. Cells of a parallel-array processor are identical, each consisting of a memory register for storing the state of a cell and a neighbourhood logic module for computing the cell's transition state. Interconnections between cells represent the cell's neighbourhood relation. Alternatively, the Cytocomputer pipelined neighbourhood processor [2] developed at the Environmental Research Institute of Michigan significantly reduces processor hardware without significantly affecting overall image-processing rates because its input and output are overlayed with processing and also because the pipeline stages have been customized for the neighbourhood operations of mathematical morphology.

2. SPECIALIZED ARCHITECTURE

The Cytocomputer pipelined image processor employs a chain of serial neighbourhood processing stages, each stage capable of generating the transformed value of a single pixel within a single clock pulse interval. The serial neighbourhood processing stage employs a neighbourhood logic module identical to its counterpart in the parallel-array processor cell, and line-delay memory for receiving a serial pixel stream from a row-by-row raster scan of the input data array and for configuring the neighbourhood window by providing

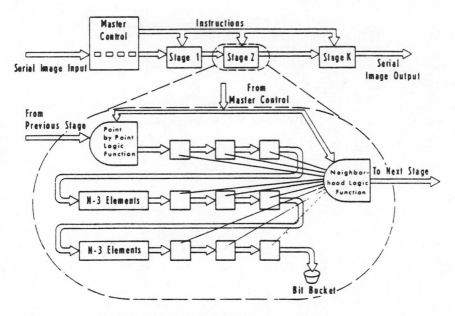

Figure 1. Cytocomputer block diagram.

the appropriate matrix elements to the neighbourhood logic module. The serialized input matrix is provided to the line-delay memory and the data bits are serially shifted through the line delays. When the line-delay memory has been filled it then contains the neighbourhood configuration for the first element to be transformed. Taps at appropriate positions in the line-delay memory provide parallel neighbourhood elements values to the neighbourhood logic module. These tapped memory elements in the line-delay memory constitute the neighbourhood window registers. A block diagram of a Cytocomputer is shown in Figure 1.

Binary image morphological transformations are implemented in a serial neighbourhood logic module which consists of a 512x1 RAM look-up table. The look-up table approach is impractical however for the grey-scale morphological transformations, which are executed in a programmable three-dimensional stage. An exclusive OR flip circuit at the input to the three-dimensional stage compliments the image data for the erosion transformation, while the bias generation and subtraction circuits permit the image data to be treated modulo 255 which increases the effective dynamic range of the three-dimensional stage. Programmable contribution values of +1, 0, −1, and X (don't care) permit grey-scale morphological transformations by three-dimensional structuring elements taken from a 3x3x3 window. The recently completed ERIM CYTO II Cytocomputer combines both binary and grey-scale processing stages on a single custom LSI chip. A block diagram of a three-dimensional stage is shown in Figure 2.

The parallel-array processor ideally executes neighbourhood image transformations at a maximal rate of one per clock pulse interval, while the serial-array processor performs the same image transformations at the rate of K/P image transformations per discrete time step, where P is the total number of pixels in image and K is the number of process-

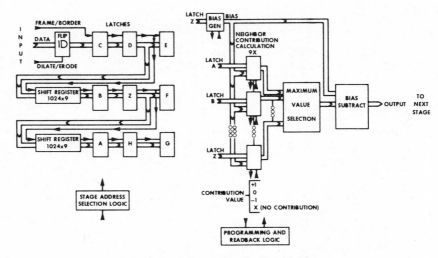

Figure 2. Three-dimensional stage block diagram.

ing stages in the serial array. For large images, the ratio of serial-array processor speed to parallel-array processor speed can be very small. Since processing speed can only be increased by increasing the ratio of neighbourhood logic modules to the total number of data elements in the matrix, the question arises as to whether it is possible to incorporate more than one neighbourhood logic module per serial processor stage, or equivalently, whether it is possible to reduce the number of line-delay memory elements associated with each neighbourhood logic module.

An alternative way of viewing this problem is in terms of the dimensionality of the picture elements transformed in a single processor cycle. In the parallel array, all pixels in a two-dimensional array are simultaneously transformed, but in a serial array a single stage transforms only a single pixel in one time step. Therefore, it is reasonable to examine architectures which, potentially, can transform one-dimensional arrays (entire rows, for example) in a single time step.

The parallel-partitioned serial-array processor [4] allows the achievement of transformation rates at a wide range of levels between the extremes presented by the parallel-array processor and the serial-array processor. This image-processor architecture takes the form of two or more serial-array processors which equally divide the task of transforming a single data matrix. Assuming a data matrix N elements wide and a serial-array processor stage laterally partitioned into M identical segments, the first N/M columns are fed to the first-stage segment, the next N/M columns to the next segment, and so forth. These individual pixel streams are staggered so that each segment receives its pixel stream delayed by one scan line with respect to its left-hand segment.

Parallel-partitioned serial-array processor segments resemble conventional serial processing stages but incorporate connections to certain of the window registers of the companion or outboard processor segment. These connections are provided by multiplexors which effectively switch either an outboard or an inboard window register to the

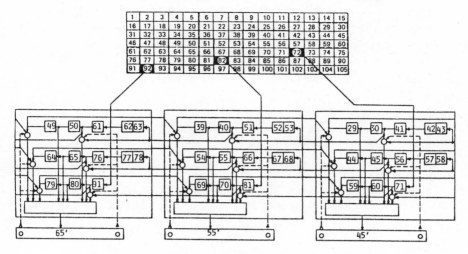

Figure 3. Triple partitioned stage.

neighbourhood logic module of the inboard processor. The switching of the multiplexor depends upon the position of the pixel in the matrix which is currently being transformed. If that pixel's neighbours are not all contained within a single matrix segment, the multiplexors switch the required pixels from the adjacent segment into the given segment. In this way image processing is partitioned among several processors but problems associated with image segment edges are avoided.

In the cyclic serial-array processor feedback is provided for selectively coupling the output of the last stage in a pipeline to the input of the first stage. Enough stages are included in the pipeline for line-delay memory to hold all the pixels in an image, including at least one blank line at the beginning of a frame. To operate, first all stages are programmed to perform a first sequence of neighbourhood transformations. The pixels are then circulated through the programmed stages. Before any nonblank pixels are recirculated through the pipeline, the first stage is reprogrammed with a new transition function. Thereafter stages are progressively reprogrammed in synchrony with the blank scan line. By this method a greater number of transformations may be accomplished than there are stages in the pipeline.

3. CONCLUSION

Pipeline processors suffer a disadvantage that array processors inherently to not have, image-processing algorithms requiring branching on measurements taken on image content cannot be implemented entirely in the pipeline. Parallel array architectures may contain a wired OR to the state register of each cell which may serve as a flag to interrupt or otherwise modify the flow of neighbourhood transition instructions. In a pipeline, stages may be performing different neighbourhood transformations concurrently, hence the need for taking the image out of the pipeline before testing for image content.

An image-processor architecture which combines advantages of both pipelines and arrays utilizes a banyon multistage interconnection network to couple logic modules to pixel memories. This arrangement permits a much greater degree of flexibility of processor performance than either arrays or pipelines yet is conservative of hardware.

REFERENCES

1. M. J. B. Duff, "Review of the Clip Image Processing System", AFIPS Conf. Proc., Vol. 47, 1978 NCC, pp. 1055–1060.
2. R. M. Lougheed, D. L. McCubbrey, S. R. Sternberg, "Cytocomputers: Architectures for Parallel Image Processing", Proc. Workshop Picture Data Descr. and Management, Pacific Grove, Calif., Aug 27–28, 1980, pp. 281–286.
3. J. P. Serra, *Mathematical Morphology and Image Analysis*, Academic Press, London (1982).
4. S. R. Sternberg, "Pipeline Architectures for Image Processing", *Multicomputers and Image Processing-Algorithms and Programs*, L. Uhr, ed., Academic Press (1982) pp. 291–305.

S. R. Sternberg
Environmental Research Institute
University of Michigan
Ann Arbor
MI 48107
U.S.A.

D. Image-processing-architecture design

Serge Castan

Image-processing architectures are functions of different parameters such as:
1. The problem for a given application.
2. The specificity of the images.
3. The processing time:cost ratio.
4. The algorithms to solve the problem, with different constraints.
5. The technical implementation constraints.
We shall briefly discuss these different points.

1. Vision problems are generally considered as two subproblems: low- and high-level problems. The lower level performing operations are, e.g., image input, output, restoration (filtering), enhancement, feature extraction segmentation and description. The higher level problem is an image understanding one. Separation between these two levels is not sharp and a big interaction exists. The higher level uses information from the lower level, but also from any source knowledge data base for a given application, using artificial intelligence techniques.

2. The specificity of the images. A digital image is determined by many parameters, such as:
 (i) Number of pixels: 4096 (64 × 64) up to more than 30 326 400 bytes (2340 × 3240) × 4 in Landsat images.
 (ii) Number of grey levels: generally 2–256.

(iii) Black and white or colour.

(iv) Multispectral in remote-sensing and spatial applications.

(v) Temporal sequences of images: compression problems for image transmission (digital TV) or temporal features for interpretation.

3. The processing time:cost ratio is specific for each application and it is a technical and financial problem. In some cases real time is expected, but what does real time mean? In industrial robotic applications video TV rate is generally used. Generally speaking image-processing techniques involve much data, but for a given price people want to get the quickest results.

4. To solve a given problem there are many algorithms working in serial or parallel mode, at pixel level or region level processing, and in each case the most convenient architecture should be adapted. Therefore, pipeline machines may be used when an image is considered, such as a one-dimensional data-flow, or when processing line-by-line, such as in TV video mode. S.I.M.D. machines are well adapted for pixel-level processing in local neighbourhood algorithms. M.I.M.D. machines are more convenient for region-level processing in semiglobal or global algorithms.

But in each case what will the best configuration be? In S.I.M.D. machines is it one- or two-dimensional arrays, and how many processors will be necessary? In multiple CPU (M.I.M.D.) systems, how to choose the number of CPUs? In this last instance, a not too bad criterion of studying such structures would be to consider the effect of the processing time in comparison to the transfer delay, and the variation of the total processing time when the different characteristics of the system vary, such as number of processors, data exchange rate, number of regions the image is divided into, memory storage capacity, using the normalized speed-up defined as the ratio of the processing time required on the considered multiprocessors divided by the number of processors.

Theoretical and simulation results show that for some given algorithms, the normalized speed-up grows rapidly with the number of processors while the ratio of the processing time over transfer delay for corresponding data is greater than the number of processors. Otherwise, speed-up is not improved significantly.

5. Technical implementation constraints. We have to choose between hard-wired systems for some specific problems, or software systems for a more general image-processing task, the number of processors and their interconnection possibilities. But the advent of VLSI and automated design tools have removed a fundamental constraint from computer architecture.

In conclusion, a parallel structure adapted to image understanding must be a multi-structure system: a multiple parallel-structure machine for the lower-level problems, with a high level of flexibility and reconfigurability, connected with a high-level language machine (List Machine for example) to solve the higher-level problems.

S. Castan
Département Informatique
Université Paul Sabatier
Avenue Rangueil
31077 Toulouse
France

E. For the round table discussion "What computer architecture?"

Per-Erik Danielsson

It is important that we make the following distinctions when we discuss computer architecture for image processing:

dedicated	*v.*	general purpose,
total system solutions	*v.*	processing part,
high-level processing	*v.*	low-level processing,
research and		
algorithm development	*v.*	end-user requirement.

Thus, we have four different more or less clear-cut choices and if we don't declare these
choices we may find we are talking about 16 different things. I will relate my discussion
to only one of these, namely, how the general purpose processing part of a low-level
image-processing architecture for end-user requirement should be optimized.

A fifth dichotomy that might be considered important is

interactive operation	*v.*	noninteractive operation,

which is probably the same distinction as between semiautomatic and automatic operation. I believe, however, that speed and other requirements could be just as demanding
for one as for the other, at least for low-level image processing.

I have spoken previously at this conference about the pros and cons for different
types of parallelism. I then concluded that although no commercial success is in sight for
parallel processor arrays there is a strong principal argument *for* the bit-serial processor
array working in SIMD-mode. It bears the promise of being cost-effective for all types
of neighbourhood operations, on large and small images, on binary as well as on greyscale or multispectral images.

Even if neighbourhood operations are important there are other operation types that
also must be performed by a general-purpose image processor. Before going into that I
want to show the general strategy we are using for our on-going study of parallel image-
processor architecture in Linköping. See Figure 1.

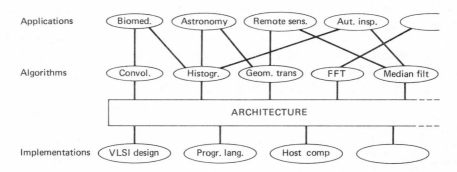

Figure 1

We are trying to do a limited top-down design in the following way. We try to understand as much as possible about different application areas. Then we examine the algorithms and operation types that are employed. This guides us when we design the architecture.

However, this is only the first step. The architecture level is unique because it is only at this level that we can clearly see the totality of the design. The different pieces fall together here and we can use our eyes to grasp the data flow, the control problems, the memory and its addressing mechanism, etc. Because of this there is a strong temptation to arrive at the architecture level rather quickly. That is fine as long as we understand that we have to *verify* the design over and over again. This is done by performance analysis (simulation by more or less sophisticated means). "Given this architecture what is the actual performance for the interesting algorithms?" "If performance is unsatisfactory, can we modify the architecture?" Or "Maybe we should check whether the applications could use another algorithm?" If we decide to modify the architecture, then we have to check all the other algorithms again, etc.

There are also similar verifications to be done downwards from the architecture level. For instance, if the partition of the architecture is impossible to do without creating an excessively large amount of pins for certain chips, then it is unsuitable for VLSI and implies a redesign.

Obviously we need an exhaustive list of algorithms, or rather operations, to match against different architectures. My list looks as follows.

- + Input/output of image data.
- + Feature extraction (counts, event coordinates).
- ++ Neighbourhood operations.
- ++ Global transforms (FFT, etc.).
- ++ Table look-up.
- + Geometry transforms (scaling, registration, rotation).
- + Propagation (recursive logic op).
- − Strongly data-dependent neighbourhood operations.
- − Data-dependent traversal (contour following).

Two plus signs means that we have full efficacy, one plus sign means reasonable but not full efficacy in our present design proposal. A negative sign means that the architecture is useless.

So far we are encouraged to go on with our project. The pros for image parallelism seem to more than counterbalance the cons. Note, however, that this situation has demanded several novelties in the design, e.g.:

(i) Distributed processor topology (one processor per subimage).
This enhances the neighbourhood operation capacity tremendously. It was suggested but not implemented on DAP.

(ii) Fast orthogonalization circuitry.
This gives us both reasonable high-speed input and output and a good means to do FFT data shuffling.

(iii) Table look-up (= index register).
This is almost indispensable for many image-processing tasks such as histogramming, logic

operations, grey-level transforms, etc.

(iv) Three variable-length shift registers.

These greatly enhance both ordinary arithmetic and special algorithms, such as median filtering.

All these points except the first one increase the complexity of the processor. We believe this is justified. The memory part of the PE-element is likely to be at least 65 Kbit and it seems unreasonable to attach a vastly increased memory to the extremely simple processors of CLIP IV, DAP or MPP.

In summary, I would like to give the processor array architecture a fair chance. It will not easily gain acceptance because the inertia of the von Neumann process is tremendous and hitherto the record of image parallel computers is far from impressive.

P.-E. Danielsson
Department of Electrical Engineering
Linköping University
S-58183 Linköping
Sweden

20 The MIMD level of the system SY MP A TI simulation and performances expected

I L Basille and S Castan

1. INTRODUCTION

Among all the numerous structures that have been studied and involved in image processing we did not choose a particular architecture, as it appears more and more obviously
relevant to consider a multistructure system. So we conceived a multilevel parallel system,
where the highest level is of M.I.M.D. type with standard processors and the lowest level
is of S.I.M.D. type. Furthermore, specialized hard-wired facilities may be connected to
the system.

In this article we first describe the structure of our system SY.MP.A.T.I. Then, for
the highest level, we study the variation of the processing time in relation with some
different features of an M.I.M.D. structure.

2. THE SY.MP.A.T.I. STRUCTURE

2.1. Introduction

The general structure of the SY.MP.A.T.I. system has been conceived in order to manage
different parallelism approaches. In such a way we designed a double-data-bus-oriented
structure:

(1) Standard processors are connected to an interprocessor bus making it possible to
    perform parallel algorithms in an M.I.M.D. mode.
(2) The second bus is a fast one used to exchange data at TV rate. On this bus are
    connected one or more image processors in which picture information is stored and
    may be processed in an S.I.M.D. mode.

Specialized modules may also be connected to this bus in order to solve specific
problems. These modules may generally be hard-wired. These two buses are connected
to each other through a transfer module. The whole system is managed by a resource
allocator (Figure 1).

2.2. The standard processors

Each standard processor has its own memory, large enough to store the image region
which is to be processed. The size of each region and the number of regions depend on
the algorithm performed and on the image analyzed. We shall see in the next part the
influence of the number of these processors on the performances.

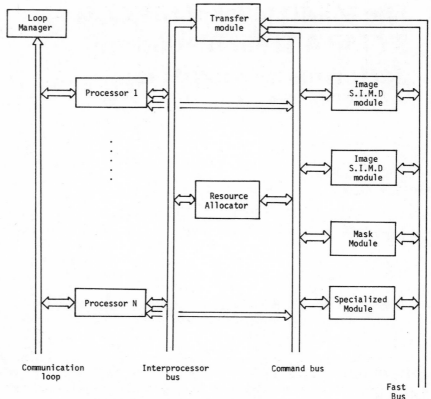

Figure 1. The general SY.MP.A.T.I.
structure.

The parallelism to be performed has to be explicitly indicated by the programmer. Different resource allocation strategies may be used in the operating system. We adapted a strategy based on the dynamic tree of resource requests.

2.3. The S.I.M.D. modules

2.3.1. The column structure

Each S.I.M.D. module may contain a 512 × 512 256 grey-level picture. The data are column structured so the neighbours of the same column are in same block and the close-line neighbours are in the neighbour blocks (Figure 2). For that, each column of the image is contained in a memory-block and to each block is attached a processing unit. All the processing units communicate with each other through a shifting loop (Figure 3). We took 16 blocks and so we work in parallel on 16 consecutive pixel line segments.

It seems it would be satisfying to have only one column in each memory-block, that is to have 512 blocks as we are working on 512 × 512 images. But for economical reasons and for a simplification purpose we took only 16 blocks. It is the minimum number in order to be compatible with TV rate considering the dynamic RAM we chose to realize a memory (1 μs cycle).

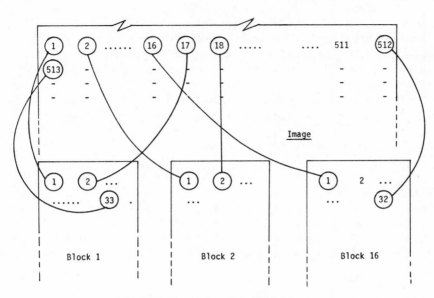

Figure 2. Image location/Memory address correspondence.

2.3.2. *The shifting loop*

We have to work modulo 16, which is made possible with the shifting loop (Figure 3). The extremity of the loop is composed of:

(1) two registers to memorize the left or right point of the considered 16 points segment,

(2) one register to initialize some or all of the 16 registers of the shifting loop, and

(3) a set of gates which make it possible to communicate with the fast bus, shifting left or right, input or output and to shift circularly.

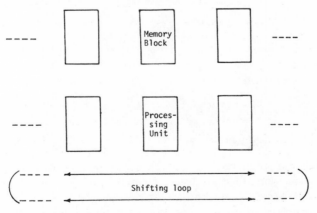

Figure 3. The column structure of one S.I.M.D. image memory of SY.MP.A.T.I.

2.3.3. The processing units structure

Each processing unit consists of the following components:

(1) an arithmetical and logical unit in order to process some plain local expressions,

(2) an 8 register scratch pad to memorize some temporary results or to avoid memory accesses, and

(3) an indicator set where the ALU indicators may be memorized. This is necessary to process conditional calculus as the 16 ALUs work in an associative way.

This structure shows that memory access and ALU processing cannot be run simultaneously. But memory access and shifting loop or ALU processing and shifting loop may be run so.

2.3.4. The command unit

It manages the 16 blocks (memory and processing unit) and the shifting loop. Procedures running on are microprogrammed with two level microinstructions:

(1) the "short" microinstructions which give the sequence of the microprogram, and

(2) the "long" microinstructions which command the different parts of the image memory.

2.4. Specialized modules

On top of the image memories, SY.MP.A.T.I. will have specialized modules for processing specific problems. For image input-output, it will have three modules:

(1) an image analyzer to digitalize an image from a TV signal,

(2) an image synthetizer to visualize image contents on a TV monitor, and

(3) an input-output module, to manage the connection with another computer for instance.

Many other modules may be thought of in order to process classical algorithms such as histograms, objects counting, etc.

3. THE LIMITS OF M.I.M.D. STRUCTURES

M.I.M.D. structures are, of course, the more general and flexible ones. However, when studying the evolution of the total processing time for an image, as the number of processors and the processing time of one region vary, we can observe some limits due to the different features of the system. In this part we shall study the effect of some of these different features of an M.I.M.D. structure.

3.1. The terms of the problem

Let us try to define more precisely the problem to be solved. We are considering region-level processing which an M.I.M.D. structure is rather devoted for. For such treatments each region has to be sent to a processor in order to be processed. So, simultaneous transfers of different regions to different processors could be thought of. But this would make it necessary to have some interconnection network or cross-bar matrix between the memory split into several banks and the processors. In fact, it is not essential to study this more general configuration as it can be considered as an improvement of a single bus. Furthermore, the complexity of such an interconnection network becomes rapidly awful as the number of processors and memory banks increases.

The whole image may be split into regions, somehow or other. The only thing required is to take into account the memory storage capacity of each processor. Once more the generality of the study is not reduced when considering that the image has been divided into regions of identical shape and area.

We shall study the evolution of the total processing time for an image, as the number of processors and the processing time of one region vary.

The number of processors will be considered in relation to the number of regions the image has been divided into and the processing time of a region in relation to the transfer delay required to send a region to a processor.

We can sum up the hypothesis, with the following parameters, as follows.

3.1.1. *The parameters*
The following features will be used in the present study:
(1) The number NR of regions the whole image is divided into.
(2) The number NP of processors.
(3) The processing time TP for one region.
(4) The transfer delay TD required to send or receive a region.
(5) The total processing time TT for the whole image.

3.1.2. *The hypothesis*
(1) All the NR regions are shape and area identical.
(2) All the regions are processed the same way.
(3) The interconnection network between the image memory and the processors is reduced to a single bus.
(4) One processor processes only one region at once.
(5) $NR \geq NP$.

3.1.3. *Definitions*
3.1.3.1. The region cycle of a region R_i of an image is the time required to send the region data to the processor, to process the data and to get back the results, that is

$$TD + TP + TD = 2TD + TP$$

3.1.3.2. The processor cycle is the time passed by between two successive treatments processed on a same processor. This processor cycle includes the time while the processor is waiting for receiving or sending data.

3.1.3.3. The processing ratio of an algorithm is the quotient

$$\rho = TP/TD.$$

3.1.3.4. The speed-up S is the quotient of the time required on a monoprocessor over the time required on a multiprocessor, therefore

$$S = TP \cdot NR/TT.$$

3.1.3.5. The normalized speed-up $\hat{S}$ is the speed-up divided by the number of processors:

$$\hat{S} = S/NP = TP \cdot NR/(TT \cdot NP).$$

It is defined in order to compare multiprocessors which have a different number of processors. It is necessarily lower than 1.

3.1.3.6. The half-time number of processors $NP_{1/2}$ is the number of processors required to get half the total processing time corresponding to NP processors.

3.2. The theoretical study

We shall consider the simple case where the timing diagram is periodical. In fact a good operating system with a good bus allocation strategy, may improve the performance and reduce the waiting delays [8].

We have to consider separately two cases:

3.2.1. $\rho \geq NP-1$

In this case the timing diagram of a processor cycle, for NP processors is as shown in Figure 4. Processor NP has to wait $(NP-1)TD$ before it gets its data and begins processing them. But, as the $(NP-1)$ data transfers, towards processor 2 to processor NP, are run out before processor 1 has processed its data, because $TP \geq (NP-1)TD$, processor 1 may send its results on the bus without waiting, if it is not busy.

One processor cycle is

$(NP-1)TD +$ one region cycle $= (NP-1)TD + 2TD + TP = (NP+1)TD + TP.$

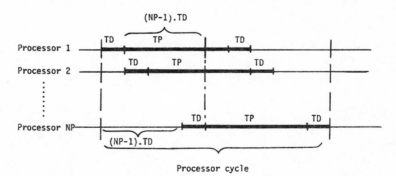

Figure 4. $\rho \geq NP-1$.

When considering all the NR regions of the image, we have to take into account whether NP divides NR or not.

3.2.1.1. NR = K·NP. We need exactly K = NR/NP processor cycles, so we get

$$TT = [(NP+1)TD + TP] NR/NP.$$

3.2.1.2. NR = K·NP + L with 0 < L < NP. We need K processor cycles and an extra cycle for the remaining L regions.

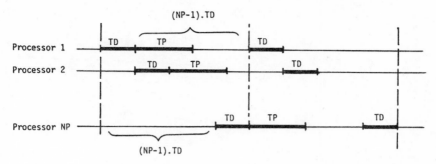

Figure 5. $\rho < NP-1$.

For these L regions, we have to consider L processors only, and the relation TP/TD $\geq L-1$ is still verified as L < NP. Therefore, we get

$$TT = [(NP+1)TD + TP] K + (L+1)TD + TP,$$
$$TT = (K+1)TP + [K+(NP+1) + L+1] TD.$$

3.2.2. $\rho < NP-1$

In this case, processor 1 has processed region data before the (NP−1) data transfers, towards processor 2 to processor NP, have all been executed. So it has to wait for (NP−1)TD − TP before sending its results. The timing diagram is as shown in Figure 5.

Furthermore, processor 1 has to wait (NP−1)TD while the other processors send their own results before processing another region. A processor cycle is

$$(NP-1)TD + [(NP-1)TD-TP] + \text{one region cycle} = 2(NP-1)TD-TP + 2TD+TP$$
$$= 2NP-TD.$$

When considering all the NR regions of the image, we have to take into account whether NP divides NR or not.

3.2.2.1. $NR = K \cdot NP$. We need exactly K = NR/NP processor cycles, so we get

$$TT = 2NP \cdot TD \cdot NR/NP,$$
$$TT = 2NR \cdot TD.$$

3.2.2.2. $NR = K \cdot NP+L$ with $0 < L < NP$. We need K processor cycles and an extra cycle for the remaining L regions. For this extra cycle, two cases have to be considered:

3.2.2.2.1. $TP/TD \geq L-1$. The processing time of the extra cycle is

$$(L-1)TD + \text{one region cycle} = (L-1)TD + 2TD + TP = (L+1)TD + TP$$

and it leads to

$$TT = K2NP \cdot TD + (L+1)TD + TP,$$
$$TT = (K2 \cdot NP+L+1)TD + TP.$$

3.2.2.2.2. $TP/TD < L-1$. The processing time of the extra cycle is

$$(L-1)TD-TP+(L-1)TD + \text{one region cycle} = 2(L-1)TD-TP + 2TD + TP = 2L \cdot TD$$

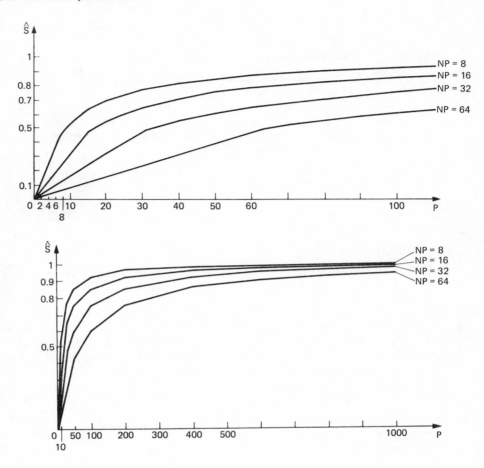

Figure 6.

and it leads to

$$TT = (K2NP+2L)TD$$
$$= 2(K \cdot NP+L)TD$$
$$= 2NR \cdot TD.$$

3.3. The speed-up

For more simplicity we shall only consider the case where NP divides NR, that is,

$$NR = K \cdot NP.$$

Therefore, $\hat{S} = S/NP = TP \cdot NR/(TT \cdot NP)$,
$\hat{S} = K \cdot TP/TT.$

And we can get $\hat{S}$ as a function of ρ:

When $\rho < NP-1$ we have
$$TT = 2NR \cdot TD.$$
Therefore, $\hat{S}$ $= K \cdot TP/(2NR \cdot TD),$
$\quad\quad\quad\;\; \hat{S}$ $= \rho/(2NP).$

When $\rho \geq NP-1$ we have
$$TT = K[(NP+1)TD + TP].$$
Therefore, $\hat{S}$ $= K \cdot TP/(K[(TP+1)TD + TP]),$
$\quad\quad\quad\;\; \hat{S}$ $= \rho/(\rho+NP+1).$

Figure 6 shows the evolution of the normalized speed-up when the ratio ρ varies. We can observe that speed-up increases quickly at first but then is not significantly improved over a certain value of ρ depending on the number of processors NP. Furthermore, for the same value of ρ, speed-up is best when the number of processors is lowest.

3.4. The half-time number of processors

If TT is the total processing time obtained with NP processors, then the number $NP_{1/2}$ of processors required to get a total processing time $TT/2$ has to verify the equation:

$$\frac{NR}{NP_{1/2}} [(NP_{1/2} + 1)TD + TP] = \frac{1}{2} \frac{NR}{NP} [(NP + 1)TD + TP],$$

which leads to

$$\frac{NP_{1/2} + 1 + \rho}{NP_{1/2}} = \frac{1}{2} \frac{NP + 1 + \rho}{NP},$$

therefore, $2NP(NP_{1/2} + 1 + \rho) = NP_{1/2}NP + 1 + \rho),$

$\quad NP_{1/2}(NP + 1 + \rho - 2NP) = 2NP(\rho + 1),$

$$NP_{1/2} = \frac{2NP(\rho + 1)}{\rho + 1 - NP}.$$

Figure 7 shows the evolution of $NP_{1/2}$ when NP varies for different values of ρ. We can observe that $NP_{1/2}$ increases quickly and grows the fastest when ρ is lowest. Therefore, it is important to have procedures with a high value for the processing ratio, and, when that cannot be done, it shows how much it costs to make the total processing time lower.

3.5. The simulation results

A simulation of the M.I.M.D. level of our system SY.MP.A.T.I. has been undertaken and has provided us with results that confirm the theoretical ones [8]. The bus allocation strategy of the operating system involved in our system is more elaborate than the plain one considered in the theoretical study. Furthermore, the time required by the operating system itself has not been taken into account in the above study.

According to both these simulation results and the theoretical study, two standpoints

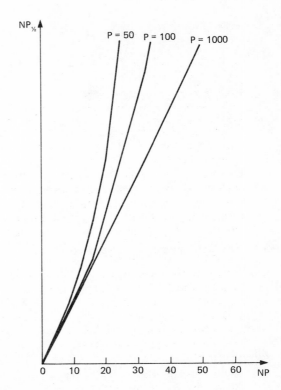

Figure 7.

have to be considered depending on whether the algorithm or the number of processors is given. If the algorithm is given, the number of processors may be increased to a limit of TP/TD+1 but no further than a certain number which, under our hypothesis, may be equal to 20 or thereabouts [8]. On the other hand, if the number of processors is fixed then such algorithms interesting enough to be processed on this structure, must have a ratio TP/TD greater than NP−1 [8].

4. CONCLUSION

The main conclusion we can draw from this study of an M.I.M.D. structure is the importance of the processing ratio ρ. Such an architecture is most interesting when this ratio is high. It confirms the intuitive feeling that it is a region-level-processing-oriented structure. So good advice would be to take care not to split a problem into too many modules but, appositely, to define procedures with a large enough processing ratio.

Furthermore, when considering speed-up, we can observe that it is best when the number of processors is low, especially when the processing ratio ρ is low. This correlation vanishes as this ratio gets large.

Finally, when one wants to get half the total processing time, one has to take N times more processors, where $N = 2(\rho+1)/(\rho+1-NP)$. So we can see that this multiplying factor becomes NP independent when the processing ratio tends towards infinity, and, in that case, N tends towards 2, a limit we can also get near to when NP is rather low.

REFERENCES

1. AL ROZZ, M.
Gestion de procédures et de processeurs sur un système multiprocesseur: SY.MP.A.T.I. Rapport de D.E.A. Toulouse 1981.
2. BARNES, G. H. et al.
The ILLIAC IV computer. Trans. IEEE on Comp. C-17, 746, 1968.
3. BASILLE, J. L., CASTAN, S., and LATIL, J. Y.
Structure logique et physique de l'information dans un multiprocesseur adapté au traitement d'images. GRETSI, 7ème colloque sur le traitement du signal et ses applications. Nice 1979.
4. BASILLE, J. L., CASTAN, S., and LATIL, J. Y.
Système multiprocesseur adapté au traitement d'images. 2ème congrès AFCET-IRIA, Toulouse 1979.
5. BASILLE, J. L., CASTAN, S., and LATIL, J. Y.
Système multiprocesseur adapté au traitement d'images. Workshop on New Computer Architecture and Image Processing, Ischia, Italy, Ac. Press, 1981.
6. BASILLE, J. L., CASTAN, S., and LATIL, J. Y.
A two-level parallel structure. SY.MP.A.T.I. Application to chromosome Analysis. IVth European chromosome Analysis Workshop. Edinburgh, Scotland, 1981.
7. BASILLE, J. L., CASTAN, S., DELRES, B., and LATIL, J. Y.
A typical propagation algorithm on the line processor SY.MP.A.T.I. The region labelling. Workshop on application of New Conventional Computers in image processing: Algorithms and programs. Madison (Wisconsin) USA, Ac. Press, 1982.
8. BASILLE, J. L., CASTAN, S., and AL ROZZ, M.
Parallel architectures adapted to image processing and their limits. Workshop on Computing structures for Image Processing. Abingdon (England) 25–28 May 1982. Ac. Press (to be published).
9. BERNSTEIN, A. J.
Analysis of programs for parallel processing. Trans. IEEE on Elect. Comp. EC-15, 757–763, 1966.
10. COFFMAN, E. G.
Operating systems theory. Prentice-Hall, Inc.
11. COMTE, D., and DURRIEU, G.
Techniques et exploitations de l'assignation unique. 5.7.75, Contrat SESORI 74.167, 1975.
12. CONWAY, M. E.
A multiprocessor system design. Proc. AFIPS Fall Joint Comp. Conf., 1963.
13. CORDELLA, L., DUFF, M. J. B., and LEVIALDI, S.
Comparing sequential and parallel processing of pictures. Proc. 3rd IJCR, Coronado USA, p. 707, 1967.
14. DUFF, M. J. B., WATSON, D. M., and DEUTSH, E. S.
A parallel computer for array processing. Proc. IFIP Congress 1974, Stockholm, Sweden, pp. 94–97.
15. DUFF, M. J. B.
Parallel processing techniques. In Pattern Recognition. Ideas in practise, Batchelor, B. G. (ed) Plenum Publishing Co., New York, 1978.
16. FISHBURN, J. P.
Analysis of Speedup in distributed algorithms. Computer Sciences Technical Report #431, University of Wisconsin-Madison, May 1981.

17. HAYNES, L. S., LAU, R. L., SIEWIOREK, D. P., and MIZELL, D. W.
A Survey of Highly Parallel Computing. IEEE Trans. on Computer, January 1982.
18. LENFANT, J.
Mémoires parallèles et réseaux d'interconnexion. Technique et Science Informatiques, vol. 1 no 2, RAIRO, 1982.
19. LEVIALDI, S., MAGGIOLO-SCHETTINI, A., NAPOLI, M., and UCCELLA, G.
Considerations on parallel machines and their languages. Workshop on high level languages for image processing. Windsor U.K., 1979.
20. LISTER, A. M.
Principes fondamentaux de systèmes d'exploitation. Eyrolles, Paris 1977.
21. SHAPIRO, D.
Theoretical limitations on the efficient use of parallel memories. IEEE Trans. on Computers, vol C-27, 5, pp. 421–428, mai 1978.
22. TIMSIT, C. and BOUDAREL, R.
PROPAL II: une nouvelle architecture de calculateur adaptée au traitement du signal. Colloque National sur le traitement du signal et ses applications. Nice 1977.
23. LEVIALDI, S., MAGGIOLO-SCHETTINI, A., NAPOLI, M., and UCCELLA, G.
Considerations on parallel machines and their languages. Workshop on High Level Languages for Image Processing 4–8 June 1979, Windsor U.K.

J. L. Basille and S. Castan
Laboratoire C.E.R.F.I.A.
Département Informatique
Université Paul Sabatier
Avenue Rangueil
31077 Toulouse
France

21 REST: a powerful element for reconfigurable processing structure

D Marino

1. INTRODUCTION

REST is a single-bit versatile module easily "concatenable", allowing implementation both of simple structures (arbitrary word-length processors) and of the most complex ones (arrays multiprocessors, fault-tolerant structures, dynamic architectures).

REST architecture derives from that developed in the SINBIT project first presented in 1979 [1–3]: the elementary cell was in SINBIT, too, a single-bit processor (SBP) *ad hoc* designed for highly reconfigurable structures. The SBP was tested during 1981 and its structure revised after an extensive implementation programme of different level "virtual" machines: F-8, Z-8000, PdP-11, P-code, Z-code, VAX-780. More about the SBP is referred to in refs. [4–6]. This "refinement" work ended with the design of REST.

The REST module, as well as the SBP, includes the memory, formed by single-bit words (also the memory concatenates). Its most remarkable feature is to be itself both *logically* and *physically* reconfigurable:

(1) *Logical reconfiguration*: REST, used *alone* can efficiently emulate up to 16-bit commercial microprocessors, exploiting its deep prefetching possibility (this feature is later clarified).

(2) *Physical reconfiguration*: REST is basically formed by only three specialized sub-modules: they can be replicated and combined in order to assemble a large variety of different type of powerful *single-bit* slices: using this facility we can expand — in the higher-level structures — the number of bus lines, the number of buses (for redundancy sake as well as for parallelism exploitation), the number of communicating processors and to speed up the arithmetic operations.

In this contribution we will discuss some implementation schemes and a fault-tolerance evaluation model.

Investigations about optimal exploitation of REST modules in "image-processors" are at present in progress.

2. BASIC SUBMODULES AND EXPANSION MECHANISMS

As mentioned above, submodules of only three kinds are present on each slice:
(1) controller (C),
(2) memory processor (MP), and
(3) communication processor (CP).

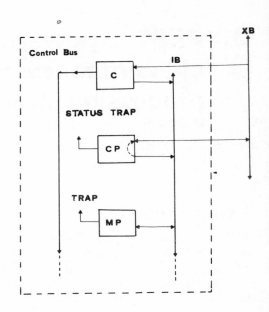

Figure 1. A simplified REST configuration. C, control unit; CP, communication processor; MP, memory processor; IB, internal bus (on the REST module); XB, external bus (outside the REST module); CB, control bus (on the REST); STATUS, TRAP, branch condition sent to C.

Figure 2. Replication of C-units on a same REST module: an application for "multiprocessor slice" implementation. C_1, C_2, C_3, different control units on a same REST; IB_1, IB_2, IB_3, internal buses controlled by the corresponding control units; CB_1, CB_2, CB_3, control buses from corresponding control units; XB_1, XB_2, XB_3, different external buses for the same "multiprocessor slice" REST.

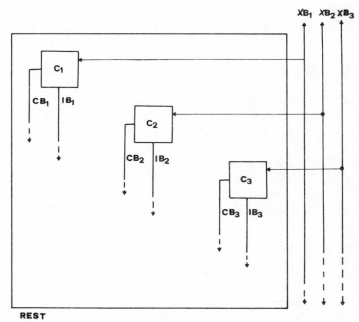

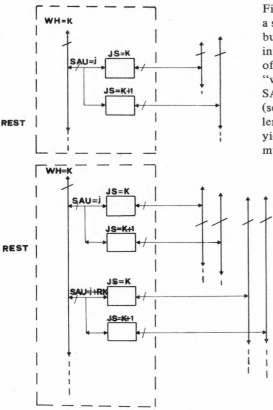

Figure 3. Using more CP submodules (on a same slice) as bus-size expanders and bus-number expanders. IB_1, XB_1, XB_2, internal and external buses; WH, "weight" of a REST module in a system; JS, "weight" of each CP on a *same REST*; SAU, address of the IB "source" line (sent to CP); RK, rank of the slice (word length of a single REST: each REST can yield on an external bus single- or multiple-bit contributions).

In Figure 1 a simplified REST configuration is shown; here IB and XB, respectively, represent an "internal bus" and an "external bus". Figure 2 shows how C submodules can be replicated in order to assemble a *multiprocessor slice*, i.e., a slice having more interconnected control units on separate buses (true parallelism).

The use of CP units both as bus-size expander and as bus-number expander is shown in Figure 3, while the replication mechanism of MP submodules is shown in Figure 4. We will note here that only the CP submodule can directly communicate via external bus (XB) and then it is the only module able to process whole words. The C submodule can only *see* the whole XB and takes from it the "machine instructions". The MP module operates using only the internal bus (IB); direct access to XB is forbidden to it.

3. DYNAMICAL BEHAVIOUR OF REST

The dynamical behaviour of REST can be rigorously described using Petri graphs. Figures 5–8 give such graphs in reference to the whole REST and to the three submodules C, CP and MP.

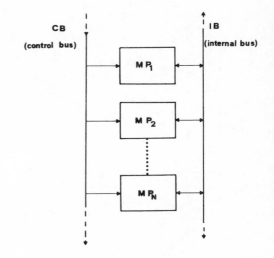

Figure 4. Interconnection scheme or more MP units on a same REST module. CB, IB, control and internal buses; MP$_1$, MP$_2$, MP$_N$, memory processors.

Figure 5. Petri graph of the REST module. TRAPS, STATUS, branch conditions sent to C unit; I, next microinstructions; WAIT, external "wait" signal; MIR, microinstruction register; CIN, COUT, carry in/out signals of the ALU; FALU, microinstruction to ALU (from MIR).

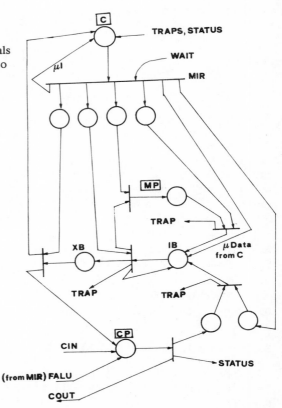

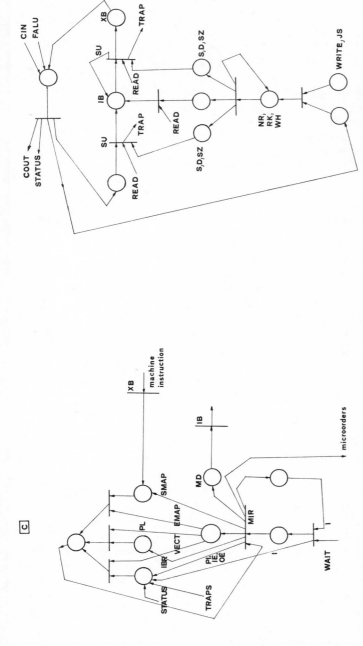

Figure 6. Petri graph of the C submodule.
I, next microinstruction to the
"microsequencer"; PI,
microinstructions to a priority encoder;
IE, microinstruction "enable", to the
priority encoder; OE, output enable of
interrupts to the "microsequencer";
VECT, vector map enable (vector "branch
address" corresponding to interrupts and
status); IBR, indirect microbranch

address; PL, "read" signal of the
microbranch address from the C unit;
EMAP, "read" the "mapped" address of
the microroutine corresponding to an
operating code of "machine" instruction;
SMAP, map-zone selection; MD, set
microdata on to IB; MIR, microinstruction
register; I, next microinstruction; micro-
orders, controls to other submodules of
the REST.

Figure 7. Petri graph of CP submodule.
COUT, CIN, FALU, STATUS, TRAP, as
in Figure 5; SU, switching units; RAU,
TIX, control of the access to IB and XB;
SAU, DAU, SIX, DIX, RK, control signals
to the two switching units SUs; JS, RK,
WH, as in Figure 3; RNR, RWR, read
SAU, DAU, SIX, DIX, RK, WH (to be
read before reconfigurations or during
diagnosis).

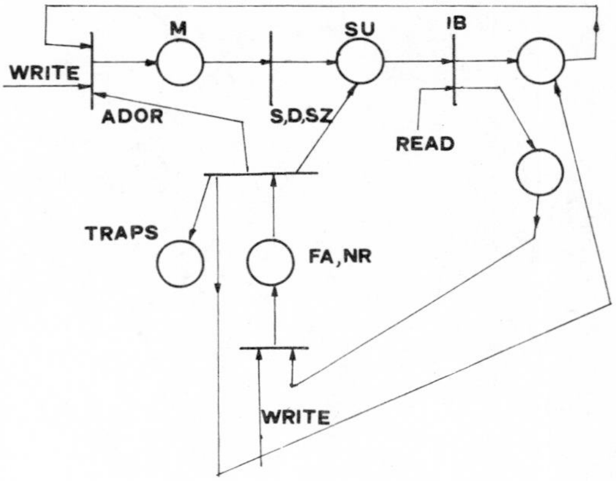

Figure 8. Petri graph of MP submodule.
M, memory blocks (16 blocks of 4096
single-bit cells); ADDR, block and cell
address (4+12 bits); R, read memory on
IB; FA, NR, computed address and
control parameter for reading and writing
of the memory.

4. IMPLEMENTATION EXAMPLES

The implementation of typical structures is here presented in order to clarify the meaning
of "physical" and "logical" restructurability.

A single N-bit processor can be assembled using N different REST modules each of
them yielding an external bus single- or multiple-bit contribution: if REST works as an
effective single-bit processor, it presents a 16 single-bit words prefetching, while if REST
works as a two-bit processor, its prefetch possibility is consequently reduced to eight
2-bit words. The *external* "processor" word length is not affected by this reconfiguration.

Let us consider, é.g., the implementation of a 16-bit processor. If the maximum
number of "immediate" operands in the machine instruction set is 1, this processor can
be implemented using only two REST slices holding true parallelism: in fact the pre-
fetched 16-bit words can be logically considered as two 8-bit words which concatenate
on the external bus forming the 16-bit word: in this way prefetch deepness is just of two
words. Obviously the same processor can be emulated by a single REST but efficiency
decreases.

The submodules present on the slice are only the three basic submodules "controller"
(C), "memory processor" (MP) and "communication processor" (CP), without any
replication (Figure 9). One more REST module can be inserted if an additional 8-bit
tag-processor is requested. Tag processing can be effected sequentially, but if, for efficiency, we overlap tag processing and instruction processing, a further bus must be provided
and then on any REST one more CP submodule has to be considered.

Let us now consider a 32-bit high-performance processor, having virtual memory.
Consider that "immediate" operations of the instruction set use no more than three
operands (see, e.g., a VAX processor). A prefetch deepness of four words is sufficient to
warrant full parallelism. To obtain them we can use 8 REST slices, each bringing a 4-bit
contribution on the external 32-line bus (XB). The total cache capacity becomes 16000
words (considering a word size of 32 bits).

The "physical reconfiguration" meaning can be clarified considering a virtual memory of 64 address lines (some of them used as tags and parity bits): three more CP units
are to be added on any REST. The total size of the line number rises to 64 lines. One

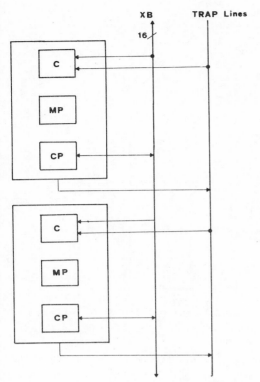

XB   TRAP Lines
16

Figure 9. C, MP, CP, XB, TRAP, as in Figure 1.

more memory submodule can be useful in order to provide fast context switching and virtual address expansion bits (Figure 10). Hardware stacks, local store and so on, are optional memory expansion to be implemented on any slice simply by adding similar memory submodules ("physical reconfiguration"). Tag manipulation can, of course, be overlapped with that of instruction without a further expansion of the external bus.

Figure 11 shows an essential scheme of a high fault-tolerance multiprocessor system. It is organized with three hardware levels. Detection schemes and reconfiguration procedures are presented elsewhere [3]. In general spare elements can be inserted at any level, but fault-tolerance and reconfiguration possibilities in REST architectures are also effective, if spare REST modules are not available, by exploiting its logical reconfiguration possibilities.

Among the implementation schemes, we will mention here the possibility of making machine structures also having an expansible number of buses at any level. This bus redundancy (allowed by the mentioned "physical" reconfiguration possibility) can be exploited for efficiency in fault-tolerance purposes in the case of bus failure.

5. FAULT-TOLERANCE EVALUATION

For fault-tolerance evaluation, we used the Markov-chain model introduced first by Ng and Avizienis [7] (Figure 12). This model has been adapted in order to include the reconfiguration mechanism introduced by us: let us suppose that the starting "active"

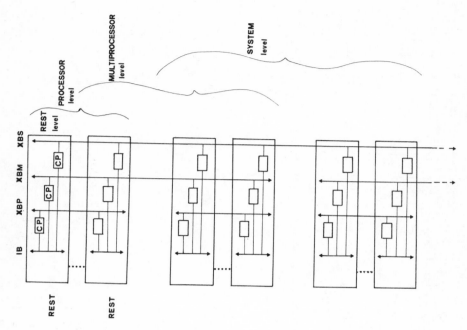

Figure 11. C, MP, CP, XB, TRAP, as in Figure 1.

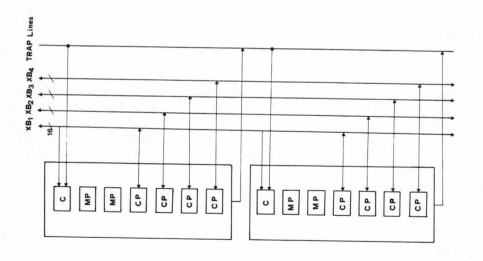

Figure 10. C, MP, CP, XB, TRAP, as in Figure 1.

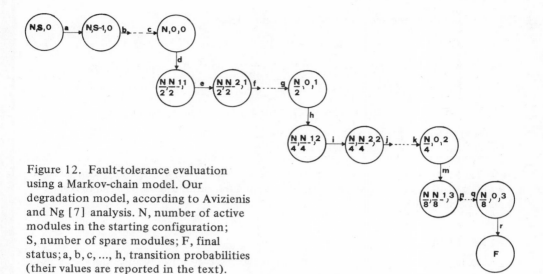

Figure 12. Fault-tolerance evaluation
using a Markov-chain model. Our
degradation model, according to Avizienis
and Ng [7] analysis. N, number of active
modules in the starting configuration;
S, number of spare modules; F, final
status; a, b, c, ..., h, transition probabilities
(their values are reported in the text).

configuration of an N-bit processor exploits N REST modules. If a failure occurs in one
of them, the new configuration will exploit N/2 slices of REST type, each yielding on
the bus 2-bit contributions. The external word length remains invaried, the prefetch
deepness reduces to height words and N/2−1 spare RESTs are produced.

As a consequence system performance can decrease but afterwards it remains invaried until the produced N/2−1 spares are "consumed". The occurrence of a further
failure brings the system in a new configuration in which the number of active RESTs
becomes N/4; the prefetch deepness is reduced to four words, each REST brings on the
bus 4-bit contributions and N/4−1 "spares" are produced. This process can continue
until the prefetch deepness reduces to a single word (maximum allowed degradation).
This graceful degradation possibility can, of course, be increased using a starting configuration with more than N REST modules. These additional spares, as is known, allow the
system to work without degradation.

Figure 12 shows our fault-tolerance model and fault-tolerance parameter evaluation.
The degradation and reconfiguration mechanism we introduced, holds the maximum
efficiency. In reference to Figure 12 the "transition probability" in a new configuration
after a failure, can be evaluated according to the following expressions:

$a = \lambda N + \lambda S,$

$b = \lambda N + \lambda(S-1),$

$c = \lambda N + \lambda,$

$d = \lambda N,$

$e = \lambda N/2 + \lambda(N/2 - 1),$

$f = \lambda N/2 + \lambda(N/2 - 2),$

$g = \lambda N/2 + \lambda,$

$h = \lambda N/2,$

$i = \lambda N/4 + \lambda(N/4 - 1),$

$j = \lambda N/4 + \lambda(N/4 - 2),$

$k = \lambda N/4 + \lambda,$

$m = \lambda N/4,$

$n = \lambda N/8 + \lambda(N/8 - 1),$

$p = \lambda N/8 + \lambda(N/8 - 2),$

$q = \lambda N/8 + \lambda,$

$r = \lambda N/8.$

For evaluation of the above expressions we considered:

$C_a = C_d = 1$ (coverage parameters for active and degraded configurations)

moreover, CY, the coverage vector for the three degraded configurations, given by

$CY[1] = CY[2] = CY[3] = 1;$

Y[i], the number of active modules in the ith degraded configuration, given by:

$Y[1] = N/2; \quad Y[2] = N/4; \quad Y[3] = N/8;$

$\lambda = \mu$ (same permanent-failure probability for active and spare modules: spare modules are "powered"); the number of spare modules $S = N/2 - 1$ in the first degraded configuration; $S = N/4 - 1$ in the second degraded configuration and $S = N/8 - 1$ in the final degraded configuration.

REFERENCES

1. D. Marino; Progetto Sinbit, 1979. Report CNR (P.F.I.) 79-P1-2-Sinbit-1- C.N.R. Roma - Italy.
2. M. De Blasi, G. Degli Antoni, M. Mallamo, D. Marino; Progetto di calcolatore Fault-Tolerant Single-Processor ed elevato grado di emulazione. Proc. AICA, Oct. 1979. Bari - Italy.
3. M. De Blasi, C. Degli Antoni, D. Marino; Proposta di architettura multiprocessor con capacità di riconfigurazione a più livelli. Proc. AICA, Oct. 1979.
4. M. De Blasi, D. Marino; The sinbit reconfigurable computer: Architectural insights. Proc. AICA, Oct. 1980. Bologna - Italy.
5. G. Aloisio, F. Corvino, M. Mallamo, O. Murro; Memory management and protection methods in the sinbit experimental computer. Submitted to Fitcs-13, 1983.
6. G. Aloisio, M. De Blasi, M. Mallamo, D. Marino; Implementazione di macchine virtuali di differenti livelli su architettura sinbit. Report INFN, 1982 (in press).
7. Y. W. Ng and A. Avizienis: "A unified reliability model for fault-tolerant computers". IEEE - TC, V.C-29, 11 (1980) 1002–1011.

D. Marino
Dipartimento di Elettrotecnica e di Elettronica
Università di Bari
Italy

22 Parallel image processing: uni- or bidimensional? Example: Romuald

B Bretagnolle, P Jutier and Cl Rubat du Mérac

1. INTRODUCTION

Let us first try to find the exact limits of architecture in general, and computer architecture in particular. It is hardly necessary to require the help of an architect when building a log cabin. If you build even a small house it may be useful. For any larger building you cannot dispense with architectural considerations, because you are confronted with the physical world, with the laws of nature, etc.

Conversely, we think that architecture stops where town planning starts, in other words when problems of long-range communication take the upper hand. Thus, in our opinion, architecture involves strong interaction of all parts, and prevalence of physical constraints, arising mainly from gravity in buildings, and electricity and time in computers.

Various factors are important: history — personal and institutional; intended use; site and environment — host computer — type of sensor/display. Building materials entail additional constraints, and may procure new opportunities, such as microprocessors in our case. Special features often mean special opportunities: segmentation of the address space is a good example, as will be seen later.

2. THE ROMUALD PROJECT

2.1. Historical

It all started in the autumn of 1978. Two software engineers (one having just produced a thesis on image processing) started looking for a replacement for their I.P. facility which had collapsed (and was anyway obsolete).

2.2. Intended use

To try interactively new I.P. algorithm if possible without any limitation.

2.3. Environment

A large regional university computer centre (Grenoble), with strong emphasis on research. The host computer would be a large main-frame computer, supporting multics.

2.4. Basic choices

The image sensor/display system should use ordinary TV technology. It appeared that for exploration of new algorithms the images to work with at the start would be mostly photographs (the cheapest image memory anyway!). Thus it was important that image acquisition could be made repetitively and, therefore, be as quick and simple as possible. Note that this situation is exactly opposite to (for example) Landsat image processing, where numerous images placed on tape have to be processed successively. This entails a *strong interaction* between the acquisition/display part and the image memory/processor part of the machine.

2.5. Definition

ROMUALD was defined as a real-time TV-type acquisition and display facility with enough image memory and computing power to take care of image preprocessing (part or all of low-level processing in Danielsson-Levialdi's definition [1]) and acting as a front end to a powerful main-frame computer. It should be fast enough for comfortable interactive use; speed *not* being the prime consideration. The master word is ease: ROMUALD was thought of as easy to conceive, to build, to program, and to operate, and the main object-ive is *flexibility*.

2.6. Easy to conceive

Four basic ideas:
(1) Image in memory is an exact replication of image at sensor level (TV camera).
(2) To avoid bus blocking, processing power (as opposed to old-fashioned multiproces-sors) is distributed in the image memory (true parallelism).
(3) For that purpose, image memory has to be partitioned; the obvious way to do it: *horizontal* slices, between raster lines (see Figure 1).
(4) *A motto*: Dynamic reconfiguration: don't move data, don't talk about it, switch it! using a general bus with appropriate control means (bus control buffers everywhere).
For details on these four points, see refs. [2–4].

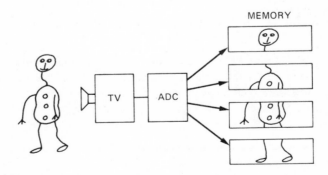

Figure 1

2.7. Easy to build

Systematic use, as building blocks of large integrated "off the shelf" circuits offering the best possible "intelligence/electricity ratio": 16-bit microprocessors, single chips, micro-computers, EPROMS, DRAMS, three state buffers, PROMS, PALS. Carefully avoiding bipolar slices (too exotic for large computer programmers . . . and too many electrical problems in sight) as we were not prepared to pay that much just for speed! For the same reasons as little TTL as possible.

2.8. Easy to program

ROMUALD *can* be configured, at system level, as a pure SIMD machine (single instruction multiple data). The supervisor is then able to make parallelism totally transparent to unsophisticated users. It is currently programmed using a cross assembler running on multics. A cross higher-level language is possible. A resident assembler is also a possibility.

2.9. Easy to operate

A command language enables easy man-machine interaction; all the more as the process-ing power (about 10 MOPS in the latest version) is large enough for ordinary neighbour-hood operators and other simple programs (histogram, etc.) to run in a matter of seconds (fractions of seconds for the simpler ones).

3. ROMUALD'S REALITY

In fact, as it now exists,* ROMUALD is a multimicroprocessor system which can be partitioned to operate as several independent SIMD and/or MIMD machines of varying sizes. Furthermore, as a piece of hardware it is practically uncommitted, as no particular issues in image processing influenced the design choices (apart from the very strong influence of the raster principle chosen for the sensor/display system).

Among various proposals in the image-processing field, its nearest parent seems to be PASM which was apparently advocated simultaneously but is a better known and much more ambitious project [6].

We will now describe the salient features supporting our previous assertion.

3.1. Basic hardware: a bus controlled by a star

3.1.1. *A unidimensional array of processors* (Figure 2)
It is a typically bus centred architecture. To a large general bus (32 data, 15 addresses, 6 general control utilities) are tied logical blocks of three different types:
(1) One general processor (GP) — the master — built around an 8001 segmented machine

---

* First described in an internal report (CICG 1978). French patent no. 2465273 (09/11/ 79) [7]. First published in Toulouse (Sept. 1979) [2] and Houston (Nov. 1979) [3]. Feasibility demonstrated (with two processors) from August 1980. First complete prototype (eight processors) in operation since August 1981 [5]. French patent applic-ation no. 8200590 (01/15/82) [8]. Colour prototype currently under tests.

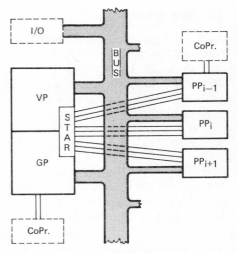

Figure 2

(originally ZILOG) with program memory (EPROM), system RAM, one duplication image memory and a memory management unit (MMU) type logic.

(2) One I/O processor or videoprocessor (VP) with A/D and D/A video speed concerters; working in DMA mode. It is centred on a single chip microcomputer.

(3) 2^n (eight in the present versions) parallel processors (PP) (8002 nonsegmented) each with its private program memory and its own full capacity memory (DRAM), all of them together constituting the main image memory.

Each block is separated from the bus by appropriate three state buffers. In addition each PP (with program memory) is separated from its own image memory slice by buffers.

Finally, the general processor (in association with the video processor) is the centre of a star: each PP receiving two control lines from the GP/VP and sending one STATUS line the GP. One control line (AA_i on Figure 4) is bidirectional. How this control star operates is a matter of soft and/or firmware as will be seen later on.

3.1.2. Why not bidimensional?

For the present version, with eight processors, a bidimensional array does not seem a practical proposition anyway. But why not a bidimensional 16 processor (4 × 4) array, dividing the image in 16 equal squares?

The answer is to be found in the properties of the strictly unidimensional output signal of TV-type sensors. As can be seen from Figure 3, the line duration is 64 μs (in the most common TV standard), enough to run a program on a modern microprocessor or computer. Conversely, only about 80 ns elapse between two points of a line (with a 512 × 512 pixel image). Of course, it is possible to actuate switches in such a short time, but only with dedicated hardware, totally inflexible.

Applying the unidimensional principle, dispatching of data can be prepared under program supervision during a full-line time (about 60 μs) and effectively executed (by actuation of bus switches) in several microseconds (duration of the black level). Thus this basic architectural choice is a logical consequence of the maximum ease/flexibility principle and the choice of TV sensors.

3.2. Open hardware structure

ROMUALD hardware is physically open in all directions.
(1) All the "back doors" of GP and PPs can receive additional hardware (coprocessors, I/O, etc.).
(2) Any new logical block type (for example, a different I/O processor) can be tied to the central bus and then gain direct image memory access (Figure 2).

3.3. Open system

All hardware/software trade-offs are pulled systematically toward soft and/or firmware for two reasons:
(1) Reliability: at the present integration level, programs — especially when frozen in PROMS — are safer than wired logic.
(2) Maximum flexibility: the uncommitted ROMUALD hardware is specified, transformed into *one* effective ROMUALD system (among many potential ones) by programs and tables loaded at all levels of organization: PAL-PROM-EPROM-RAM (the last only temporarily).

Note that, thus organized, the ROMUALD system is something much more general than just an image-processing machine: a straight-forward freely partitionable parallel computer equipped with image capture/display facilities. It will start being a real image-processing machine the moment appropriate programs are loaded in GP and/or PPs (see later).

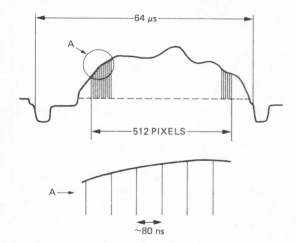

Figure 3. Video-type signal.

3.4. Delegation of power . . . and feed back

Maximum simplification of the control system is achieved by systematic delegation of power from the master to the slaves (Figure 4):
(1) Global in the case of the video processor (VP), which then becomes a "master in second" substituting itself to the GP in the centre of the control star. This is obtained via an asynchronous link (L).

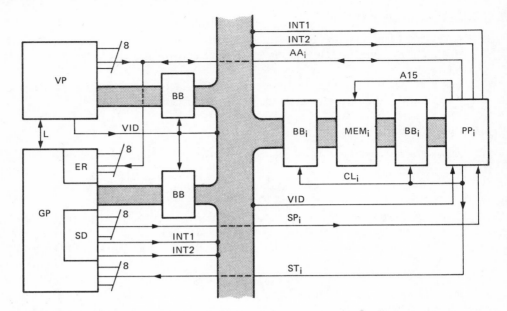

Figure 4. Relationship between GP/VP and a current parallel processor. SD, segment decoder, originates interrupt lines INT1, INT2 (bus lines) and selection lines SP$_i$ (star lines); ER, emergency requests incoming from parallel processors; VID, indicates which of GP or VP controls the bus.

(2) Individual — by a combination of mail box messages and general commands sent through the bus to the PP$_i$ and selective activation of PP$_i$ by the control star (SP$_i$) — the GP monitoring execution by polling of the STATUS line (ST$_i$).

In this manner each block can be made responsible for the control of its own bus buffers BB$_i$ through appropriate command lines CL$_i$. In addition, and as an alternative to polling, the bidirectional star lines (AA$_i$) can be used by PP$_i$ for urgent requests of service to the GP.

Note that this very simple mechanism is exactly what is needed to allow (i) arbitrary partitioning and (ii) SIMD (uniform commands and mail box messages) and/or MIMD operation (specified commands and messages).

3.5. Concept of generalized address-command space

It is common to extend memory address space to I/O. It may be less common to extend this principle to commands. It is what is done in ROMUALD, taking full advantage (in a nonorthodox way) of the segmentation.

Segment bits (seven of them) are transformed and decoded:
(1) to address various parts of the internal GPs memory space (normal use);
(2) to activate any arbitrary partition of the control star in order to reconfigure the machine;

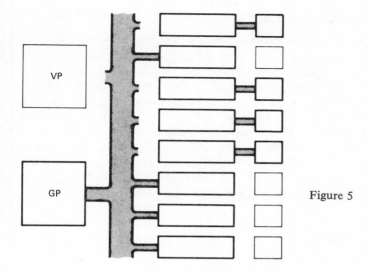

Figure 5

(3) to address (using (2)) any part of the external addressing space of the GP; coinciding with the image memory considered as a whole;

(4) to generate various housekeeping and reconfiguration commands.

All this happens in one single normal memory transaction cycle, taking advantage of the fact that the segment number is obtained well in advance of the regular address.

3.6. Dynamic reconfiguration

It is thus possible to create instantly under program control five different types of machine:

(1) Acquisition machine, ADC feeding the memory, slice after slice, from any TV-type source.

(2) Display machine, the other way around.

(3) A group of totally separated processors, each operating on its memory. This concerns, of course, the eight local processors, but also possibly the global processor, if need be.

(4) A memory intensive computer (monoprocessor). Power can be added to the global processor, by various means, in view of that.

(5) Any combination of (3) and (4). Some local processors are at work independently, the global processor using the part of the memory left at its disposal by the idling local processors (Figure 5).

3.7. Hardware-software homogeneity

The segmented GP (8001) and unsegmented PP (8002) are in fact exactly the same chip, differing only by the package type and by the fact that the segmented one can be made to act in unsegmented mode (but not the contrary of course!). Thus a nonsegmented code can be tested in system environment on the GP and then transferred to the PPs.

It is also possible to write a universal code, using a restriction of the assembler, which then runs indifferently in both modes, and consequently on all alike processors.

3.8. No transversal connection between PPs

This would involve considerable hardware complications without real benefits. It has been amply demonstrated in a previous paper that for typical neighbourhood algorithms, data transfers between image-memory slices through the GP has very little effect on speed [3].

Moreover, arbitrary large overlap can be obtained under control of the VP between adjacent image-memory slices suppressing in most cases the necessity of data transfers. This is done using the bidirectional lines AA_i to send, from the VP to the PP_i, individual values of address A15 (which discriminates between the upper and lower halves of the image memory). Thus image data can be written simultaneously in the upper half of memory slice M_i and in the lower half of slice M_{i+1}. This is, of course, only one of many different possibilities, all under program control by the VP. Note that the PASM's interconnect network is different; it is apparently equivalent to our control star.

4. ROMUALD'S PRESENT AND FUTURE

4.1. Colour

A colour prototype is now nearing completion. It uses exactly the same GP and PPs as the black and white machine. The I/O processor is of course different, even if the general principles are maintained. It will process either a $256 \times 256 \times 16$ bit $(6 + 5 + 5)$ image or a $512 \times 512 \times 8$ bit $(4 + 2 + 2)$ image, the last particularly for pseudocolouring of black and white originals.

4.2. Short-term upgrading (with current technology)

The present image memory of less than 512 Kbytes in PPs and 256 Kbytes in GP may quickly appear as too small. There are two ways out:
(1) Step from 8 to 16 processors, thus doubling the speed and the memory capacity. This might be done without too much trouble. To go over to 32 would raise important electrical problems on the bus side (FANOUT and reflections). Moreover, there are also physical limits to the "star system"!
(2) Replacing the 16000 DRAM in PPs by 64000.
The ones coming out these days are fast enough and prices are going down. This means replacing the 8002 (unsegmented processors) by 8001 (segmented). Then, all that is needed is a partial redesign of the PVC board of the PPs to accommodate the different pinout. Due to the software homogeneity principle the bulk of the program can be kept as it is. This means quadrupling memory size (with speed unchanged) for a very moderate price increase.

4.3. Long-term upgrading

In a few years time it will be possible to put on a single chip the whole of a present-time PP (memory-processor-buffers). It will totally modify timing and electrical constraints. By then, the high-definition solid-state camera will probably be a common piece of hardware. The application of a matrix image on a matrix memory, on a bidimensional array of ROMUALD-type parallel processors might then look a very sensible proposition.

REFERENCES

1. P. E. Danielsson, S. Levialdi, Computer architectures for pictorial information systems, Computer (November 1981) 53–67.
2. B. Bretagnolle, Claire Rubat du Mérac, ROMUALD: Un multiprocesseur interactif pour la saisie et le traitement d'images, in: *Reconnaissance des formes et intelligence artificielle* (Tome III) 2ième Congrès AFCET-IRIA, Toulouse 12–14 septembre 1979.
3. B. Bretagnolle, Claire Rubat du Mérac, Jean Seguin, Architecture of a multi-micro-processor for pictures capture and processing: ROMUALD, 1979 *International micro and minicomputer conference*, Houston (14–16 November 1979) IEEE catalog no. 79CH1474-6.
4. B. Y. Bretagnolle, C. Rubat du Mérac, P. Jutier, Looking for parallelism in sequential algorithms: an assistance in hardware design, in J. C. Simon, and R. M. Haralick (eds), *Digital image processing*, Reidel (1981) 95–103.
5. Presentation at 3ième Congrès AFCET: Reconnaissance des formes et intelligence artificielle, Nancy, septembre 1981.
6. H. J. Siegel, PASM: A reconfigurable Multi-micro computer System for Image Processing, in: M. J. B. Duff and S. Levialdi (eds), *Languages and architectures for image processing*, Academic Press (1981) 257–265.
7. B. Y. Bretagnolle, Claire Rubat du Mérac, Multiprocesseur interactif pour la saisie et le traitement d'images, french patent no. 2465273 (11 septembre 1979).
8. P. Jutier, B. Bretagnolle, Cl. Rubat du Mérac, Dispositif de saisie et de restitution en temps réel d'une image formée de trames successives de lignes de balayage, french patent application no. 8200590 (15 janvier 1982).

P. Jutier and Cl. Rubat du Mérac
Inserm U 194
91 Boulevard de l'Hôpital
75013 Paris
France

B. Bretagnolle
CICG
Grenoble
France

23 A computer aided system for interactive definition of digital image interpretation

U Cugini, M Dell'Oca, D Merelli and P Mussio

1. INTRODUCTION

In many and different fields of science and technology, images are used as a tool for the analysis of physical events and as a means of communication among different experts. For example, photographs taken by astronomers carrying out observations, are spread out in order to be interpreted by different members of the scientific community. This work of interpretation is often ambiguous, not univocal, strictly connected to the cultural background and personality of each interpreter.

Once the problem is recognized, the necessity arises of defining procedures that, taking into account the processes of production of images, allow an interpretation as uniform as possible and connected to the physical reality represented by the image. Automatic computation tools, properly used, can help define this procedure.

In opposition to the great effort in automatic classification [1] and machine image interpretation [2], users specialized in other fields point out their doubts on the use of automatic tools [3, 4]. These doubts seem to be connected to the low control, which the user can obtain, of the procedures leading to classification and interpretation.

The aim of this paper is the definition and implementation of tools useful to the experimenter willing to create a controlled and controllable procedure for interpreting digital images. The main characteristic of such a procedure is maintaining the language, conventions, and operating modalities typical of the group that has to manage those images. The second characteristic is allowing the user to check, step by step, the procedure and the consequences, even very specific, of the decisions taken. This implies the necessity of giving to the user in every moment, the synthesis of the results reached up to then by analysis.

These are the motivations that have led to a uniform approach to the problems of image analysis and synthesis, as the basis to the realization of a system for the interactive definition of procedures for the interpretation of classes of images.

2. PRELIMINARY DEFINITIONS

ISIID (Interactive System for Image Interpretation Definition) is an interactive tool designed to help an experimenter to create procedures for the interpretation of non-laboratory (real-world) images [3]. By image interpretation we mean the application of a set of rules which map patterns present in an image into objects present in a scene,

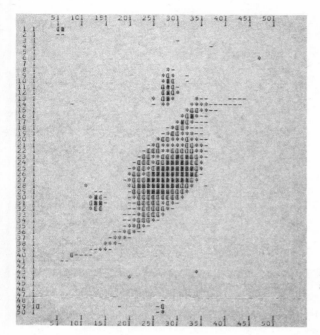

Figure 1. A "coloured" digitized astronomical image.

from which the image was drawn. A pattern is a set of picture elements (pixels) which are associated together by one or more descriptions, a description being a mono- or bidimensional string in a language; the description is computed by a function whose variables are features and relations among features [2]. A description, relating to a set of pixels, points out hints about the nature of the object in the scene, from which the pattern in the image was generated.

ISIID is designed to help the definition of a procedure for a well-defined class of images: that is, for images that represent homogeneous events and have to be interpreted for the same goal, according to a precise point of view. Let us think, for instance, of images taken from a plane or a satellite with the aim of individuating certain physical characteristics of the earth or to sky images taken by astronomers (Figure 1).

According to our experience, in most cases the production of an image interpretation is often an ill-defined problem. In these situations, before proposing a solving procedure, it is necessary to state clearly the problem. Therefore, the process of defining an interpretation is divided into three stages. In the exploration phase, hints about the considered set of patterns in the experiment in hand, are collected from a set of images, drawn from known scenes. This first set of images are *a priori* interpreted by the user, that is, before any automatic procedure is applied. The experimenter knows the meaning (corresponding object in the scene) of each blob in the image. He studies these correspondences so that hints and clues are found about those structures in the data that are meaningful for interpretation. In this way a set of descriptions and related interpretations may be proposed and translated into programs to extract them from a digitized image. The experimenter tests these tools on the first set of data. Then he checks the correspondence between the obtained results and the *a priori* interpreted data. The results of the checks often compel the experimenter to change his *a priori* point of view about the interpretation of the whole class of images [5].

In the second phase, the confirmation, the experimenter's point of view, descriptions and programs are first used on a new set of data, whose interpretation is unknown. The results obtained are then checked against the interpretation, performed by a disciplinary expert, who does not know the results of the procedure. If the check is positive according to some predefined criteria, the new point of view, descriptions and programs are accepted. If the check is not satisfactory, they are rejected to a new refinement or redefinition step. If a redefinition or refinement step is required the redefined tools have to be tested on a new set of data of the same class which were never previously used in the definition procedure.

At last, when a satisfactory problem definition and derived solution are reached, they have to be communicated to different experimenters. In such a way, definitions and solutions can be checked and criticized following a different experience and against different sets of data.

3. REQUIREMENTS FOR A PICTORIAL DATA-INTERPRETATION SYSTEM

A pictorial data-interpretation system has to allow the experimenters:
(1) To explore digitized data both from a numerical or structural point of view.
(2) To propose the description and interpretation of objects of their interest.
(3) To obtain an algorithmic description of the proposed solution.
(4) To implement quickly some tools and to check them with a new set of data.

In the two steps of exploration and confirmation, automatic analysis needs to be overlapped and supported with visual image interpretation, so that at each stage of the definition process, the experimenter is able to control, check, and change, if necessary, the solution being proposed.

These requirements suggest that a system designed to help an experimenter in defining his interpretation of some kind of pictorial data, has to allow:

(1) A graphic person-machine interaction, which may be achieved if tools for image analysis can systematically interact with tools for image synthesis (generation).
(2) An easy adaptation of the strategy and tactics to new situations.
(3) A friendly person-machine communication. This means that not only the user interacts with the system through a subset of its specialized natural languages but also understands and checks each action performed by the system. This last requirement is achieved using well-formalized tools for describing the activities, the strategy to perform, and the actions followed by each activity.
(4) A formal description of each action performed (pattern description and definition).

4. DESCRIPTION AS A TOOL FOR IMAGE INTERPRETATION

To match these constraints:

(1) Descriptions are used as common tools to fulfil both the analysis and the synthesis

(a)

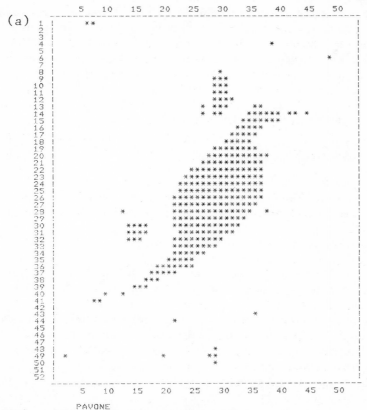

PAVONE

(b) NUCLEO ELLITTICO CON ESTREMI NEI PUNTI (9,24) (17,25) (23,15) (13,14)
BRACCIO DA (23,13) A (23,18) CON ESTREMO IN (29,5)
BRACCIO DA (9,19) A (9,23) CON ESTREMO IN (3,25)

Figure 2. (a) Black and white sketch
identifying the set of "interesting" pixels
in Figure 1. (b) Verbal description of the
main object in the astronomical image of
Figure 1 (braccio stands for arm, nucleo
for nucleus, punti for points, estremo for
extremity).

tasks. In this view, a description is the goal of the analysis, and the starting point of
the synthesis. In the machine environment, the description is a relational data structure, which is built step-by-step by the analysis tools and at any stage of the process
may be used by the synthesis device to show to the experimenter the obtained results.
In a person-to-person or machine-to-person communication environment a description may assume the form of a verbal description (e.g., a set of user-defined words
and sentences identifying structures and relations) or a graphic form (Figure 2). A
well-defined map exists between the two forms of descriptions.

(2) Activities are described by a suitable technique [6], while strategies are formalized as production systems [7, 8] and tactics are displayed by the $\alpha-\omega$ notation [9, 10]. In such a way each algorithm, interpreted by a machine, is formally described in a way which can unambiguously be understood by every person interested in the problem.

5. SYSTEM IMPLEMENTATION DESCRIPTION

The core of the system is a data base (in the sense of Hayes-Roth [8]) which can be accessed by set of table-driven, data-driven programs. The data base is composed of five sections (Figure 3):

(1) n-Level data structure. The user describes the image by means of n categories of graphic structures. For example, an astronomer may describe his image identifying the position and type (star or galaxy) of each blob in it. Each object is then described by its parts: a galaxy by the presence of a nucleus, arms, bar, etc.

Each part is recognized by some typical subset of pixels, such as a more or less linear structure in the boundary, a convexity, etc. At the last level, the pixels recognized of interest in the experiment are stored. This point of view is reflected in an n-level data structure.

At each level a relation is stored, and its elements are descriptions of sets of pixels, belonging to a well-defined category. Each description is characterized by the name of the specimen or prototype recognized (category and identifier within the category) and by the values assumed by an m-tuple of variables.

The variables describing a category are defined by the user in the exploration phase. Their values are evaluated from the data by the hint-recognizer and are used by the synthesizer to produce sketches or verbal descriptions. The descriptions of

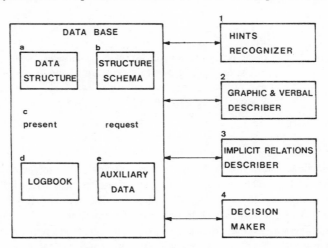

Figure 3. The schema of the proposed system.

the r-level are associated to a set of pixels, which are subsets of the set of pixels described in the r+1 level.

(2) The data structure schema. The number of categories and of levels in the data structure, the number of variables (degree of relation) defining the description at each level, and the domains of each variable are typical of each experiment. They are described in the user-defined data-structure schema. It is worth while to notice that the schema resumes the point of view of the experimenter about the patterns to be researched in the image. At the beginning of the exploration phase the point of view is based on his *a priori* knowledge. Very often it has to be changed during this first phase.

(3) The present request. It is the statement which is being interpreted by one of the programs.

(4) The logbook. The aim of the logbook is to allow a reconstruction of how the user has reached the present results. All the decisions that the user has taken during the section (even the defaults) are stored in the logbook. Some information stored in the logbook is even automatically associated to the output as headings of the different images, sketches, or descriptions.

(5) Auxiliary data. In this area each program stores auxiliary-data structures, which are necessary to its computation, but need not appear in the final results of its activity.

The programs, which have access to the data base, have been designed and partially implemented as coroutines. They are all production systems; an interpreter controls their activities. Given a set of rules [8], the interpreter looks for some antecedent in the data. If some antecedents are present, the interpreter fires the actions, whose names are the consequence in the set of rules.

These programs are:

(1) The hints recognizer. Its activity is the recognition of features in the image which may be hints for its interpretation. Hints may be predefined or defined interactively by the user. The experimenter may collect hints about grey-level behaviour in the image and grey-level classes or about the structures formed by sets of pixels of some grey-level class, such as the presence of concavity, convexity on their contour, or of protusions or inclusions in their body. These hints are described and stored in the different levels of the data structure (Figure 4).

(2) The graphic and verbal describer (synthetizer). Its activity consists in synthetizing recognized structures as sketches or as verbal description, which have to be meaningful to the user. Given the request of synthetizing an image (or object) at a given level of description, the user can require the representation of any sublevel of description (Figures 5 and 6).

(3) The implicit relation describer. The relations, which are described in the data structure, are generally not exhaustive of the possible relations of interest to the experimenter. These programs allow the evaluation of some features which are implicit in the structure.

(4) The decisionmaker. The decisionmaker classifies on the basis of the collected hints by means of multivalued logic tools [11].

```
                    FAIDISEGNO
          VUOI IL COLLOQUIO ESTESO O SINTETICO?
              (RISPOSTE POSSIBILI: ESTESO/SINTETICO)
          .SINTETICO
          DAMMI IL NOME DELLA SCENA
          .ASTR1
          LIVELLO DA CUI TRARRE LE STRUTTURE DA DISEGNARE
          .PARTI
          VUOI DISEGNATE TUTTE LE STRUTTURE  O ALCUNE ?
          .TUTTE
          LIVELLO DI SCOMPOSIZIONE DELLA  DESCRIZIONE
          .VERTICI
          VUOI UN UNICO SIMBOLO O PIU' SIMBOLI PER IL DISEGNO?
          .PIU'
          LIVELLO DA CUI TRARRE LA COLORAZIONE
          .PARTI
          VUOI UN SIMBOLO PER OGNI STRUTTURA INDIVIDUATA,
          O UN SIMBOLO PER OGNI TIPO DI STRUTTURA?
          .TIPO
          SONO PRESENTI NELLA DESCRIZIONE I SEGUENTI
          TIPI DI STRUTTURE:
          PROTUBERANZA
          PROTUBERANZA ESTESA
          ANSA
          VUOI I SIMBOLI SCELTI AUTOMATICAMENTE O MANUALMENTE?
          .MANUALMENTE
          DAMMI 3 SIMBOLI PER IL DISEGNO
          .*xo
          OCCORRE UN SIMBOLO PER I PUNTI NON CLASSIFICATI
          VUOI I SIMBOLI SCELTI AUTOMATICAMENTE O MANUALMENTE?
          .MANUALMENTE
          DAMMI 1 SIMBOLI PER IL DISEGNO
```

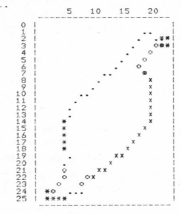

```
SIMBOLO   STRUTTURA

   *         PROTUBERANZA
   x         PROTUBERANZA ESTESA
   o         ANSA
   .         NON CLASSIFICATI
```

Figure 4. The hints recognizer classifies low-level monodimensional structures in the main object in the astronomical image (protuberanza stands for protusion, estesa stands for large, ansa stands for cove and non classificato for not classified).

```
        FAIDISEGNO
VUOI IL COLLOQUIO ESTESO O SINTETICO?
   (RISPOSTE POSSIBILI: ESTESO/SINTETICO)
.SINTETICO
DAMMI IL NOME DELLA SCENA
.ASTR1
DAMMI IL NOME DEL LIVELLO DA CUI TRARRE LE STRUTTURE DA DISEGNARE
.OGGETTI
VUOI DISEGNATE TUTTE LE STRUTTURE   O ALCUNE ?
.TUTTE
DAMMI IL NOME DEL LIVELLO DI SCOMPOSIZIONE DELLA  DESCRIZIONE
.PARTI BIDIMENSIONALI
VUOI UN UNICO SIMBOLO O PIU' SIMBOLI PER IL DISEGNO?
.PIU'
DAMMI IL NOME DEL LIVELLO DA CUI TRARRE LA COLORAZIONE
.PARTI BIDIMENSIONALI
VUOI UN SIMBOLO PER OGNI STRUTTURA INDIVIDUATA, O UN SIMBOLO PER
OGNI TIPO DI STRUTTURA?
.TIPO
SONO PRESENTI I SEGUENTI TIPI DI STRUTTURE:
NUCLEO
BRACCIO
VUOI I SIMBOLI SCELTI AUTOMATICAMENTE O MANUALMENTE ?
.MANUALMENTE
DAMMI 2 SIMBOLI PER IL DISEGNO
.2*
```

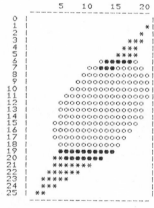

```
SIMBOLO   STRUTTURA

   o      NUCLEO
   *      BRACCIO
```

Figure 5. The synthetizer is required to sketch bidimensional structures using prototypes. An arm (braccio) is interpreted as a triangle, a nucleus (nucleo) as an ellipse.

6. CONCLUSIONS

The motivation and requirements which were at the base of ISIID design have been presented. The system is partially implemented in APL under the virtual machine operating system VM/SP CMS on a Siemens 7.865 and it is being used in the analysis of real data.

The procedure and the tools have been used in real cases and the results can be found in the literature [5, 12, 13]. These cases have pointed out how the proposed hints recognizer is useful in the exploration phase, but they have also shown that it would be useful to extend the automatic analysis to more complex structures. But their definition is often connected to the single experiment: it seems impossible to define a set of complex structures useful for every experiment. It would instead be interesting to build a device able to infer a specialized generating system from examples. In this way, it would be possible to implement an (explorative) specialized hints recognizer for each experiment.

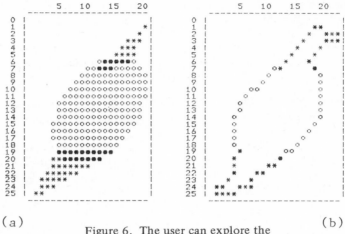

(a)   (b)

Figure 6. The user can explore the
description at the same level through
information of different levels. Here we
see the verbal description of Figure 2(b)
sketched using bidimensional prototypes
(a) or a monodimensional specimen of
the contour (b).

A second point has evolved, when dealing with a large amount of descriptions. It is desirable to manage them automatically. We plan to face it by the use of IRS-APL [14]; an information retrieval system embedded in the APL system, which is able to handle graphic and verbal descriptions.

REFERENCES

1. N. Ahuja, B. Schacter, Image Models, ACM Computing Surveys, Vol. 13 No 4 (1981) 373–397.
2. M. Brady, Computational Approach to Image Understanding, ACM Comp. Surv., Vol. 14 No 1 (1982) 3–71.
3. F. Leberl, J. Soha, A. Meijerink, Digital picture processing and its impact on image interpretation, in: G. Hildebrandt and H. J. Boehmel (eds), Proc. Intern. Symposium on Remote Sensing (1978) 245–260.
4. G. A. Thorley, J. M. De Noyer, Remote Sensing from space, experience from the seventies — programs for the eighties, as viewed by the Eros program, in: G. Hildebrandt and H. J. Boehmel (eds), Proc. Int. Symposium on Remote Sensing (1978) 51–60.
5. S. Bianchi, G. Gavazzi, P. Mussio, P. Ossola, An application of ISIID: automatic interpretation of digitized astronomical images, presented at: 2nd Int. Conf. on Image Analysis and Processing, Selva di Fasano (1982).
6. Architect's manual ICAM definition method "IDEF 0" April 1980, CAM-I Report n.DR-80-ATPC-01.
7. H. F. Ledgard, Production Systems: A Notation for Defining Sintax and Translation, IEEE Trans. on Soft. Eng. Vol. 3 No 2 (1977) 105–124.

8. F. Hayes-Roth, D. A. Waterman, B. Lenat, Principles of pattern directed inference systems, in: D. A. Waterman and F. Hayes-Roth (eds), Pattern Directed Inference Systems (Ac. Press, 1978).

9. K. E. Iverson, Algebra: an algorithmic treatment, (APL Press, New York, 1977).

10. S. Bianchi, A. Della Ventura, M. Dell'Oca, P. Mussio, A. Rampini, An APL pattern directed module for bidimensional data analysis, API Quote-Quad Vol. 12 No 1 (1981) 54–61.

11. U. Cugini, M. Dell'Oca, A. Mirioni, P. Mussio, An Interactive drafting system based on bidimensional primitives, IEEE 78CH1289-8C, Interactive Techniques in Computer Aided Design (1978).

12. U. Cugini, G. Ferri, P. Micheli, P. Mussio, M. Protti, An Application of ISIID: arcs fillets and tangent points restoration in digitized line drawing, presented at: 2nd Int. Conf. on Image Analysis and Processing, Selva di Fasano (1982).

13. A. Della Ventura, R. Rabagliati, A. Rampini, R. Serandrei Barbero, An application of ISIID: Remote-Sensing observation of glaciers, presented at: 2nd Int. Conf. on Image Analysis and Processing, Selva di Fasano (1982).

14. F. Naldi, IRS-APL Descrizioni Generale, Rapporto SIAM 005/82, Milano (1982).

U. Cugini
Politecnico di Milano
Dipartimento di Meccanica
Piazza Leonardo da Vinci 32
20133 Milano
Italy

M. Dell'Oca
SAE SpA
Via Fara 26
Milano
Italy

D. Merelli and P. Mussio
Istituto di Fisica Cosmica del CNR
Via Bassini 15/A
Milano
Italy

24 Iconics: computer aided visual communication

Z Kulpa

The term "iconics" proposed here is to denote the whole field of image manipulation (encoding, processing, analysis, interpretation, generation, synthesis, etc.) done with the aid of computer techniques. A term "picture engineering" has been proposed recently to denote this discipline [5]. The term "iconics" was chosen here to put the emphasis on the connection between this area and the one of *visual communication*, i.e., the exchange of information by means of visual images.

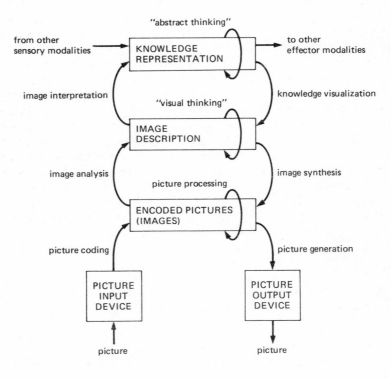

Figure 1. The general structure of an iconic system.

Undoubtedly, efficient means of communication between people are indispensable for the development of human culture and civilization, and the human himself. The prevailing method used so far has been the phonetic (spoken) language and its visible record: writing. However, its one-dimensional structure does not correspond too well to the essentially many-dimensional structures of the world and our knowledge of it. This fact negatively affects the speed and accuracy of language communication.

As was recognized long ago, pictures — with their two-dimensional structure and our excellent abilities to analyze them in parallel by our eyes — are potentially more suitable as a medium of communicating complex knowledge. The picture is worth a thousand words! However, use of pictures in everyday communication is badly hindered by our lack of an efficient pictorial effector.

Recent developments of computer technology and artificial intelligence make us hope the construction of such an effector will be possible. The research and practical design efforts oriented toward construction of effective means of visual communication, based on computer technology, constitute just the contents of iconics. Recently, this field divided into two major disciplines: image analysis and computer graphics. In an "iconic system", i.e., a system capable of visual communication, these two aspects can, and should, be unified (Figure 1).

Symptoms of this unification become perceptible now, e.g., on the hardware level we have growing use of raster-graphics devices and frame-buffer memories serving both fields; in image analysis there is a growing tendency to use structural image description approaches and model- (and knowledge-)driven picture interpretation. Probably the best examples are provided by pictorial data-base systems as well as advanced knowledge-base expert systems (which can be easily interpreted using the above scheme).

The proliferation of personal computers endowed with good-quality graphics, development of computer networks, and services such as Teletext, computer film animation, and many other phenomena of this sort, signify the growing influence of iconics techniques on both development of our civilization and shaping of our everyday lives.

The most challenging research problems here seem to be the visualization problem (design of appropriate knowledge representation, design of "pictorial languages" and "intelligent" translators of knowledge into the pictorial language), and the interface problem (how to interface a man with the "iconic computer", allowing efficient and natural flow of information to be visualized). Also of great importance is the construction of new parallel architectures of image-manipulation processors.

REFERENCES

1. W. M. Turski, *Nie sama informatyka . . . (Not Only with Computers . . .*, in Polish), WNT (Science and Technology Publ.), Warsaw 1980.
2. D. R. Clark, ed., *Computers for Imagemaking*, Pergamon Press, London 1980.
3. A. Blaser, ed., *Data Base Techniques for Pictorial Applications*, Lect. Notes in Computer Science, vol. 81, Springer-Verlag, Berlin 1980.
4. S.-K. Chang, K.-S. Fu, eds., *Pictorial Information Systems*, Lect. Notes in Computer Science, vol. 80, Springer-Verlag, Berlin 1980.
5. K.-S. Fu, T. L. Kunii, eds., *Picture Engineering*, Springer-Verlag, Berlin 1982.
6. J. F. Blinn, Image processing: the future vital link?, In: C. White, ed., *Computer*

Graphics, Infotech State of the Art Report, series 8, No. 5, vol. 2, Infotech Ltd., Maidenhead 1980.

7. R. Williams, Image processing and computer graphics, *Computer Graphics and Image Processing*, vol. 10, 1979, pp. 183–193.

8. N. J. Nilsson, *Principles of Artificial Intelligence*, Tioga Publ., Palo Alto 1980.

Z. Kulpa
Iconics Laboratory
Institute of Biocybernetics and Biomedical Engineering
00-818 Warsaw
Poland

25 Artery detection and tracking in coronary angiography

M Kindelan and J Suarez de Lezo

1. INTRODUCTION

Angiographies are made up of a set of time-sequenced X-ray pictures, and constitute for the physician an important tool in order to analyze the structure and function of different human organs. An immense amount of visual data is obtained using this technique, including blood flow and the evolution with time of the shape and dimensions of different elements.

However, only a very small fraction of this information is derived by the physician when he observes the passage of the contrast medium through the organs. Furthermore, this information is mostly qualitative and highly dependent on the observer.

The need for quantitative analysis was felt very early, but this type of analysis is often tedious and laborious, so that its development has only been possible since the beginning of the seventies with the application of digital-image-processing techniques, which had been very useful in the study of other types of radiographic images.

From that time several research groups have been working on the application of digital techniques to the analysis of angiographic image sequences, either to improve the quality of the images (digital substraction angiography [1-4]) or to extract quantitative information (videodensitometry [5-8], ventricular volume [9-12], ventricular motion [7,13,14], contraction patterns with radiopaque markers [15,16], myocardial perfusion [3,17], evolution of coronary tree geometry [18,19], parametric images [8,20,21]).

A particularly interesting case is that of coronary angiography, which provides the physician with information regarding the morphology of the coronary arterial tree, and, therefore, may be used to study the presence and importance of coronary disease.

The procedure generally used to extract this information is by simple visual inspection of the film, with the drawbacks of inter- and intraobserver variability [22-24], inaccurate quantitative results, and missing a considerable amount of information. It appears that digital-image processing is a very promising technique of overcoming these drawbacks. However, it should be pointed out, that the use of some of the successful techniques in other types of angiograms, such as enhancement by digital substraction [3,4,17] and elaboration of parametric images [8,21], is rendered difficult by the motion associated with heart contractions and even with respiration.

One of the first attempts to perform quantitative angiographic analysis of the coronary arteries was the measurement of the "critical" stenosis [25,37]. The procedure starts by the selection of frames that show the lesion most clearly. These frames are projected and magnified on a large screen, where the contours of the arteries are traced by an observer and digitized. A computer program calculates from these data the vessel diameter, cross-sectional area and percent diameter and area stenosis. Thus, digital processing was only used for computational purposes, but the subjectivity and labour associated with manual tracking of the arteries was still present.

A further step in the direction of automating the quantification process is described by Starmer and Smith [19]. These authors were interested in the representation of coronary trees in order to study the contraction patterns of the heart and the relationship of these patterns with the presence of coronary disease. To this end, digital-image-processing techniques were used to recognize and codify the skeleton of the arteries in the tree, assuming that the skeletons in the previous frame are known. The coronary tree in the first frame of the sequence is traced manually using a digitizing tablet, and is used to locate the tree in the following frames, on the assumption that the motion of the tree between two frames is small.

Recently, Gerbrands *et al.* [16] developed a semiautomatic procedure to analyze and quantify lesions in the coronary arteries, in order to minimize variations in the interpretation of coronary angiograms. In their approach, the user traces a tentative centreline of the artery using a digitizing tablet, and a one-dimensional first derivative operator is applied perpendicularly to this centreline in order to detect the edges of the artery. From these data the percentage diameter and area narrowing are computed.

An automatic method of detecting and tracking the contours of coronary arteries has been reported by Fukui *et al.* [27]. Their approach is based on the location of cord-like patterns of unimodal grey-level distribution, although very few details are given of the exact procedure used.

In the present paper an interactive image-processing approach to the analysis and quantification of coronary obstructions is described. The user selects with a cursor on an image display terminal, the origin and end points of an artery. Then, a Hueckel-based algorithm automatically traces its contour and overlays it on the original image. The user can accept the contour traced or reject it, in which case he may change some of the parameters of the algorithm and repeat the procedure until the artery is correctly tracked. Once accepted, the coordinates of the middle line, the width of the artery, and the average grey level at each point, are stored on a file for later processing with a graphic editor. This editor may be used to display the coronary tree, magnify it, link arteries, eliminate portions of arteries, plot the width or grey-level distribution, perform average filtering of those distributions, locate candidate points for coronary obstructions, or compute the percentage diameter or grey-level narrowing on those candidate points.

2. IMAGE-ANALYSIS SYSTEM

From the original 35 mm cine-angiograms the best frames in each projection are selected for later processing. These frames are digitized with a Perkin-Elmer 1010 A flat bed microdensitometer, using a square digitizing window of 70 m x 70 m, thus obtaining images of 270 x 360 pixels quantized in 256 grey levels, which have an adequate resolution for our purposes.

These images are visualized with a RAMTEK 9351 image display terminal having 512 x 512 pixels and 256 grey levels. To process and analyze them, a set of interactive programs using full-screen menus, have been developed. These programs, written in FORTRAN and ASSEMBLER, run in a CMS environment under a VM 370 operating system.

Once the middle line, width and grey-level distributions are obtained, the arterial tree is displayed and manipulated on a Tektronix 618 graphics monitor, using a graphic editor running on the graphics attachment for the IBM 3277 Display Station.

3. ARTERY-DETECTION ALGORITHM

In order to locate and track the arteries, an interactive approach has been chosen. The user selects with a cursor the initial and final points in the artery, and the algorithm tries to locate the best line-shaped object linking these two points. The algorithm has to overcome the difficulties associated with high noise, low contrast, nonuniform grey-level distribution in the artery, and complexity of the coronary tree structure.

To this end, it is necessary that the algorithm incorporates in an efficient manner, the main characteristics of the objects it tries to locate. In the present case, its line shape in which an elongated region of high grey level is separated from a noisy, lower grey-level background, by a couple of approximately parallel lines.

Typically, contour or boundary detection is performed in two steps: firstly, local operators are applied in order to locate edge points, and, secondly, edge-following algorithms are used to connect these edge elements and build continuous contours [28–30]. Edge-following algorithms generally use heuristic global information about the structure of the objects of interest.

The artery-detector algorithm designed for this application, combines a simple one-dimensional local detector, with a global detector based on the Hueckel operator [33,34], which incorporates heuristic information to describe the main characteristics of the arteries shapes.

A similar approach combining a local and a regional detector has recently been described [38] in relationship with the problem of locating line-shaped objects (roads, rivers) from aerial images. However, the local and regional operators chosen for that application are different from those used in the present case.

4. LOCAL DETECTOR

In order to simplify the problem and to take full advantage of the characteristics of the arteries that we try to locate, a one-dimensional local detector approach is used [31]. Figure 1(b) and (c) shows typical one-dimensional grey-level profiles taken perpendicularly to the artery. It is apparent that the arteries may be distinguished from the background by looking for regions of high grey-level profile with rather steep transitions to the surrounding lower grey-level regions. To perform this identification in an efficient manner, a sliding t-test operator is used. This operator, described in ref. [32], takes a one-dimensional window of $2n+1$ pixels centred at P, where $x_1, x_2, ..., x_n$ are the n pixel values to the left of P, and $y_1, y_2, ..., y_n$ are the n pixel values to the right of P. If P is

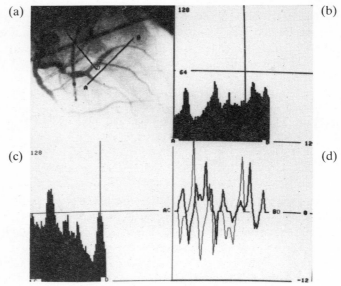

Figure 1. (a) Angiographic frame
showing selected lines. (b) – (c) Grey-level
profiles. (d) Profiles of parameter t.

located in the border of the artery a significant difference in the average grey level of
x's and y's is to be expected. Thus, t is defined as

$$t = (\bar{y} - \bar{x})/s,$$

where

$$s^2 = (\Sigma\, x_i^2 - nx^{-2} + \Sigma\, y_i^2 - ny^{-2})/n(n-1).$$

The centre of window P is slid across the line perpendicular to the artery in order to
locate the points, where t is maximum and minimum, which correspond to the edges of
the artery separating regions with significant difference in grey levels. Figure 1(d) shows
the result of applying the t-test to the grey-level distribution of Figure 1(b) and (c) for
n=4. The edges of the arteries are easily identified by looking at the maximum and
minimum values of t.

To apply the t-test, it is therefore necessary to know the direction of the artery and
the approximate width of the object one tries to locate. This information is provided by
the global detector which is described in the following section.

5. GLOBAL DETECTOR

The objective of the global detector is to link candidate artery points, as obtained by the
local detector, into continuous arteries, and to discard other candidates which cannot be
fitted into the global coronary artery structure. Furthermore, the global detector is used
to guide the search of the artery by selecting the next point where the local detector is to

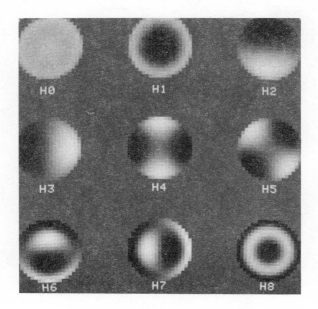

Figure 2. Hueckel basis functions.

be applied, and by updating the direction and width of the artery which is being followed.

For instance, in the case of ref. [26] the tentative centreline traced by the user can be considered as the global detector which guides the search of the artery by means of a one-dimensional local detector, similar to the one described in the previous section.

In the present case, a detector based on the Hueckel operator [33,34] has been chosen. This operator is a regional edge detector which is generally considered as superior to most other detectors [35,36], and consists essentially in optimally fitting an ideal edge-line pattern to the image intensity values in a small circular window. To this end, the values in the circular window are projected onto a set of nine orthonormal basis functions $H(x, y)$, and a minimum mean-squared error criterion is used to find the parameters of the ideal edge which best fit the grey-level distribution on the window. A computationally efficient algorithm has been described in ref. [36] to perform the minimization. Figure 2 shows the discretized basis functions for a window of 293 pixels.

In particular, the operator may be used to locate only pure lines [34], in which case the ideal line is defined by its direction, the location of the centreline with respect to the centre of the window, the grey level on the line, and the grey level on the background. This implementation of the operator is very well suited to the task of locating the arteries, although it still presents problems, especially when the borders of the arteries are not parallel (for instance in the presence of an occlusion), or in the intersection of crossing of two arteries, or when the artery is located in the boundary of two regions with significant grey-level differences. Figure 3 shows the result of applying the Hueckel operator in a window of 293 pixels, in order to locate clear lines on a dark background in a typical angiographic image. For this example, the operator is applied every 12 rows and 12

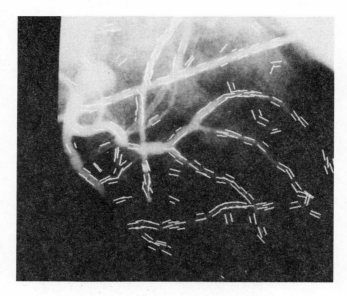

Figure 3. Lines detected by Hueckel
operator.

columns, and the parameters CONFF and DIFF, defining the noise level of the image
(see ref. [34]), are assigned the values 0.7 and 6, respectively.

An adequate combination of the local operator, previously described, to track the
contours of the arteries, and the Hueckel detector to guide the global search, together
with heuristic information such as low curvature of the arteries and continuity of the
width distribution, provides a rather efficient algorithm to analyze the structure of the
coronary tree.

The following section describes the actual operating procedure.

6. OPERATION PROCEDURE

The user selects a starting point in an artery. The Hueckel detector is then applied in a
window of 137 pixels centred at the starting point. If no line is detected, a new starting
point is requested, otherwise the width and direction of the line are computed and passed
to the local detector. The one-dimensional operator is then applied at the starting point
in a direction perpendicular to that returned by the Hueckel detector, and the borders of
the artery at that point are computed. From these data, the coordinates of the centre of
the artery, its width and the average grey level are calculated and stored for later use.

A new skeleton point is predicted by moving a distance d from the point of the
artery just detected, in the direction returned by the Hueckel detector. Its exact location
is refined by applying again the local operator to obtain and store the information on a
new point of the artery. This procedure is repeated a certain number of times nd (usually
three), and then a new call to the Hueckel detector is carried out in order to update the
direction and width of the artery.

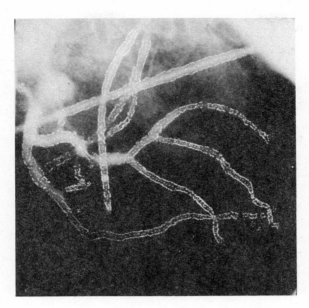

Figure 4. Result of artery-detection
algorithm.

When the Hueckel detector fails to detect a line, the frequency of its application is
increased (nd=1) and the direction of the artery is computed from the direction of its
two borders at the three last valid artery points. If the two borders are not approximately parallel, the direction of the artery is not updated.

In addition, the artery detector incorporates in an *ad hoc* manner, functions to
ensure continuity in the width distribution, to prevent very high curvature points and
to neglect lines with a width higher or lower than some user-selected thresholds. Finally,
when the detector is unsuccessfully applied a certain number of times in a row, an attempt
to recover the lost artery is carried out by extrapolating from the last valid point.

The contours of the artery being traced, as well as the points where the Hueckel
operator is successfully applied, are overlaid on the display terminal. Figure 4 shows a
typical result of the application of the artery detector. The white dots located approximately in the middle line of the arteries, correspond to points where the global detector
has successfully located a clear line. The white dots located approximately at the borders
of the arteries, correspond to the edges detected by the local operator. It may be observed
that the detector usually encounters difficulties at points of intersection of two arteries,
although sometimes it is capable of correctly crossing over such points.

The algorithm ends either by reaching the user-marked end point, or by being unable
to find the artery going from the origin to the end point, or by a user-generated inter-
ruption when he observes that the detector is following an erroneous path. Once stopped,
the user may accept or reject the artery traced. When accepted, a file with the coordinates
of the centreline, its width, and grey level at each point, is written and stored, and a new
artery may be traced. When rejected, a new attempt to track the artery may be carried

out by changing origin or end point, or some of the parameters of the Hueckel detector, such as the expected noise of the pattern, its amplitude, or the maximum distance from the line to the centre of the window.

7. ARTERY EDITOR

The structure of the coronary arterial tree, resulting from the application of the artery detector, is processed with an interactive graphic editor that runs on the graphics attachment for the IBM 3277 Dispaly Station. This editor is used to visualize, correct errors resulting from the artery detector, and finally to extract quantitative information regarding the presence and degree of coronary lesions.

The functions presently available may be divided into the following categories:

(1) Functions to visualize and store the coronary tree, such as:
    (a) Display of the coronary tree with arbitrary magnification.
    (b) Display of a certain artery with arbitrary magnification.
    (c) Store in a file the coronary tree structure modified with the graphic editor.

(2) Functions to modify and correct errors in the coronary tree resulting from the artery detector, such as:
    (a) Link two arteries.
    (b) Erase an artery.
    (c) Erase an artery from a certain point until its end point.
    (d) Erase an artery from its origin until a certain point.
    (e) Divide an artery into two different arteries, breaking them at a user-selected point.
    (f) Extrapolate an artery until a certain point of another artery.

(3) Function to extract quantitative information, such as:
    (a) Average width on the vicinity of a certain point.
    (b) Average grey level on the vicinity of a certain point.
    (c) Width distribution along an artery (Figure 5).
    (d) Grey-level distribution along an artery.
    (e) Average filtering of the width or grey-level distribution and display of the resulting distribution (Figure 6).

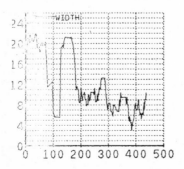

Figure 5. Width distribution.

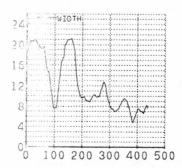

Figure 6. Average filtered width distribution.

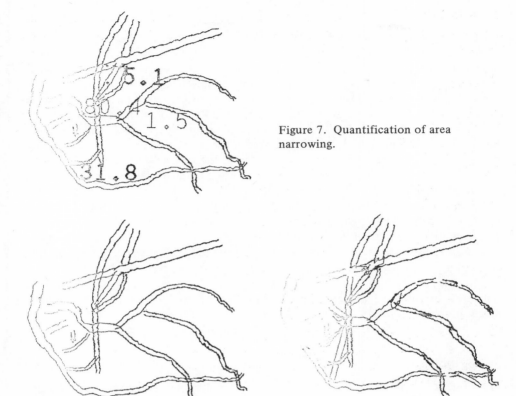

Figure 7. Quantification of area
narrowing.

Figure 8. Comparison of artery editor
input and output.

   (f) Average filtering of the width or grey-level distribution and substitution of the
original values by the filtered ones.

(4) Functions to locate and quantify candidate points for coronary occlusions, such as:

   (a) Location of points where the average filtered width or grey-level distribution
exhibits a local minimum, above a user-selected threshold.

   (b) Quantification of the degree of diameter or area narrowing, on the points of
local minimum (Figure 7).

Figure 8 compares a typical coronary tree structure, obtained by the artery detector, with
the same coronary tree after being manipulated by the graphic editor. It should be pointed
out that at the present time, the results of the artery detector contain too many errors, so
that it is necessary to carry out a considerable amount of work to correct them with the
editor, before it is possible to derive quantitative results. Therefore, it will be necessary to
improve the performance of the detector, in order to make the application of the system
practical. At the present time, however, the system is being used on experiments to
quantify and analyze the information contained on a coronary angiography.

8. DISCUSSION

Digital-image-processing techniques have been applied to coronary angiographies in order to extract quantitative results, not dependent on the observer. To this end, a line-detection algorithm is applied in an interactive manner, in order to locate the main arteries visible in the angiography. This algorithm may also be of interest in other applications, where the objective is to locate lines in a noisy background, such as location of roads in aerial images [38,39], location of rivers in satellite images [40], recognition of DNA molecules in electron micrographs [41,42], or analysis of the structure of neuron cells. However, the heuristic information should be changed according to each particular case. The interactivity, which appears to be very useful in the case of coronary angiographies, could be eliminated in other cases, by incorporating an algorithm to locate promising starting points [38].

The artery detector computes the coordinates of the centreline of the arteries, its width distribution and its grey-level distribution, thus producing an adequate description of the coronary tree structure. This information is analyzed and manipulated by means of a graphic editor which, for instance, may be used to plot the width distribution in a given artery, or to locate and quantify candidate points for coronary occlusions.

At the present time, the line-detection algorithm produces several errors which can sometimes be corrected with the graphic editor. It is believed that to make this approach practical, those errors have to be reduced, either by improving the artery detector, or by increasing the quality of the image with some type of digital enhancement.

REFERENCES

1. Chow, C. K., Hilal, S. K. and Niebuhr, K. E., X-Ray Image Substraction by Digital Means, IBM J. Res. Develop., 17, 206–218, 1973.
2. Ran Vas, Diamond, G. A., Forrester, J. S., Whiting, J. S. and Swan, H. J. C., Computer Enhancement of Direct and Venous-Injected Left Ventricular Contrast Angiography, American Heart Journal, 102, 719–728, 1981.
3. Heintzen, P. H., Brennecke, R. and Bursch, J. H., Digital Cardiovascular Angiography. In "Digital Image Processing in Medicine", Ed. K. H. Hohne, Springer-Verlag, Berlin, 1–14, 1981.
4. Mistretta, C. A. and Crummy, A. B., Diagnosis of Cardiovascular Disease by Digital Substraction Angiography, Science, 214, 761–765, 1981.
5. Wood, E. H., Sturm, R. E. and Sanders, J. J., Data Processing in Cardiovascular Physiology with Particular Reference to Roentgen Videodensitometry, Mayo Clin. Proc., 39, 849, 1964.
6. Bursch, J. H., Heintzen, P. H. and Simon, R., Videodensitometric Studies by a New Method of Quantitating the Amount of Contrast Medium, Europ. J. Cardiol., 1, 437, 1974.
7. Heintzen, P. H., Brennecke, R., Bursch, J. H., Lange, P., Malerczyk, V., Moldenhauer, K. and Onnasch, D., Automated Video-Angiocardiographic Image Analysis, Computer, 8, 55–64, 1975.
8. Boehm, M. and Hoehne, K. H., The Processing and Analysis of Radiographic Image Sequences. In "Digital Image Processing in Medicine", Ed. K. H. Hoehne, Springer-Verlag, Berlin, 15–41, 1981.
9. Chow, C. K. and Kaneko, T., Automatic Boundary Detection of the Left Ventricle

from Cineangiograms, Computers and Biomedical Res., 5, 388–410, 1972.

10. Chow, C. K. and Kaneko, T., Boundary Detection and Volume Determination of the Left Ventricle from a Cineangiogram, Comput. Biol. Med., 3, 13–26, 1973.

11. Falsetti, H. L. and Carroll, R. J., Single Plane Angiography: Current Applications and Limitations. In "Cardiovascular Imaging and Image Processing", Eds. D. C. Harrison, H. Sandler and H. A. Miller, Society of Photo-Optical Instrumentation Engineers, Palos Verdes, 123–127, 1975.

12. Smalling, R. W., Skolnick, M. H., Myers, D., Shabetai, R. and Johnston, D., Digital Boundary Detection, Volumetric and Wall Motion Analysis of Left Ventricular Cine Angiograms, Comput. Biol. Med., 6, 73–85, 1976.

13. Leighton, R. F., Rich, J. M., Pollack, M. E. and Altieri, P. I., Clinical Applications of a Quantitative Analysis of Regional Left Ventricular Wall Motion. In "Cardiovascular Imaging and Image Processing", Eds. D. C. Harrison, H. Sandler and H. A. Miller, Society of Photo-Optical Instrumentation Engineers, Palos Verdes, 203–208, 1975.

14. Stewart, D. K., Dodge, H. T., and Frimer, M., Quantitative Analysis of Regional Myocardial Performance in Coronary Artery Disease, In "Cardiovascular Imaging and Image Processing", Eds. D. C. Harrison, H. Sandler and H. A. Miller, Society of Photo-Optical Instrumentation Engineers, Palos Verdes, 217–224, 1975.

15. Carlsson, E., Experimental Studies of Ventricular Mechanics in Dogs Using Tantalum-Labeled Heart, Fed. Proc., 28, 1324–1329, 1969.

16. Gerbrands, J. J., Booman, F. and Reiber, J. H. C., Computer Analysis of Moving Radiopaque Markers from X-Ray Cinefilms, Computer Graphics and Image Processing, 11, 35–48, 1979.

17. Smith, H. C., Robb, R. A. and Wood, E. H., Myocardial Blood Flow: Roentgen Videodensitometry Techniques. In "Cardiovascular Imaging and Image Processing", Eds. D. C. Harrison, H. Sandler and H. A. Miller, Society of Photo-Optical Instrumentation Engineers, Palos Verdes, 225–232, 1975.

18. Starmer, C. F. and Smith, W. M., Problems in Acquisition and Representation of Coronary Arterial Trees, Computer, 8, 36–41, 1975.

19. Starmer, C. F. and Smith, W. M., Computer Storage and Retrieval of Coronary Trees. In "Cardiovascular Imaging and Image Processing", Eds. D. C. Harrison, H. Sandler and H. A. Miller, Society of Photo-Optical Instrumentation Engineers, Palos Verdes, 195–199, 1975.

20. Hoehne, K. H., Boehm, M. and Nicolae, G. C., The Processing of X-Ray Image Sequences. In "Advances in Digital Image Processing, Theory, Application, Implementation", Ed. P. Stucki, Plenum Press, 147–163, 1979.

21. Boehm, M., Obermoeller, U. and Hoehne, K. H., Determination of Heart Dynamics from X-Ray and Ultrasound Image Sequences, Proceedings 5th International Conference on Pattern Recognition, I.E.E.E., 403–408, 1980.

22. Detre, K. M., Wright, P. H., Murphy, M. L. and Takaro, T., Observer Agreement in Evaluating Coronary Angiograms, Circulation, 52, 979–986, 75.

23. Zir, L. M., Miller, S. W., Dinsmore, R. E., Gilbert, J. P. and Harthorne, J. W., Interobserver Variability in Coronary Angiography, Circulation, 53, 627–632, 1976.

24. De Rouen, T. A., Murray, J. A. and Owen, W., Variability in the Analysis of Coronary Arteriograms, Circulation, 55, 324–328, 1977.

25. McMahon, M. M., Brown, B. G., Cukingnan, R., Rolett, E. L., Bolson, E., Frimer, M. and Dodge, H. T., Quantitative Coronary Angiography: Measurement of the "Critical" Stenosis in Patients with Unstable Angina and Single-Vessel Disease without Collaterals, Circulation, 60, 106–113, 1979.

26. Gerbrands, J. J., Reiber, J. H. C. and Booman, F., Computer Processing and Classification of Coronary Occlusions. In "Pattern Recognition in Practice", Eds. E. S. Gelsema and L. N. Kanal, North Holland Pub. Co., 223–233, 1980.

27. Fukui, T., Yachida, M. and Tsuji, S., Detection and Tracking of Blood Vessels in Cine-Angiograms, Proceedings 5th International Conference on Pattern Recognition, I.E.E.E., 383–385, 1980.

28. Montarani, U., On the Optimal Detection of Curves in Noisy Pictures, Comm. ACM, 14, 335–345, 1971.

29. Martelli, A., An Application of Heuristic Search Methods to Edge and Contour Detection, Comm. ACM, 19, 73–83, 1976.

30. Ashkar, G. P. and Modestino, J. W., The Contour Extraction Problem with Biomedical Applications, Computer Graphics and Image Processing, 7, 331–355, 1978.

31. Enrich, R. W. and Schroeder, F. H., Contextual Boundary Formation by One-Dimensional Edge Detection and Scan Line Matching, Computer Graphics and Image Processing, 16, 116–149, 1981.

32. De Souza, P., Edge Detection Using Sliding Statistical Tests, Computer Graphics and Image Processing, to appear.

33. Hueckel, M. H., An Operator which Locates Edges in Digitized Pictures, J. ACM, 18, 113–125, 1971.

34. Hueckel, M. H., A Local Visual Operator which Recognizes Edges and Lines, J. ACM, 20, 634–647, 1973.

35. Nevatia, R., Evaluation of a Simplified Hueckel Edge-Line Detector, Computer Graphics and Image Processing, 6, 582–588, 1977.

36. Shaw, G. B., Local and Regional Edge Detectors: Some Comparisons, Computer Graphics and Image Processing, 9, 135–149, 1979.

37. Brown, B. G., Bolson, E., Frimer, M. and Dodge, H. T., Quantitative Coronary Arteriography Estimation of Dimensions, Hemodynamic Resistance and Atheroma Mass of Coronary Artery Lesions using the Arteriogram and Digital Computation, Circulation, 55, 329–337, 1977.

38. Groch, W. D., Extraction of Line Shaped Objects from Aerial Images using a Special Operator to Analyze the Profiles of Functions, Computer Graphics and Image Processing, 18, 347–358, 1982.

39. Fischler, M. A., Tenenbaum, J. M. and Wolf, H. C., Detection of Roads and Linear Structures in Low-Resolution Aerial Imagery using a Multisource Knowledge Integration Technique, Computer Graphics and Image Processing, 15, 201–223, 1981.

40. Montoto, L., Digital Detection of Linear Features in Satellite Imagery, Proc. of the International Symposium on Image Processing, Graz, 149–153, 1977.

41. Lipkin, L., Lemkin, P., Shapiro, B. and Sklansky, J., Preprocessing of Electron Micrographs of Nucleic Acid Molecules for Automatic Analysis by Computer, Comp. Biomed. Res., 12, 279–289, 1979.

42. Speck, P. T., Automated Recognition of Line Structures on Noisy Raster Images Applied to Electron Micrographs of DNA, Proceedings 5th International Conference on Pattern Recognition, IEEE, 604–609, 1980.

M. Kindelan
IBM Scientific Center
Madrid
Spain

J. Suarez de Lezo
Ciudad Sanitaria Reina Sofia
Cordoba
Spain

26 Electromagnetic effect evaluation by Markovian texture analysis of nucleated cells

L Bertolini and G Vernazza

1. INTRODUCTION

The biological effects of low-frequency electromagnetic exposure are strictly connected with the modifications induced by the field in the electrochemical cell microenvironment, and with the related variations in the concentrations of cell macromolecules and ions. The spatial conformations of chromatin-DNA itself can be affected so that electromagnetic control of cell functions becomes feasible [1, 2].

In order to measure *in situ* chromatin conformational changes induced by electromagnetic exposure inside single cells, Markovian texture analysis has been performed. The behaviours of these parameters, as well as the geometric and densitometric ones, allow one to distinguish effects due to different field waveforms, and to enhance a dedifferentiation process from the first stage.

As sample population, frog erythrocytes have been considered as they are nucleated cells which undergo morphological changes when exposed to a suitable electromagnetic (e.m.) field in a chemically potentiated microenvironment. According to Becker and Murray, erythrocytes dedifferentiate toward a totipotent stage from which they differentiate along an osteoblastic line [3].

The purpose of this report is to check if Markovian texture parameters can be employed as a quantitative tool for measuring frog erythrocytes chromatin conformation, associated with the previously mentioned electromagnetically induced morphological changes *in vitro*.

2. MATERIALS AND METHODS

Experiments have been described previously [4]. Petri dishes with frog erythrocytes in solution were inserted between the exposure coils. A particular electric field waveform was applied [1], with 10 mV/cm as average amplitude at 2.5 cm off-centre of the coils.

In Figure 1 two different signal waveforms are shown: (a) for upper waveform $T = 18$ ms and $\tau = 0.3$ ms; (b) for the lower waveform $T = 65$ ms and $\tau = 5$ ms; a single shot period $= 200$ μs. For the experiments described in this report, waveform (b) was applied.

100 cells after 24 hours of exposure time (treated cells) and 100 cells at time t=0 (control cells) were acquired by our automated image-analysis system (ACTA 500) [5]. Every cell was converted into about 100×100×8 pixels, with a spatial sampling at about the allowed limit of optical resolution, i.e., 0.1 μm.

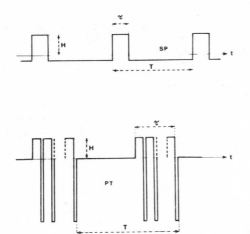

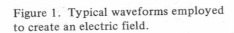

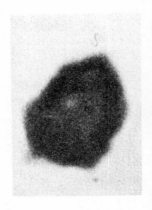

Figure 2. Image acquired before the calibration step.

Figure 1. Typical waveforms employed to create an electric field.

After a masking phase, a preprocessing technique was followed to calibrate optical and electronic systems. One dark image and one background image were periodically acquired: the dark image was employed to subtract the offset, while the white image was employed to adjust nonuniform Plumbicon target responses. In Figure 2 one image is shown that was acquired before the calibration step.

In Figure 3 four typical images relevant to the cell dedifferentiation process are presented: at the beginning, nuclear chromatin assumes a compact structure with low spatial frequencies, while, under external agents, successive modifications are induced toward a mitosis phase, with phenomena of chromatin relaxation and high spatial frequencies. Texture parameters should characterize appropriately this nuclear structure evolution.

3. MARKOVIAN ANALYSIS

Various approaches and model investigators can be used for texture analysis [6, 7]. We adopted the Markovian method, which is based on the estimation of the second-order joint conditional probability density functions $p(i, j/d, \theta)$. Each element $p(i, j/d, \theta)$ presents the probability of finding a grey level i and a grey level j at the spatial distance d and in the direction θ; d also stands for "Markov's step size". These probability values can be written in matrix form, called a "transition matrix":

$$\Phi(d, \theta) = p(i, j/d, \theta),$$

where the indices i and j go from 0 to the maximum number of grey levels (N). Simple relationships exist among matrices computed with θ and $\theta + 180°$; consequently, in our investigation, we only considered

$$\theta = 0°, 45°, 50°, \text{ and } 135°;$$

with $\theta = 0°$, TV raster scan was followed.

Referring to grey levels, an equalization of the grey-level histogram (pixel number v. grey level) is needed, in order to apply the following property:

$$p(i, j) = p(i/j)p(j) = p(i/j) \frac{\text{number of pixels at grey level } j}{\text{image area}}.$$

The most frequent equalized grey levels turned out to be N=4, 8, 16, 32, while the starting number of grey levels in the picture after preprocessing was 256. Figure 4 shows a picture (a) before and (b) after the equalization step, with N = 8.

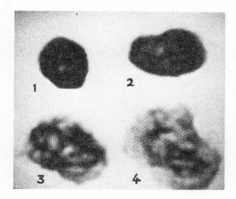

Figure 3. Four typical images of a cell dedifferentiation process. The upper left cell is an initial configuration, followed by the upper right cell, the lower left cell, and the lower right cell.

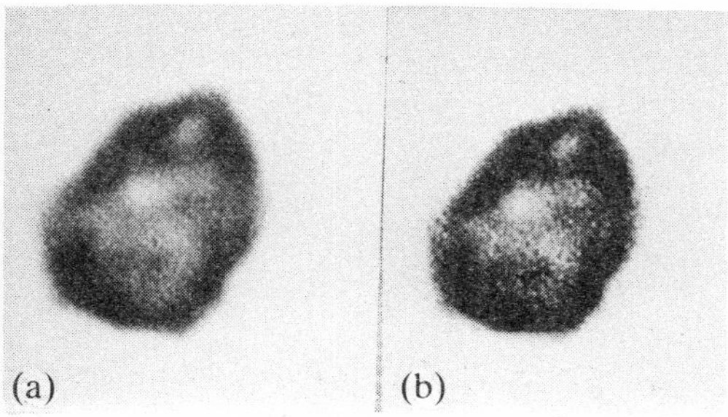

Figure 4. A cell (a) before and (b) after the equalization process with N = 8. Figure 4 corresponds to Figure 2 after the calibration step.

The above equalization procedure is not only required for statistical reasons but it is also useful since it allows successive computer elaborations with reduced-size matrices, e.g., 4x4 or 8x8, etc.; besides, for a first comparative texture analysis, this approach does not modify the basic texture structure, as will be seen in the next section.

From Markov's transition matrix (Φ), different texture features can be computed, as reported in the literature [8]. We considered 20 parameters and we recall the most usual ones:

(1) Energy or angular second moment

$$E(d, \theta) = \sum_{i=0}^{N-1} \sum_{i=0}^{N-1} [p(i, j/d, \theta)]^2.$$

(2) Entropy

$$H(d, \theta) = -\sum_{i=0}^{N-1} \sum_{j=0}^{N-1} p(i, j/d, \theta) \log p(i, j/d, \theta).$$

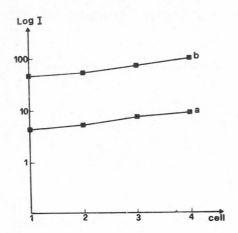

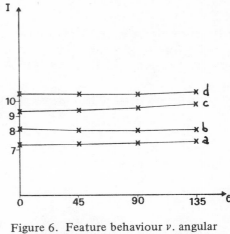

Figure 5. Feature behaviour with two different equalization levels: N=4 (lower curve) and N=8 (upper curve) (d=1; $\theta=0°$).

Figure 6. Feature behaviour v. angular direction θ (d=1; N=8). (Ref.: Figure 3; a=1, b=2, c=3, d=4.)

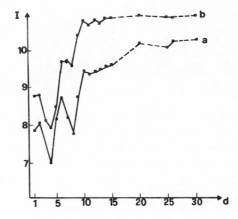

Figure 7. Feature behaviour v. Markov's step size d ($\theta=0°$; N=8). (Ref.: Figure 3; a=2, b=4.)

Figure 8. Feature behaviour v. mask size ($\theta=0°$; N=4; d=1). (Ref.: Figure 3; a=2, b=4.)

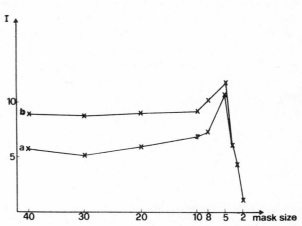

(3) Inertia or difference moment

$$I(d, \theta) = \sum_{i=0}^{N-1} \sum_{i=0}^{N-1} p(i, j/d, \theta)(i-j)^2 .$$

(4) Local homogeneity or inverse difference moment

$$L(d, \theta) = \sum_{i=0}^{N-1} \sum_{i=0}^{N-1} \frac{1}{1+(i-j)^2} p(i, j/d, \theta).$$

4. DISCUSSION

Five different tests were implemented to analyze texture behaviour; in particular, we considered only one variation in each test, to better evaluate the effect of the selected parameter.

Test A: feature value v. number of equalization grey levels N.

Test B: feature value v. angular direction θ.

Test C: feature value v. Markov's step size.

Test D: feature value v. mask size.

Test E: feature values for control and treated cells.

For Tests A,B,C,D, the four cells shown in Figure 3 were considered, and as texture feature, reference was made to the inertia moment (I) (defined in the previous section at (3)).

The results obtained by Test A are presented in Figure 5 where, on the vertical axis, a logarithmic scale has been adopted. Even if a marked absolute difference was noticed when using eight equalized levels as compared with four, these higher values are maintained during the cell modification sequence so that for relatively homogeneous comparisons, that is, with the same number of equalized levels, no discrepancy occurs. Anyway, a larger number of equalized grey levels are required whenever very slight variations in texture must be detected. Pressman [9] suggests the employment of eight grey levels.

Test B shows a substantial invariance value of the selected texture feature for four different angular directions (θ). These results are illustrated in Figure 6.

Test C. If we keep on incrementing d, a correlation loss is obviously achieved; this fact is also brought into evidence in Figure 7 where, after starting oscillations, a more or less delayed saturation effect appears in every cell.

Test D. In this case a quadratic smaller mask is considered step by step. After a nearly constant behaviour at the beginning, i.e., for large mask dimensions, a strong variation occurs for a mask size of about 20x20 pixels. The relevant results are shown in Figure 8.

Test E. By this test, a fundamental analysis of the basic biological problem is carried out. Twenty texture parameters have been computed for control and treated cells, assuming $\theta=0°$, d=1, N=4, and a maximum mask size consistent with cell dimensions.

Histograms of some selected features defined in the previous section, with related cumulative probabilities, are shown in Figures 9–12. In every figure (a) shows the curve of control cells and (b) shows the curve of treated cells. In Figure 9 the energy feature

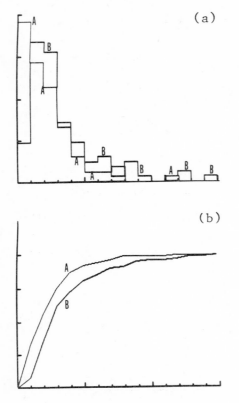

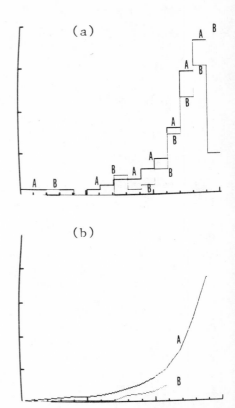

Figure 9. Energy for (a) control and
(b) treated cells (histograms and
cumulative curves).

Figure 10. Entropy for (a) control and
(b) treated cells (histograms and
cumulative curves).

shows an average decrease for treated cells, while in Figure 10 the entropy feature shows an average increase for treated cells. Similarly, in Figure 11 the inertia feature increases in treated cells; in Figure 12 local homogeneity decreases in treated cells. These behaviours are related to cell structure evolution; in fact, during chromatin relaxation, higher spatial frequencies appear, and so elements of the Markovian matrix are spread everywhere from the main diagonal.

5. CONCLUSIONS

Markovian texture analysis can be a useful approach to enhance morphological modifications in nuclear chromatin structure; for this purpose, some texture parameters can be employed. Geometric parameters can also be used to detect such modifications but texture parameters exhibit higher sensitivity. To better evaluate experimental results, texture analysis should be implemented by some further method (e.g., grey-level run-length method, grey-level difference method, power spectral method, etc.). In the near

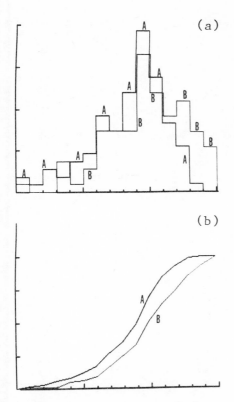

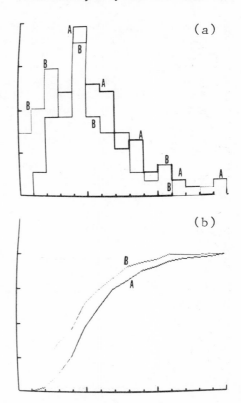

Figure 11. Inertia for (a) control and (b) treated cells (histograms and cumulative curves).

Figure 12. Local homogeneity for (a) control and (b) treated cells (histograms and cumulative curves).

future, we shall investigate these images, taking a bidimensional Markovian approach for comparing results.

REFERENCES

1. F. Beltrame, A. Chiabrera, M. Grattarola, P. Guerrini, G. Parodi, D. Ponta, G. Vernazza, R. Viviani, Electromagnetic control of cell functions, *Alta Frequenza* XLIX-2 (1980) 101–114.
2. A. Chiabrera, R. Viviani, G. Parodi, G. Vernazza, M. Hinsenkamp, A. A. Pilla, J. Ryaby, F. Beltrame, M. Grattarola, C. Nicolini, Automated absorption image cytometry of electromagnetically exposed frog erythrocytes, *Cytometry* 1-1 (1980) 42–48.
3. R. O. Becker, D. G. Murray, The electrical control system regulating fracture healing in amphibians, *Clin. Orthop.* 73:169 (1970).
4. A. Chiabrera, M. Hinsenkamp, A. A. Pilla, J. Ryaby, D. Ponta, A. Belmont, F. Beltrame, M. Grattarola, C. Nicolini, Cytofluorometry of electromagnetically controlled cell de-differentiation, *J. Histoch. and Cytochem.* 27 (1979) 375.

5. F. Beltrame, A. Chiabrera, M. Grattarola, P. Guerrini, G. Parodi, D. Ponta, G. Vernazza, R. Viviani, ACTA: automated image analysis system for absorption, fluorescence and phase contrast studies of cell images, *IEEE 1980 Frontiers of Engineering in Health Care* (1980) 58–60.
6. R. W. Conners, C. A. Harlow, A theoretical comparison of texture algorithms, *IEEE Trans. on Pattern Analysis and Machine Intelligence* PAMI-2, 3 (1980) 204–222.
7. R. Haralick, Statistical and structural approaches to texture, *Proc. of the IEEE* 67-5 (1979) 786–813.
8. N. J. Pressman, Markovian analysis of cervical cell images, *J. Histochem. and Cytochem.* (1976) 138–144.
9. N. J. Pressman, Optical texture analysis for automatic cytology and histology: a Markovian approach, Lawrence Livermore Laboratory, Univ. of California, UCRL-521555 (1976).

C. Bertolini and G. Vernazza
Istituto di Elettrotecnica
SIBE
Viale F. Causa 13
16145 Genova
Italy

27 Detectability enhancement of thermographic data

A R Fiorini, R Fumero and M Pisarello

1. INTRODUCTION

Cardioplegy, during heart surgery by extracorporeal circulation, is one of the latest outstanding techniques in patient safety. According to this technique, a 4°C physiologic solution is directly injected into coronary arteries: the amount of injected physiological solution is proportional to patient weight. The cold cardioplegic solution transport, in coronary arteries, cools myocardium; therefore myocardium cooling and coronary flow efficiency are strictly related. Myocardial effect monitoring by this technique is generally coarse; in fact, a sensor makes a single and localized temperature measurement. Information about myocardial metabolic activity and coronary artery characteristics, i.e., localized flow, dimensions, and possible occlusions, can be made available by a technique capable of enhancing temperature change dynamics all over the myocardium.

Thermographic monitoring during the infusion of cardioplegic solution and digital-image processing allows continuous and quantitative assessment of rapid temperature changes and shows exactly the extent of myocardium involved. Hence, during a cardio-surgical operation, thermocardiography monitoring can be useful in three ways:

(1) It allows the monitoring of localized temperature values all over the myocardium; therefore it is possible to evaluate cardioplegy efficiency and to decide the exact time for treatment renewal.
(2) It can enhance the detectability of possible coronary occlusions.
(3) Continuous assessment of rapid temperature distribution changes is particularly useful in by-pass efficiency evaluation immediately after operation.

Digital-image-processing techniques allows the computation of quantitative data and the facility to have them easily evaluated for the considered area of the inspected heart.

The recording subsystem contains:

(1) A front surface reflecting mirror.
(2) AGA 1010 reference surface.
(3) AGA 680 thermographic equipment and related control unit.
(4) OSCAR digital recording system on magnetic tape.

The digital-image-processing subsystem uses an HP 1000/E computer with 512 Kbytes central memory, 140 Mbytes of storage memory, TESAK VDC 501 interface for high-

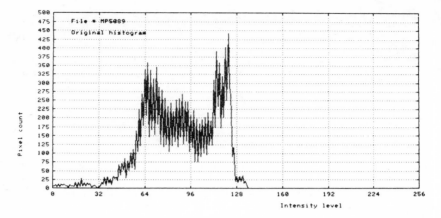

Figure 1. MP5089 original image
histogram.

resolution colour monitor, a high-resolution (256×256) colour monitor **BARCO**, and
a videographic terminal HP 2648A with a hard copy unit HP 2631G.

Considering that our prefixed goal is to study the optimal features for a real-time
monitoring system which is capable of offering the largest amount of visual information
as soon as possible to a human observer, it is necessary to develop a study to improve
image intelligibility.

2. IMAGE PROCESSING

The luminance histogram of a typical image, in matricial notation $A(i, j)$ is computed.
Figure 1 is an MP5089 original histogram. A tight distribution can be observed; this is
usual in a compressed range image: it is necessary to redistribute all brightness levels

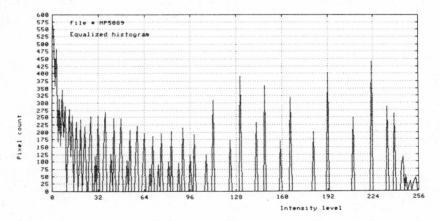

Figure 2. MP5089 hyperbolic equalized
image histogram.

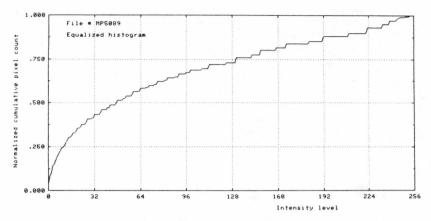

Figure 3. MP5089 hyperbolic equalized
image cumulative histogram.

over the available dynamic range so that the new histogram can follow some desired form.

Frei has suggested a histogram hyperbolization procedure [1]. By this technique the input image histogram is modified by the following transfer function

$$g = ([g_{max}^{1/3} - g_{min}^{1/3}] [P_f(f)] + g_{min}^{1/3})^3 \qquad (1)$$

so that the output image probability density will be of a hyperbolic form. In essence, histogram equalization is performed after the rods and cones of the retina. Figure 2 is the hyperbolic equalization of the original histogram of Figure 1 and Figure 3 is the related cumulative histogram.

Numerous experiments have indicated that although the human eye can differentiate only perhaps thirty or forty brightness levels in an image, it is able to separate thousands of various colours [2]. Therefore colour mapping of grey levels in an image can improve intelligibility. In general terms the pseudocolour mapping for a data matrix $A(i, j)$ is defined as:

$$R(i, j) = F_R \{A(i, j)\}, \qquad (2)$$

$$G(i, j) = F_G \{A(i, j)\}, \qquad (3)$$

$$B(i, j) = F_B \{A(i, j)\}, \qquad (4)$$

where $R(i, j)$, $G(i, j)$, and $B(i, j)$ are display tristimulus values and F_R, F_G, and F_B are linear or nonlinear functional operators. The three display tristimulus values $R(i, j)$, $G(i, j)$, and $B(i, j)$ can be considered to form the three axes of a colour space, so a mapping defines a path in three-dimensional colour space parametrically in terms of the data matrix $A(i, j)$. Now we have to build a path, in colour space, able to emphasize peculiar image characteristics in order to offer the largest amount of visual information as soon as possible to a human observer. It is necessary to define the information of typical interest to us. We are studying fast temperature distribution changes over a myocardium surface, so temperature change edges and their associated changing rates are useful; in fact, cardioplegic cooling and distributed coronary flow efficiency are strictly related [3-5]. Our optimum path should be able to enhance, by means of the greatest chromatic contrast of the display, a restricted number of different temperature areas.

Figure 4. Four paths in colour space for maximum chromatic contrast of continuous colours.

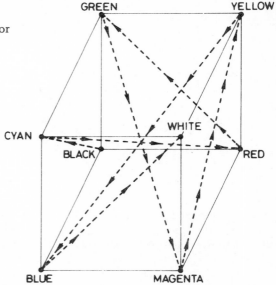

A restricted number of different temperature areas is rapidly intelligible: up to now eight is thought to be a satisfactory compromise. We can use the eight constant saturated colours which are on the vertices of the RGB colour cube shown in Figure 4. In order to maximize chromatic contrast we have to use a path able to maximize the chromatic distance of contiguous colours. Furthermore, the chromatic path is divided geometrically in equidistant intervals between colours.

3. DISCUSSION

It is interesting to observe that in a steady state from a thermal point of view, the complete occluded coronary artery thermal distribution is the same as that of the myocardium surface. Active coronary arteries are immediately enhanced with their associated cooling rate. This information correlated with coronary artery geometry allows careful monitoring of myocardium metabolic activity and possible coronary artery occlusions. Certain limitations and concerns are associated with thermocardiography. Only the anterior and the lateral aspects of the heart can be approached by the thermocamera without difficulties. In spite of these limitations, thermographic data processing, described in the present work, can be an irreplaceable tool in extracorporeal circulation operations.

4. CONCLUSIONS

The features, for a real-time monitoring system, which are capable of offering the largest amount of visual information as soon as possible to a human observer, can be summarized by observing that rapid temperature distribution change characteristic time is usually in a second range in coronary arteries. Hence image acquisition and processing time must be comparable in the same range. According to previous and present work experience, a real-time thermographic image recording and processing system could consist of:

(1) A high-speed digital image acquisition unit, 128x128 pixels image per second.
(2) A digital processing unit, able to achieve one processed image per second.
(3) A colour display unit and display image update capability every 2 s.

Essential features of the system are:

(1) Image difference display of the last two acquired images.
(2) Ability to emphasize possible occlusion points.
(3) Warning signals for cardioplegic treatment renewal.

5. ACKNOWLEDGEMENTS

The authors express their gratitude to Dr. R. Marchesi and G. Sensolo, Dipartimento di Energetica, Politecnico di Milano, for their collaboration.

REFERENCES

1. W. Frei, Image Enhancement by Histogram Hyperbolization, Journal of Computer Graphics and Image Processing, June 1977.
2. G. Wyszecki and W. S. Stiles, Color Sciences, Wiley, New York, 1967.
3. D. Tzivoni, A. Cribier, K. Kaumatsuse, C. Chew, G. Totten and W. Ganz, Epicardial Blood Flow during Coronary Occlusion Studies by High Speed Cardiothermography, Am. J. Cardiol., Vol. 41, pp. 393–394, 1978.
4. F. Robicsek, J. N. Masters, R. H. Svenson, W. G. Daniel, H. K. Daugherty, J. W. Cook and J. G. Selle, The Application of Thermography in the Study of Coronary Blood Flow, Soc. for Vascular Surgery, 32nd Annual Meeting, Surgery, June 1978.
5. F. Robicsek, T. N. Masters, R. H. Svenson and W. G. Daniel, Thermographic Study of Coronary Blood Flow during Open Heart Surgery, 11th Internat. Congr. of Audiology, Prag, Abstractbook n. 412, July 1978.

A. R. Fiorini and R. Fumero
Dipartimento di Energetica
Politecnico di Milano
Piazza Leonardo da Vinci 32
20133 Milano
Italy

M. Pissarello
Dipartimento di Elettronica
Politecnico di Milano
Piazza Leonardo da Vinci 32
20133 Milano
Italy

28 Detection of agricultural areas in tropical regions

M Dosso, J Kilian and G Savary

1. INTRODUCTION

IRAT and the IBM (France) Scientific Centre have undertaken a joint research. The aim is to evaluate the use of the future satellite SPOT in detecting the potential agricultural areas in tropical countries.

A simulation flight was realized in October 1981 in Upper Volta. The recorded image covers a surface of 10 x 24 km. Image processing consisted of two steps:

(1) Obtaining images classified in functions of the "surface status".
(2) Simplifying the classified images in terms of "regions".

The present paper describes the methods and tools which have been used and their use by agronomists.

2. METHODS AND TOOLS

2.1. Image classification

The method used to classify the image is supervised and interactive; it assumes that:

(1) Two bands are used for classification.
(2) Radiometric values in each band are coded from 0 to 63.
(3) Each class can be identified on the image by at least one training field.

The principle of classification consists in representing in a square (64 x 64) the different possible combinations of radiometric values in the two selected bands (generally the two principal components in the Karhunen Loëve transform are used). The classification is obtained by partitioning the square of radiometric values. This partition of the square induces a partition of the image. Statistics obtained from the training field and, principally, two-dimensional histograms are used to help in determining the optimal classification.

The method is interactive because the final classification is the result of successive trials with comparison between histogram, partition, and visualization of the classified image.

Such a process assumes a very fast exchange between the machine and the user and, especially, quasi-immediate histogramming of the training field, modification of partition, and visualization of the classified image after each modification. That was made possible thanks to the special image processing terminal IBM 7350 (HACIENDA), using a conversational image-classification package written under APL.

The main characteristics of HACIENDA are:
(1) A 1024 × 1024 point colour screen.
(2) A 12 plane refresh memory.
(3) A 4096 entry colour look-up table which offers a choice of 32,000 colours.
(4) A 6 million byte auxiliary memory allowing one to store images and to display them instantly.
(5) A high-speed specialized image processor which allows image transformation, combination, interpolation, histogramming, and other statistical operations.

The classification package uses the special features of HACIENDA with a view to providing the maximum interactivity through a very short response time. The classification process is made up of four phases:
(1) Computation of the two bands which will be used for classification.
(2) Visualization of the two bands.
(3) Definition of classes.
(4) Classified image creation.

(1) Computation of the two bands. In the case of SPOT simulation, the two bands were the first two components of the Karhunen Loëve transform, they represented 90% of the total variance. They were calculated directly in HACIENDA from the three original bands.

(2) Visualization of the two bands. For visualization the bands have to be recoded so that the values are in the range 0–63, this is done immediately by the HACIENDA processor using translation tables. Then the bands are sent to the refresh memory, six planes for one band, six planes for the other. The image is visualized according to a hue intensity coding (one band is used to define the brightness, the other one defines the colour). This coding is realized through the colour look-up table.

(3) Definition of classes. The 64 × 64 square representing the radiometric values is displayed on a conversational screen, preferably in colour (the IBM 3279 can be used). Each class is given a graphic symbol, e.g., an alphabetic character, which represents it. Special edition functions have been developed to define and easily modify the domain of each class (see Figure 1).

Three types of information are used which help decisionmaking during the classification process:
(a) Two-dimensional histograms of the training fields.
(b) Class distribution inside the training fields.
(c) Visualization of the classified image.

(a) Two-dimensional histogramming. The histogram is displayed on the conversational screen using digits 0 to 9, proportionally to the values obtained. Thus it is possible to draw class limits which match the statistics obtained from the training fields.

(b) Distribution of classes in the training fields. At the same time the histogram of a training field is computed, the percentages of the different classes in the same training field is displayed and the result is updated when the limits are changed. Thus it is possible to check at any time whether classification is adequate.

(c) Visualization of the classified image. The classification results can be viewed instantly at any time using the colour look-up table. This table has 4096 entries which corresponds to the 64 × 64 possible spectral-value combinations. Each class is given a colour and the colour look-up table is loaded to reflect the square partition.

(4) Classified image creation. Once a satisfactory classification is obtained, a classified image, where the radiometric values are replaced by class numbers, is generated by HACIENDA using another look-up table: the output look-up table. This table has 4096 entries, the same as the colour look-up table, but the colour code is replaced by the class number. The classified image is stored in the HACIENDA memory, from which it can be sent to a permanent storage device, disk or tape.

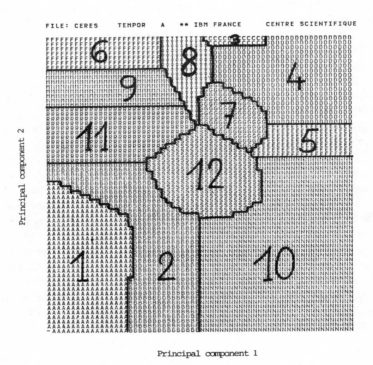

Figure 1. Partition of the principal components plane which induces the classification of the SPOT image.

2.2. Creation of regions

A region is defined as an area in which the local distribution of classes is constant. This method was developed at the IBM (France) Scientific Centre in 1979 and 1980 and was the object of several publications which are mentioned in the bibliography.

3. UTILIZATION BY AGRONOMISTS

3.1. Classified image

3.1.1. *Global study of the landscape*

The classified image is an interpretation of the spectral information. On the ground the agronomist defines homogeneous areas regarding land occupation, i.e., the surface state. On the image, to each area corresponds a spectral domain. These domains define the different classes of surface states.

For the rural planner who aims to find agricultural areas, the first task is to study the landscape. A classification of the different surface states is a good tool for such a study.

For the SPOT image, 12 spectral classes have been defined:

A 1:	Bush fires.
B 2:	Fire margins.
C 3:	Cotton.
D 4:	Various cultures and fallow lands.
E 5:	Intermediate (cultures and bare soils).
F 6:	Thick gallery-forest.
G 7:	Cultivated lowlands.
I 8:	Savanna with high grass.
M 9:	Light gallery-forest.
N 10:	Bare and eroded soils
R 11:	Savanna with low grass.
S 12:	Shrub savanna.

The classified image obtained is the basic document which can be exploited in different ways.

For the rural planner, two themes are important: the river system and the eroded soils. The river system expresses and induces the evolution of the landscape; the evaluation of the eroded soils gives the intensity and the extension of the erosion phenomenon, corresponding at the time of the satellite flight. The map of the surface states makes it possible to study both of these themes; to visualize them better, we extracted each of them from the classified image.

3.1.2. *The hydrographic system*

Seen from space, in the studied zone, the river system is revealed by the associated vegetation. So by selecting the spectral values corresponding to the vegetation associated to the river, we extracted the actual river system from the image. The principal streams were drawn using the two classes "thick gallery-forest" and "light gallery-forest". The secondary streams include parts of the two classes "savanna with high grass" and "cultivated lowlands". The image of the river system shows:

(1) The principal drainage axes and a set of "initial" circulations; this image gives an idea of the way slopes are evolving, and thereby of the dynamic aspect of the landscape.

(2) The action of cultivation on the river system. In the south, less cultivated, the flows are more linear.

3.1.3. The eroded soils

We extracted the class "bare and eroded soils" from the image of the surface states; it is a global evaluation of the eroded soils. Comparing with the classified image, we notice that two different eroded zones can be distinguished: in the north, erosion is related to cultures which modify the water circulation; erosion exists everywhere and is mainly a surface process. In the south, the erosion process is equally intense but deeper, concentrated on linear areas; the influence of cattle must be important.

These two examples show that a first study of the landscape is possible with visual exploitation of the information extracted from the image.

3.2. "Regional" analysis

We have just seen that for a given theme, relations with other classes are important. That is what a "region" takes account of. In a given landscape, a "region" is defined by a certain association of different classes. Describing a landscape with "regions" is to take into account the different associations of classes. It is no longer a spectral study of the landscape but a study of the spatial distribution for the different classes obtained. Visual analysis of the surface state image makes it possible to distinguish 10 "regions" which are defined as follows:

    1: Cultivated areas.
    2: Cultivated lowlands.
    3: Dominant thick gallery-forest.
    4: Dominant light gallery-forest.
    5: Savanna 1 (related to the rivers).
    6: Savanna 2 (related to bare soils).
    7: Savanna 3 (related to cultivated lowlands).
    8: Dominant bare soils.
    9: Dominant fires.
    10: Dominant fires margins.

We obtain such an image, using theory and programs written by Rogala (1982). It presents two advantages:

(1) It is easier to look at this new image than to read the previous one; it represents the second level of interpretation for the initial information.

(2) This document can be very useful for the rural planner; it is a synthesis of the description of the surface states. This is the first step in the search for agricultural areas; planners will also have to take into account relations between surface states and other characteristics of the environment (bedrock, topographical features, etc.); indeed, as has been observed, a spectral class, which characterizes a surface state, may occupy a slope, such as the top of a hill, or a plain.

4. CONCLUSION

Extracting information useful for the rural planner, from a satellite image, can be achieved in two steps:

(1) The surface-state classified image corresponds to a first level of interpretation; it is a

spectral study, lying on field observation data. From this image we can study different themes.

(2) The "regions" image corresponds to a second level of interpretation; it consists of the spatial study for the association of the different classes which has been defined on the first step. The resulting image is easier to study; this document is useful in the search for agricultural areas, in which rural planners have to consider external information.

The originality of the method is to give priority to field knowledge: the software interactivity allows an easier dialogue between the computer and the agronomist.

REFERENCES

1. L'Agronomie Tropicale. Aménagement écologique; réflexions méthodologiques; exemples pratiques. Série 11–111, vol. XXIX – 1974 – n 2 et 3. Fevrier–Mars.
2. Guillobez: Etudes morphopédologiques – Projet Bagre, Rapport géneral – IRAT 1977.
3. Lebart et Fenelon: Statistique et informatique appliquées. Dunod ed. 1975.
4. Raunet: Région de Bagre-Zabre (Volta Blanche), Paysages végétaux et morpho-pédologiques. Etude jointe IRAT–IBM France (1981).
5. Rogala: Approche numérique de l'espace agricole. Thèse Doc. Ing. 1982. Institut National Agronomique. Paris.
6. Tricart et Kilian: L'Ecogéographie et l'aménagement du milieu naturel. (Francois Maspero 1979).

M. Dosso and G. Savary
IBM (France) Centre Scientifique
Paris
France

J. Kilian
Institut de Recherches Agronomiques
Tropicales et des cultures vivrières
Paris
France

29 Radar image processing for remote sensing

P Y Nguyen and G Stamon

1. INTRODUCTION

Radar signals can be processed to give microwave responses of the earth surface in an image format suitable for use in remote sensing. For other sources of data, radar images possess some artefacts which users need to remove or to minimize their effects.

We limit this study to the following identified artefacts:

(1) Speckle, granular image texture resulting from the interference of dephased but coherent energy while using coherent monochromatic waves (microwaves in this case) to illuminate a surface.
(2) Foreshortening, high intensity area on the image which reflects off a surface having its centre coincident with the source of the radar.
(3) Shadows from areas not illuminated.
(4) Ranging, which causes a gradual shift in the radar response across the image due to the varying distance between the radar source and the illuminated surface.

Different methods of image preprocessing were proposed to deal with these artefacts:

(1) mean and variance correction [1],
(2) mean and median filter [1],
(3) adaptive filter [2], and
(4) fast Fourier transform filter [3].

In this paper we apply a method of selective filtering, described in ref. [4] for radar image, to reduce the speckle and render the data suitable for classification.

2. COMPARISON OF FILTERING TECHNIQUES

The above-mentioned filtering methods and two proposed ones are used for a Seasat image.

2.1. Image segmentation

A method of image segmentation as described in ref. [5] is used to divide the image into segments defined by their contours. All pixels in different segments are then replaced by their mean value. The speckles are, hence, reduced.

2.2. Segment-dependent image smoothing

A selective smoothing technique proposed in ref. [4] is used. This method is a compromise between the above method and smoothing by averaging over a moving window. However, the contour is preserved by averaging only those pixels:

(1) which are within the moving window, and
(2) which belong to the same segment of the image.

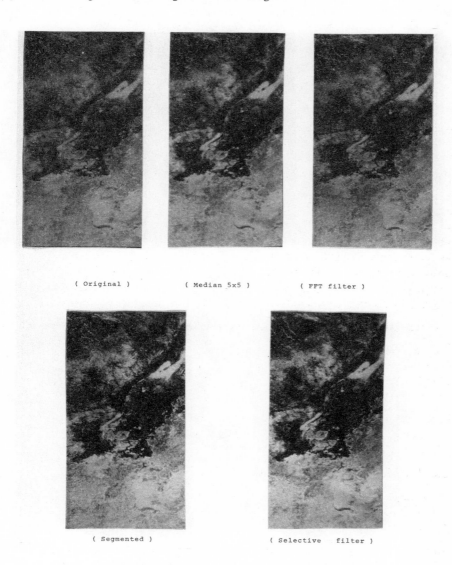

(Original)　　　　(Median 5x5)　　　(FFT filter)

(Segmented)　　　　(Selective　filter)

Figure 1. Seasat images.

The results of this filtering together with:

(1) a median filter with a window of 5×5, and
(2) a Fast Fourier Transform filter where the cut-off frequency was 90 for a transform window of 256.

These images are shown in Figure 1. The median filter, segmentation, and the selective filter all reduce speckle effectively. The FFT filter leaves some speckles, but reducing the cut-off frequency further will give an unacceptable loss in resolution. The median

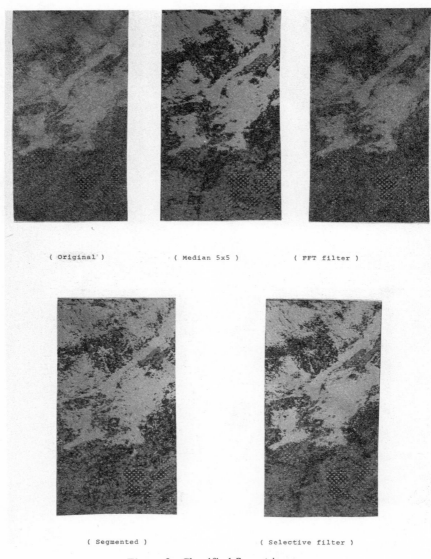

(Original)   (Median 5x5)   (FFT filter)

(Segmented)   (Selective filter)

Figure 2. Classified Seasat images.

filter also causes some loss in resolution. Segmentation completely reduces the speckles, but variation in the radar responses is lost. The best technique was the selective filter, where the speckles are eliminated while the contours and finer variations within image segments are preserved.

The effect of these filtering techniques are shown much clearer in Figure 2, where the resulting images are classified by a maximum likelihood classifier.

The study area is in Algeria and the terrain consists of:

(1) two chotts (marin deposits area) which seems to have different surface roughness,
(2) a sandy area, and
(3) a pliocene area.

The effect of speckle on the original Seasat data renders classification almost unexploitable. The two chotts are scattered with random unclassified points. The sand and pliocene are mixed together and also filled with unclassified points. The FFT reduced the speckle but its effect is still seen on the classified image. Only with the median filter, segmentation, and the selective filter are the classified images useful. In these last three classified images the chotts, sandy, and pliocene areas are well distinguished.

3. TWO-DIMENSIONAL HISTOGRAM CLASSIFIER

As explained in the previous section, speckle can be reduced and further improvement can be achieved by attempting to separate these speckles. This is done by subtracting the filtered image from the original image. Classification can then be performed on the filtered and "speckle" images.

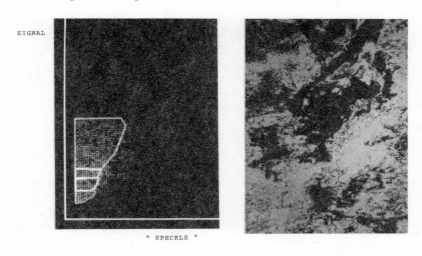

Two-dimensional histogram partitions   Classified image

Figure 3. Classified image by two-dimensional histogram partition.

One problem was encountered with the exponential nature of the "speckle" image. The maximum likelihood classifier requires the input data to be normally distributed, which is not true for the "speckle" image. A method of supervised classification allows users to divide a two-dimensional histogram according to the domain of the different classes to be discriminated. This free partitioning of the observation plane into polygons is not dependent on a hypothesis on the distribution of the class. The two-dimensional histograms, its partition, and the classified image are shown in Figure 3.

4. DISCUSSION

The results obtained are encouraging as presented, improvement of the classification was considerable but the algorithms, at their present state, required heavy computing time. The results warrant further development and the improved method can be done as follows:

(1) The segmentation process can be by-passed, selective filtering can be done by including the central point of the window and including points within increasing distances until a high gradient point is met.

(2) The mean value used can be replaced by the median mode of the points included in the smoothing to cater for different types of data, including classified image.

(3) The two-dimensional histogram partitioning can be extended to three-dimensional, using CAO CAM techniques of three-dimensional object representation.

ACKNOWLEDGEMENTS

The authors express their sincere thanks to Drs. Elachi and Rebillard, JPL, for providing the Seasat data, Mr. G. Savary for his precious help while using his APLIAS (APL Image Analysis System), Mr. L. Asfar for his works on image segmentation, and all members of the Image Processing Group at the IBM Scientific Centre who have contributed to the development of the IBM 7350 Image Processing System used during this study.

REFERENCES

1. Bloom R. G. and Daily M.
   Radar image processing for rock type discrimination. IEEE Transactions on Geoscience and Remote Sensing, Vol GE-20, No 3, July 1982, p 343–351.

2. Frost V. S. et al.
   An adaptive filter for smoothing noisy radar images. IEEE Proceedings Vol 69, No 1, January 1981, p 133–135.

3. NASA
   Synthetic aperture Radar/Landsat MSS image registration. Nasa Ref pub no 1039, p 110–112.

4. Nguyen P. T.
   Selective image enhancement and restoration. Int Geoscience and remote sensing symposium, Washington June 1981 p 291–297.

5. Asfar L.
   A method for contours detection, segmentation and classification of Landsat images. Etude F-029, Dec 1981, IBM France Scientific Centre, Paris.

P. T. Nguyen
IBM (France) Scientific Centre
36 Ave Raymond Poincaré
75116 Paris
France

G. Stamon
Université de Besançon
rue Engel Gros
90016 Belfort
France

30 An application of ISIID: remote sensing observation of glaciers

A Della Ventura, R Rabagliati, A Rampini and R Serandrei Barbero

1. INTRODUCTION

The identification of glacier surfaces and the observation of their temporal variations from remote-sensed satellite images is of great interest, both practical and scientific. Although alpine glaciers are a primary energy source and act as medium- and long-term climatic indicators, they are presently observed with traditional methods in only a small percentage ($\sim$10%), since their inspection is both difficult and costly. On the other hand, Landsat satellites offer a sufficient spatial resolution to detect the annual surface fluctuations of the majority of Italian glaciers. A series of Landsat images of a group of glaciers in the central Alps (Disgrazia group) has therefore been examined with the aim of defining a method for monitoring iced and snow-covered surfaces. IGM map 1:25,000, a series of stereo-aerial photographs (approximate scale 1:12,000), and the reports of the Comitato Glaciologico Italiano campaigns have been used as ground-truth sources. Digital-image processing has been carried out by the ISIID system which allowed completion of the exploratory analysis of this data set and definition of:
(1) the specific algorithms for the glacier surfaces identification, and
(2) the morphological structures (accumulation basin, tongue, front) and the algorithms for their recognition.

2. DEFINITION OF THE ALGORITHMS

The ISIID system is here used as an instrument for the exploration and algorithm definition [1]. It can be seen as a box having as input a first set of digital images and as output the interpretation criteria with the related personalized programs. In this phase the system allows the use of exploratory programs driven by *a priori* interpretation criteria. The reference data set, in this case the ground-truth data, allows immediate verification of the results achieved.

2.1. Surfaces identification

The multispectral images in three different years (September 13, 1975, August 28, 1978, September 4, 1980) were the first input data set.

The *a priori* interpretation criteria were based on the knowledge of the surface composition to be identified, i.e., ice and/or snow, and on instrumental performance,

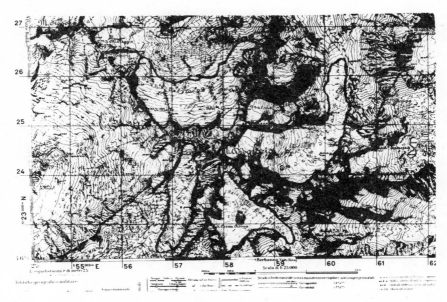

Figure 1

i.e., spectral ranges of the MSS, spatial resolution, sensor characteristics. Applying the exploratory analysis techniques to the images, the interpretation procedure was found and the specific programs were constructed. It was verified that band 5 (0.6–0.7 μm) and band 7 (0.8–1.1 μm) give information on the glacier external boundaries and internal structures [2, 3]. It was also observed that in the glacier radiance range the image obtained by multiplying band 5 by band 7 allows better discrimination of the interesting surfaces against the background.

This reference image was then divided into regions for which statistical parameters have been computed. The thresholding functions selecting the iced or snow-covered surfaces have been interactively defined on the basis of the computed parameters. Figure 1 shows the IGM map of the studied area where the approximate boundaries of the five main glaciers belonging to the group are drawn.

As an example of the application of this method, the iced surfaces identified on the September 13, 1975 image are sketched in Figure 2. The five main glaciers are easily recognizable. The computed surfaces are in good agreement with the data reported in the Catasto dei Ghiacciai Italiani [2].

2.2. Morphological structures recognition

The morphology of the objects shown in Figure 2 is complex. However, in the majority of cases it is possible to distinguish a central body and some elongated structures protruding from it. Among the largest protrusions, the glacier "tongue" is identifiable because of its direction toward the lower altitudes, while the protrusions toward the higher ones are ramification of the accumulation basin. Other small protrusions constitute side flows. The tongue is a variable structure, while the boundaries of the accumu-

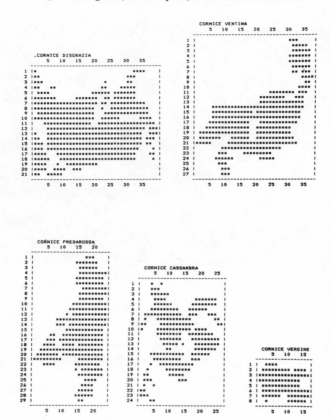

Figure 2

lation basin are time invariant. Then glacier monitoring can be done using the time invariant structures as a reference system for image registration, allowing the study of the variable features.

The input of the system used as a pattern recognizer are the silhouettes of the individual objects identified in the previous step and shown in Figure 2. The output is

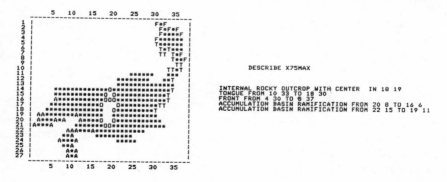

Figure 3

Figure 4

the interpretation procedure with the related algorithms. The recognition process consists of the identification of particular sequences of contour points that can be associated with meaningful structures. It is worth noting that the sequences of contour points are inter-actively defined and easily modifiable.

After protrusion recognition, the most relevant ones are selected and associated, through integration with the altitude data, to the corresponding glaciological terms. A recognized internal hole is interpreted as an internal rocky outcrop. For example, Figure 3 shows two descriptions of the Ventina glacier observed in 1975 and a silhouette of which is shown in Figure 2(b). The relevant structures (A, accumulation basin; T, tongue; F, front; O, outcrop) are now clearly recognized by the system and are shown graphically in Figure 3(a) and listed verbally in Figure 3(b).

With the fixed structures as a reference system the different images have been regis-tered. The map shown in Figure 4 is obtained through the superposition of the 1975 and 1980 images after their registration. Four classes of pixels have been identified:

blank: background in both images,
    ∗: glacier surface only in 1975,
    O: glacier surface only in 1980,
    ө: glacier surface in both images.

3. CONCLUSIONS

The explorative analysis of a first set of multitemporal images has allowed the definition of the glacier surface identification procedure, as well as the morphological structure recognition algorithms. Monitoring of the glacier variant structures is achieved using their fixed structures as reference data. Interpretation procedures yield graphical and verbal image descriptions at several syntheticity levels. Such descriptions can be usefully inserted in a data base which includes the traditional data. A subsequent confirmation analysis on a different data set is the next step in the definition of the interpretation procedure and the implementation of the specific programs.

REFERENCES

1. U. Cugini, M. Dell'Oca, D. Merelli, P. Mussio, A computer-aided system for inter-active definition of digital-image interpretation, presented at: 2nd Int. Conf. on Image Analysis and Processing, Selva di Fasano (1982).
2. A. Della Ventura, R. Rabagliati, A. Rampini, R. Serandrei Barbero, Remote-Sensing observation of glaciers toward their monitoring, presented at the Seventeenth International Symposium on Remote Sensing of Environment, Ann Arbor, Michigan (1983).
3. M. Pagliari, A. Zandonella, Utilizzazione delle informazioni Landsat per lo studio del manto nevoso, presented at Symposium A.E.I. (1982).

A. Della Ventura and A. Rampini	R. Rabagliati	R. Serandrei Barbero
Istituto di Fisica Cosmica del CNR	IBM Italia	ISDGM CNR
Via Bassini 15/A	Mestre	1364 S. Polo
Milano	Venezia	Venezia
Italy	Italy	Italy

31 An application of ISIID: arcs, fillets and tangent points restoration in digitized line

U Cugini, G Ferri, P Micheli, P Mussio and M Protti

1. INTRODUCTION

ISIID [1] has been used for defining and implementing a device following a structural approach for the restoration and vectorization of digitized line drawing. This approach [2] is based on the knowledge of the graphic and nonautomatic procedures used by the draughtsman in the creation and realization of the drawing. The technical drawing is a complex iconic message, written in order to communicate specifications about solid objects. The language used for this communication has well-defined syntactical and orthographic rules. From the orthographic point of view the basic elements are straight lines, arcs, and other well-defined curves. These curves are the strokes of this specific graphic language. When digitized, each stroke of the original drawing appears as a connected set of pixels of similar grey level, which we call a path. A segment is digitized into a nearly linear path, whose edges are nearly linear, circumference is transformed into a path similar to an annulus, and tangent points are transformed into different areas (Figure 1).

Analysis of these paths and areas allows identification of hints about the kind of original strokes. These hints are stored and used both for a proper classification of the structure or for definition of geometrical and topological properties of the classified stroke. This restoration activity is based on the knowledge of the rules of the specific graphic language.

In this paper the case of arcs, fillets, and tangent points restoration is discussed. The ability of managing these structures allows matching of the requirements of a technical drawing, representing mechanical parts whose shapes are usually represented by complex contours. Other cases, such as schematic drawings of electrical or electronic circuits or plans for drawing of civil engineering applications, are usually constructed by means of straight line tokens [3,4].

An earlier prototype [5], based on a thinning technique, showed that when dealing with data collected from a noisy line drawing, the structures are often misclassified because the restoration procedure may introduce gaps or other geometrical artefacts and this was judged a weak point for this approach [5,6]. The proposed procedure avoids these kinds of flaws, by means of a technique of classification which progressively refines the results obtained.

In a first step, the grey-level features which characterize the drawing strokes are

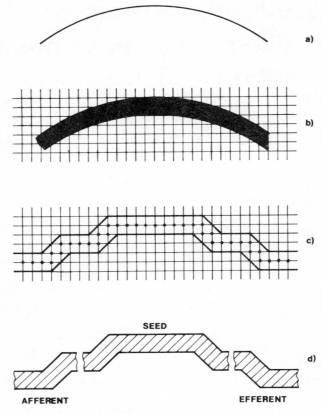

Figure 1. (a) An arc, (b) the arc and the
lattice, (c) digitized arc, and (d) parts of
the digitized arc.

determined by an adaptive device. In a second step morphological features are searched
by a data-driven, table-driven system.

2. IDENTIFICATION OF DIGITIZED ARCS

An arc when digitized looks like a sector of an annulus (Figure 1) whose edges are a noisy
approximation of the stroke. The proposed identification technique is based on the
successive accumulation of more and more complex hints that allow classification of the
considered set of pixels.

First, edges will be examined in order to check if there are sufficient hints to be
identified as rectilinear or parts of a curve. Then the classified segments of an edge will be
examined and compared in order to get an interpretation of the considered set of pixels,
identifying the stroke from which it has been generated.

Here we will examine the first step, that is the edge description and we will show
how hints can be algorithmically accumulated in order to decide if a segment of an edge

is an arc or not. To do this we will use a particular algebraic notation: the α-ω notation by Iverson [7].

3. THE IDENTIFICATION PROCEDURE OF AN ARC

The accumulation of hints, used to justify the classification of an edge as an arc, starts, of course, from the search for the simpler hints. The first kind of hints are called seeds and made of a configuration of edge points, always present when there is an arc (but sometimes also when there are oblique lines). Stronger hints, clues, and evidences are found by a function which is started by an interpreter as a consequence of the previous finding [1]. In practice, a numeric code N(P), used for the elaboration and an alpha-numeric one C(P), used for the person–machine communication [3], is associated to each possible edge point. A seed is codified as a sequence of codes, whose length corresponds to the number of points of the configuration.

The whole edge of the figure is described as a sequence of codes whose length corresponds to the number of points that form the edge. The research of a seed is reduced to the finding of a given substring in a string of codes. The seed is a weak hint: the existence of an arc depends on the presence of other points, in a predefined sequence.

An arc is made of a seed, preceded by an afferent part and followed by an efferent one. There are many configurations of the possible afferent and efferent parts; the strings of associated codes represent a defined language and they can be formally described.

Collecting other hints about the existence of an arc, means checking the codes (configurations) that surround a seed and verifying that this sequence forms a string belonging to this language. In practice, the codes associated to all the edge points do not have to be stored, only the points of the so-called contour extremes (whose more explicative subset is that of the vertex points).

The contour extremes of a figure are stored in a matrix, named SXY (Figure 2), with dimensions m x 3, where m is the number of the considered contour extremes. A line in the matrix is associated to each one of these points: in the first column the associated N(P) is maintained, in the other two are its coordinates. For easier use, the m alpha-numeric C(P) symbols associated to some points are stored, in the same sequence, in a variable GSTRINGA. The possible seeds of an arc are stored in the variable GARC. In this variable a flag is stored and it identifies the seed as "exterior" or "interior" to the annulus and the quadrant where the seed is.

The judgement of the existence of an arc cannot be based only on the existence of contour extremes of a proper configuration, but also on their geometric disposition. This

ARC Δ FROM SEED

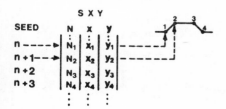

Figure 2. Relations among the content of the vector SEED, the matrix SXY and a stroke.

means that the arc has to be well proportioned. The length tolerances of the afferent and efferent parts have to be defined, they are stored in the variable TLA (too long for an arc) and MEA (minimal extension to be an arc). The same has to be done for the seed, and the tolerances are stored in TMN (threshold of minimal neighbourhood) (see the Appendix).

4. GEOMETRIC CONSIDERATIONS

When the possible nature of the strokes is defined, geometric relations need to be checked and dimensions and positions of the strokes have to be definitely estimated. Paths analysis allows individuation of the regions where a path joins and leaves another one. These regions correspond to the points (in the original drawing) where different geometric figures are tangent or cross. The codes of the extremes of contour are used to determine the kind of relation (tangency, cross) existing between the two original strokes.

The nature of the point, even if it does not allow a certain classification, makes clear at this step of the process that in that region the estimated strokes interact. All this information together allows analysis of the algorithms that define the geometric properties which are probably nearest to the original ones. For instance, in the case of two segments connected by an arc, the proposed algorithm is based on two empirical observations (which up to now are under check on the basis of new outcoming cases). The first observation is that the position and direction estimate of a rectilinear segment starting from the coordinates of the extremes of contour with all the used methods, is reasonably careful and precise. The second observation is that the estimate of the extremes, especially in the case of tangency of an arc, is usually not very precise, while its radius estimate is very precise.

In order to rebuild the connection, we will take advantage of the knowledge, obtained by the process, of the radius of the arc, the position of the two segments, and the *a priori* knowledge that the draughtsman defines the connection points making a perpendicular line from the centre of the arc to the segments. Some examples of $\alpha - \omega$ notation [7], description of the used algorithms, are given as a tool for their synthetic and strict communication.

5. CONCLUSIONS

A set of testing figures has been created to check the algorithms. Figure 3 represents a test where a set of stages of different dimensions has been drawn. The aim is to test the algorithm on a set of fillets with variable radii and with variable thicknesses of the line. The digitization step has been chosen in order that at least two fillets of minimal radius give origin to a pattern which is under the minimal identification threshold. All the other arcs should have been recognized.

The result after the identification phase is shown in Figure 4(a), where only the surely identified structures are displayed. Some observations can be made:

(1) One of the fillets, with a radius just superior to the determination threshold, has not been recognized.
(2) Dashed lines are not properly recognized because the dimension of the short segment is too small in comparison to the digitizing step.

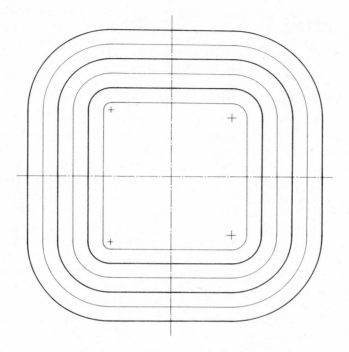

Figure 3. Original drawing.

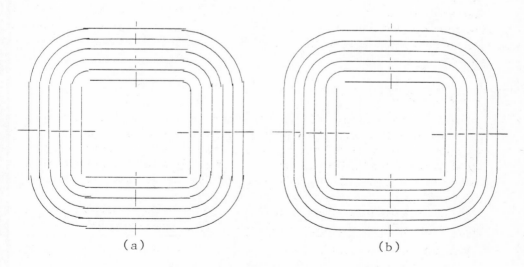

(a)   (b)

Figure 4. (a) First step of the restoration.
(b) Second step of the restoration.

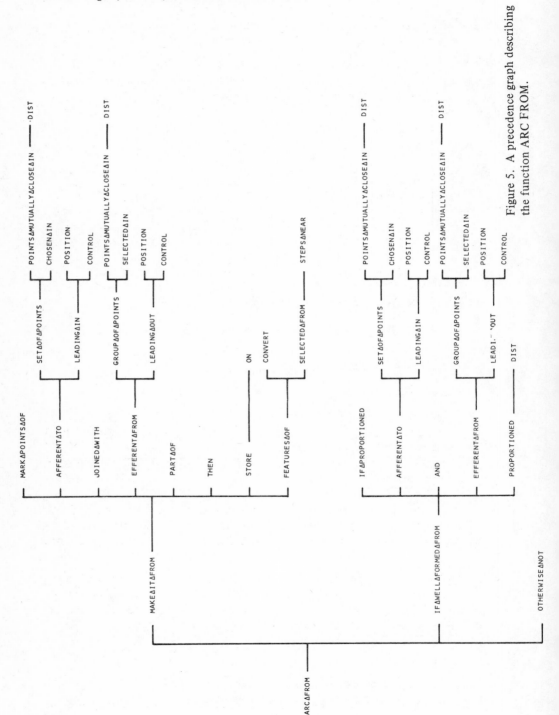

Figure 5. A precedence graph describing the function ARC FROM.

(3) It has been noticed how it is difficult to individuate the point of tangency.

(4) Information relating to the line thickness is not used in the output phase.

Results of the phase of geometric correction are shown in Figure 4(b).

Flaws in the determination of point of tangency were overcome by applying geometric considerations relating to the continuity characteristics of the different tracts of the contour.

APPENDIX

The $\alpha-\omega$ notation

The $\alpha-\omega$ notation is a synthetic one, introduced by Iverson [7] in order to describe algorithmic activities in a univocal way. This notation allows the building of formal demonstrations and every $\alpha-\omega$ well-defined formula corresponds to a performable APL program and can be interpreted in a similar, well-defined way. The notation uses APL symbolism to define and utilize the functions and primitive operators. But it is possible to define and give a name to other functions beside those defined by APL symbols. A function is defined by giving the name that identifies it, followed by a colon and then by the expression that defines it. For example, the expression

ARC△FROM: MAKE△IT△FROM α:
            IF△WELL△FORMED△FROM α:
            OTHERWISE△NOT

interpreted in $\alpha-\omega$ notation, means:

A function called ARC△FROM is defined.

This function depends on a single parameter; the function syntax will be ARC△ FROM SEED where SEED is a suitable variable.

The function is conditional: so the proposition IF△WELL△FORMED△FROM, that is defined elsewhere, will be carried out before.

If the proposition is true (and the name supposes that in this case the arc construction is possible) the left segment of the proposition, that is MAKE△IT△FROM, is carried out and this has to be the name of a function elsewhere defined.

If the proposition is false, the function OTHERWISE△NOT, niladic, is carried out.

A function of just one argument is defined through an expression and the argument is α or ω. A function of two arguments is defined through an expression and the left argument is α and the right one is ω. A function with no argument is called niladic. Functions with different numbers of arguments are not defined.

The definition of a conditional function follows this scheme:

name: left expression: proposition: right expression

A proposition is an expression that can only assume the values "false" or "true".

As a convention the function value will be the value of the left expression if the proposition gives "true" as the result, and the value of the right expression if the proposition gives "false" as the result.

The set of functions required to check the existence of an arc, after a seed is recognized, is listed below:

```
ARCΔFROM SEED

∏IO←0

ARCΔFROM : MAKEΔITΔFROM α : IFΔWELLΔFORMEDΔFROM α : OTHERWISEΔNOT

MAKEΔITΔFROM : (MARKΔPOINTSΔOF (AFFERENTΔTO α) JOINEDΔWITH (EFFERENTΔFROM α)

JOINEDΔWITH (PARTΔOF α)) THEN STORE FEATURESΔOF α

IFΔWELLΔFORMEDΔFROM : (IFΔPROPORTIONED AFFERENTΔTO α) AND (IFΔPROPORTIONED EFFERENTΔFROM α)

AND PROPORTIONED α

OTHERWISEΔNOT :→ 0

MARKΔPOINTSΔOF: ((ρGSTRINGA)ρ0)[α]←1

AFFERENTΔTO : SETΔOFΔPOINTS LEADINGΔIN α

JOINEDΔWITH : α, ω

EFFERENTΔFROM : GROUPΔOFΔPOINTS LEADINGΔOUT α

PARTΔOF : α[1], α[2]

THEN : α

STORE : VET ON α

FEATURESΔOF : CONVERT SELECTEDΔFROM α

IFΔPROPORTIONED : (ρα)>MEA

AND : α∧ω

PROPORTIONED : (+/(SXY[α[0];1 2] DIST SXY[α[1];1 2]) . (SXY[α[2];1 2] DIST SXY[α[3];1 2])>TMN

LEADINGΔIN : α[1]-POSITION φ CONTROL (lα[1], α[0 1])

SETΔOFΔPOINTS : (POINTSΔMUTUALLYΔCLOSEΔIN α) CHOSENΔIN α

LEADINGΔOUT : A[3]+POSITION CONTROL (((α[3]+1)+1ρGSTRINGA), α[2 3])

GROUPΔOFΔPOINTS : (POINTSΔMUTUALLYΔCLOSEΔIN α) SELECTEDΔIN α

CONVERT :⁻1+'0123456789'l(αz' ')/α

SELECTEDΔFROM : (STEPSΔNEAR α)≠GARC[;4+15]

ON : '' CONFORM AND CATENATE ANY STRUCTURES ''

DIST : '' COMPUTES ORDERED DISTANCE BETWEEN ELEMENTS OF
            TWO SETS OF THE SAME NUMBER OF POINTS WHOSE
            COORDINATES ARE IN α AND IN ω RESPECTIVELY ''

POSITION : (∧\α)/lρ α

CONTROL : ∨/(GSTRINGA, [0.5]⁻1 φ GSTRINGA)[⁻2+α] ε GSTRINGA[⁻2+α]

POINTSΔMUTUALLYΔCLOSEΔIN : (SXY[α;1 2] DIST SXY[α+1;1 2])≤TLA

CHOSENΔIN : (φ∧\φα)/ω

SELECTEDΔIN : (∧\α)/ω

STEPSΔNEAR : (+/GARC[;1 2] ε GSTRINGA[α[1 2]])=2
```

REFERENCES

1. S. Bianchi et al.: "A computer-aided system for interactive definition of digital-image interpretation", 2nd Conference on Image Analysis and Processing, Bari, 1982 (not included in this volume).
2. U. Cugini, P. Micheli, P. Mussio: "A structural approach to the restoration and vectorizing of digitalized engineering drawings", Pattern Recognition and Image Processing Workshop.
3. M. Gangnet, S. Coquillart, J. C. Haiat: "Numerisation par camera de plans de bati-ments", proceedings of MICAD 1982, Paris, Sept. 1982, pp. 100–115.
4. S. Kakumoto, Y, Fujimoto, J. Kawasaki: "Logic diagram recognition by divide and synthesize method", in AI and PR in CAD, Proceedings of the IFIP Working Conference, Grenoble, March 1978, pp. 457–476.
5. U. Cugini, A. Della Ventura, P. Mussio, A. Rampini: "A system for the automatic digitization of technical drawings", 1st Conference on Image Analysis and Processing, Pavia, Oct. 1980, Proceedings, pp. 101–108.
6. J. F. Harris, J. Kittler, B. Llewellyn, G. Preston: "A modular system for interpreting binary pixel representations of line-structured data", Proceedings of the NATO Advanced Study Institute, Oxford, March 1981, pp. 311–351.
7. K. G. Iverson, "Elementary Analysis", APL Press, 1977.

U. Cugini
Politecnico di Milano
Dipartimento di Meccanica
Piazza Leonardo da Vinci 32
20133 Milano
Italy

G. Ferri, P. Micheli, P. Mussio, and M. Protti
Istituto di Fisica Cosmica del CNR
Via Bassini 15/A
Milano
Italy

32 Bidimensional Fourier transform of very long baseline interferometry data

C. De Marzo, C Fanti-Giovannini, C Nuovo and G Sylos-Labini

1. INTRODUCTION

Aperture synthesis is a powerful method used by radioastronomers to obtain high-resolution maps of the radio sky. The use of aperture synthesis in linked interferometers is well tested and permits angular resolutions unreachable with single dish telescopes.

In the early 70s, however, a nonlinked interferometry technique was first used. It was named Very Long Baseline Interferometry (VLBI) because of the long distance between different antennas of the array. With this technique, resolutions even higher than those permitted by standard arrays were reached, but the fantastic limit of 1 m.a.s. (10^{-3} arcsec) must have a price. In fact, in VLBI observations all the principal deficiencies [1] of the aperture synthesis are enhanced. Some of these are intrinsic in the method, e.g., the spotty sampling of the U-V plane, that is well controlled in the layout of a linked array but is quite uncontrollable in VLBI observations, because of the random positioning of the antennas. Also, the phase instability, related with electronic fluctuations and variations of ionosphere state, becomes a harder problem to solve in VLBI observations. In conclusion, more attention is needed in the choice of the procedure to be used, in order to reduce the effects of the above-described drawbacks on the final map.

This paper concentrates on the choice of the right procedure. In the next section we shall describe limits and advantages of the standard fast Fourier approach to this problem. Then we shall discuss the solution we propose, and last, but not least, we shall present an extensive cross test of the two procedures.

2. THE FFT APPROACH

In order to perform an FFT on a set of VLBI data one must first override the problem of a nonregular spaced data set. In fact the sampled data are disposed on irregular paths on the U-V plane. This first problem is closely connected with the above discussion and could have dramatic effects on VLBI observations.

A standard way of resolving this problem is first to convolve the U-V coverage with a regular Gaussian-shaped grid and then to perform the transform on this resampled data set. The main disadvantage of this operation is the lowering of the $V(u, v)$ at the edges of the field of view. Any solution to this problem will lead to a lower S/N ratio at the edges or to an increasing of map dimensions.

The most annoying effects of an FFT operating on undersampled data are perhaps the grating sidelobes caused by the presence of spatial frequencies in the sampling plane (i.e., the U-V plane). It is possible to reduce — but not completely remove — this unwanted effect by oversampling the x-y plane (i.e., by filtering the unwanted spatial frequencies), at the expense of map dimensioning. A way to obtain a better function, which permits a more presentable map with lower sidelobes, is to introduce an apodizing function which permits a different weight for different points in the map; but even this operation could have uncontrollable effects on successive elaborations of the map.

As explained above, all the limits of the FFT approach (described fully in ref. [2]) are mainly enhanced by the poor U-V coverage in the case of the VLBI observations. These limits can be overridden by the use of a cleaner inversion procedure, such as the one we propose in the next section.

3. A PRACTICABLE SOLUTION: A DFT ALGORITHM

The only algorithm that can be applied on the visibility function $V(u, v)$, without a previous gridding of the data, is the direct Fourier transform (DFT).

In fact:

(1) DFT does not need a regular spaced grid in the U-V plane.
(2) DFT, because of the random geometry of the sampling path, does not produce grating sidelobes.

The disadvantage of DFT as opposed to FFT is in the computing time. Therefore, to reduce the time-consumption of DFT without any loss in the cleanness of the map, we propose this accurate but slow algorithm to work on a fast computing device such as an array processor. In this way, the execution times of the two procedures become comparable. DFT takes about 1 min to operate on a 50x50 pixel map, whereas FFT takes only 30 s. We hope to improve DFT performances further by optimizing I/O operations between the host computer, a VAX 11/780, and the array processor, an AP-120 B.

4. THE TESTS: DFT *v.* FFT

To test the two procedures we had two orders of problems:

(1) How to have a reference data set to compare the performances of the two procedures?
(2) The choice of objective parameters to evaluate the two methods.

The first problem was resolved by the use of a simulation program, that artificially — but in a realistic way — added random noise on the $V(u, v)$ data. This program also provides an easy way of changing:

(1) The shape of the source emission;
(2) the S/N ratio; and
(3) the U-V coverage (i.e., telescopes, locations, etc.).

The answer to the second question is more difficult. The reason for these difficulties is linked to the complexity of the point spread function (dirty beam) of a VLBI network. By direct transforming the $V(u, v)$, one obtains the function $I'(x, y)$ (dirty map) where:

$$I'(x, y) = I(x, y)B(x, y) \tag{1}$$

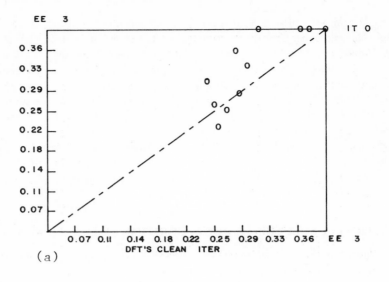

(a)

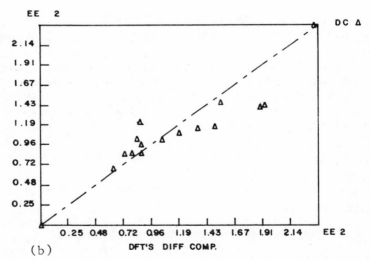

(b)

Figure 1. (a) Number of iterations of CLEAN using FFT v. DFT as function of source morphology. (b) Number of delta components used by CLEAN, in FFT v. DFT versions, as a function of morphology. (c) Differences in flux reconstruction, in per cent, among FFT and DFT versions of CLEAN, for different tests.

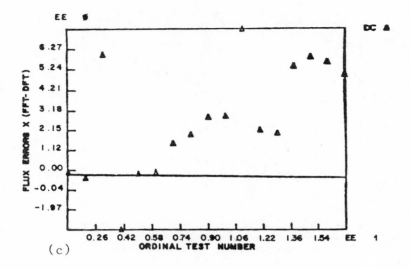

(c)

Therefore, in order to obtain the brightness distribution of the source, I(x, x), one has to deconvolve from (1) the P.S.F. of the array B(x, y). Otherwise, due to the high lobe level in the I'(x, y), judging the merit of one of the procedures would have been impossible.

To produce the I(x, y) from I'(x, y), the radioastronomers use an iterative decon-volution procedure named "CLEAN" [3, 4]. The estimation criteria we used during our test all referred to the CLEANed map. They are essentially:
(1) The accordance of the shape of the sources produced by the DFT and FFT by the start model.
(2) The value of total flux residuals.
(3) The number of iterations.
(4) The number of different components found.
The last two points are in one sense a measure of the difficulties afforded by CLEAN to reduce the maps.

All these criteria were applied in two different series of tests. The first was performed, varying the morphology of the model source, with a fixed U-V coverage; the second, varying the U-V coverage.

5. TESTS ON SOURCE MORPHOLOGIES

During these first series of tests we did not notice any great differences in the reproduction of the shape of the model by the two procedures. On the other hand some real advantages for DFT appeared regarding the other features described above. Figure 1(a) clearly shows a sensibly smaller number of iterations in CLEAN than for the maps produced by the DFT. Looking at the number of different components (i.e., the number of delta functions used by CLEAN to describe the model) it is important to point out (Figure 1(b)) that in few cases the FFT has a smaller number of components, but in this case the flux esti-mation performed by DFT is much better attained (Figure 1(c)).

(a)

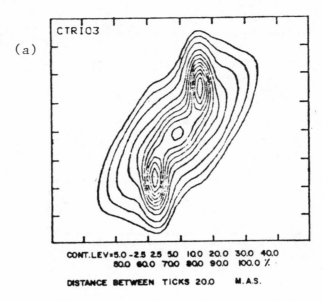

CONT.LEV=5.0 -2.5 2.5 5.0 10.0 20.0 30.0 40.0
50.0 60.0 70.0 80.0 90.0 100.0 %

DISTANCE BETWEEN TICKS 20.0   M.A.S.

(b)

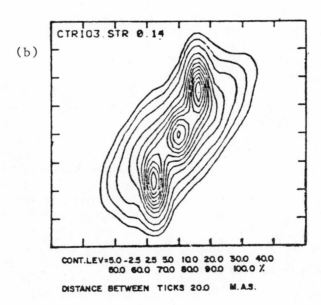

CONT.LEV=5.0 -2.5 2.5 5.0 10.0 20.0 30.0 40.0
50.0 60.0 70.0 80.0 90.0 100.0 %

DISTANCE BETWEEN TICKS 20.0   M.A.S.

Figure 2. (a) Shape of the input model.
(b) Reconstructed model by DFT. (c)
Reconstructed model by FFT.

(c)

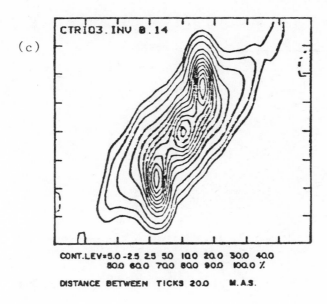

CONT.LEV=5.0 -2.5 2.5 5.0 10.0 20.0 30.0 40.0
50.0 60.0 70.0 80.0 90.0 100.0 %

DISTANCE BETWEEN TICKS 20.0 M.A.S.

6. TESTS ON THE U-V COVERAGE

As pointed out in the first section of this paper, there is very poor control on U-V coverage for VLBI experiments. Moreover, in observations involving antennas spread on more than one continental base, the enhancement in resolution results in big holes in the sampling path. This situation was fully simulated during our tests. In critical conditions — a network consisting of all European stations including Haystack and Crimea but excluding Madrid — DFT exploits a better reconstruction of the shape of the source model (Figures 2(a)–(c)), more evident with a lower S/N ratio (Figures 3(a) and (b)). In all cases CLEAN works better with maps produced by DFT. In this case it uses less iterations to converge (Figure 4(a)) and less different components to describe the final map. In the case of very bad U-V coverage both procedures generally fail in flux estimation, but the errors on the flux are smaller in the maps produced by DFT (Figure 4(b)).

7. CONCLUSION

At this moment we can say that we have confirmed all the advantages of DFT compared with FFT. Furthermore, the worse the data, the bigger the advantages for DFT. In future we hope to optimize the running time of DFT and extend our tests to well-known real sources.

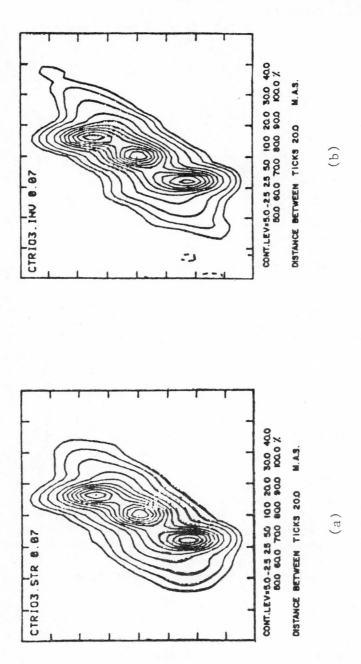

Figure 3. (a) Reconstructed model by DFT with a signal-to-noise ratio lower than Figure 2(b). (b) Reconstructed model by FFT with a signal-to-noise ratio lower than Figure 2(c).

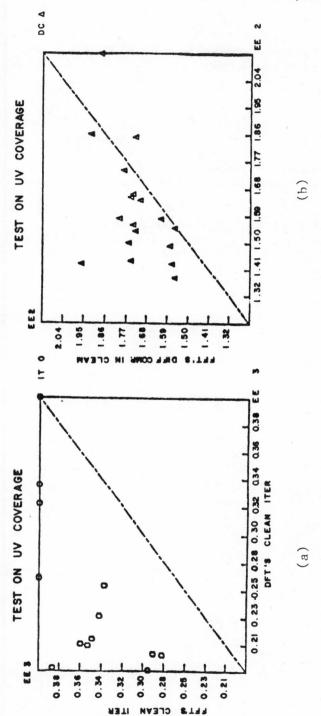

Figure 4. (a) Number of iterations of CLEAN using FFT *v.* DFT as a function of U-V plane coverage. (b) Number of delta components used by CLEAN, in FFT *v.* DFT versions, as a function of U-V plane coverage.

ACKNOWLEDGEMENT

We would like to thank Maria Massi of the Physics Department of Florence University who supplied the procedure for simulating U-V data.

REFERENCES

1. Fomalont E. B.* "Fundamental and deficiences of aperture synthesis".
2. Fomalont E. B., Proc. IEEE, 1973, 61, 9: "Earth-rotation aperture synthesis".
3. Scawartz V. J.* "The method CLEAN: Use, misuse and variations".
4. Hogböm J. A., 1974: Astron. Astrophys., Suppl. 50, 19.

* From Proc. Conf. on "Image formation from coherence functions in astronomy" Reidel Pub. Co. Dordrecht, 1979.

C. De Marzo, C. Nuovo, and G. Sylos-Labini   C. Fanti-Giovannini
Dipartimento di Fisica   Istituto di Radioastronomia del CNR
GNCB-CNR   Bologna
Università di Bari   Italy
Italy

33 A project for faint object discrimination in astronomy

M L Malagnini, M Pucillo and P Santin

1. INTRODUCTION

In astronomy, different methods for image classification are currently being sought, so as to face the problem of handling large amounts of digital data coming either from space-borne instrumentation or from photographic plates (plates of standard formats may contain images of thousands of objects, each of which require different kinds of information such as positions, magnitudes, categories). The tendency is to implement the procedures in overall systems which enable one to perform all the main operations, from the data preparation phase up to the final classification, under the control of a central computer [1–3].

The project we present here refers to the overall process, and is under development at the Astronet pole of the Trieste Astronomical Observatory. The automatically controlled digitization of the input images will be actuated as soon as the PDS microdensitometer is operated under the control of the main computer system.

The methodology relative to the phases of data analysis and classification is derived from previous works [4–6], which led to the development of a fast procedure. This procedure is already running, in Batch mode, on the CDC Cyber 720/170 computer of the Computer Centre of the University of Trieste. The main advantages provided by the implementation of this procedure in the Astronet environment lie in the improvements in the whole process made possible by interactive and pictorial facilities available there. These facilities are especially helpful during the phase of feature extraction, as different choices can be tested very quickly by interactive graphic representation of the results.

The classification procedure is based on the evaluation of suitable parameters, extracted from the modified co-occurrence matrix [5]. The overall procedure is organized in modules, linked in a sequence of interactive and iterative subprocedures, each of which is general enough to be applied to different kinds of analysis and to different classes of images. To illustrate how the method works, we present an application of it to the discrimination of faint astronomical objects (stars and galaxies).

2. THE SYSTEM

A block diagram of the system under development at the Astronet pole of the Trieste Astronomical Observatory is shown in Figure 1. It refers to both the hardware environment and to the software structure, which will be described in the following subsections.

344 *M L Malagnini, M Pucillo, P Santin*

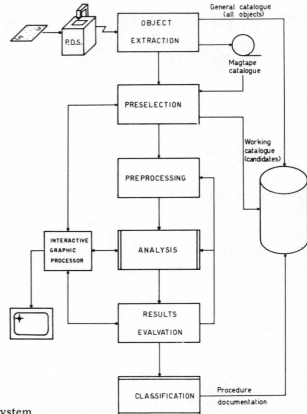

Figure 1. Block diagram of the system.

2.1. Hardware environment

The configuration of the Astronet pole of the Astronomical Observatory of Trieste is based on a VAX-11/750 processor connected to a standard set of peripheral devices [7]. It offers some special facilities, including a PDS 1010 A microdensitometer, together with many interactive graphical and pictorial systems, as shown in Figure 1.

The PDS 1010 A microdensitometer will be connected to the VAX system through an intelligent interface, under development at the local laboratories, built around a dual processor architecture provided with local memory and processing power. This approach will enable the application programs, operating in the VAX system, to avoid having to deal in any way with PDS operations. At the same time it will enable them to use the interface processing power to execute some preprocessing on the raw data, such as local filtering, thresholding and simple image segmentation.

The final version of the discrimination procedure will make full use of the micro-densitometer power during the initial stages, i.e., during the phases of object detection, preselection, and acquisition.

2.2. Software structure

The procedure has been designed in a highly modular fashion, and is run under the control of a monitor-like main program that allows the user to be fully involved in every step of the procedure. This is accomplished by making great use of the graphical and pictorial interactive capabilities of the processing system, which provides a complete and updated information on the procedure status in any moment at user request.

Moreover, when applicable, this precise view of the status of the procedure may be followed by an interactive session, which enables the user to fine tune the parameters concerning the module under examination. A log of all user and system operations is maintained in a general catalogue, which is preset at procedure start with the initial parameter values for each object, and keeps track of all intervening modifications, thus allowing the user to reconstruct his interactions with the procedure and revert to it at any point.

3. DATA PREPARATION

This phase, preliminary to the subsequent analysis and classification steps, includes both the acquisition and digitization of the raw data and, to some extent, image preprocessing used to enhance and/or restore the pictures, to perform geometric corrections, and so on. The first of these operations is standard for astronomical photographic images, while preprocessing is an application to specific astronomical problems of general processing techniques.

3.1. Input data

The physical objects to be recognized are galaxies in a very distant cluster, whose images on the photographic plate look like those of nearby stars projected on the same area of the sky. Nanni *et al.* [8] scanned a 4 cm × 4 cm area of such a plate, using a PDS micro-densitometer with a window´of 10 μ × 10 μ. From the array of 4000 × 4000 pixels (10 bits per pixel), they extracted about 2000 objects. They performed this detection by thresholding above the local background, taking into account only the regions of at least 10 connected pixels. From their general catalogue, a subset of 450 objects has been selected and kindly provided to us on magnetic tape. Each object has been extracted with its surrounding background as an array of 49 × 49 pixels, centred on the centroid of the object itself (the total image dimensions are about three times larger than the average extension of the objects of this sample).

An example of what the objects look like is given in Figure 2, where four images are shown as they appear on the video display, after compression of the original 1024 transparency levels into 64. The two upper frames contain good-quality images of a star (left) and of a galaxy (right), respectively. The lower right-hand frame shows a streak on the plate, and the lower left-hand one contains almost entirely background, in the sense that there are too few significant pixels to permit any reliable classification.

At the present stage of our project, we have chosen to delete all spurious images (such as streaks) from the working catalogue by visual preselection, using a colour video display. Visual inspection of original pictures is also used to select the training sets that

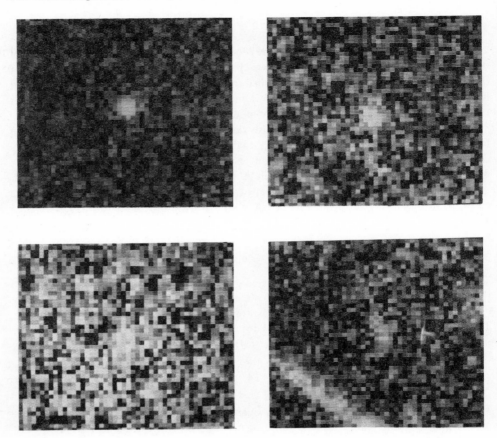

Figure 2. Examples of images of a star
(upper left), of a galaxy (upper right), of
background (lower left), and of a streak
on the plate (lower right).

are representative of the classes to be defined. Unfortunately, unlike in the case of remote-sensing imagery of terrestrial regions, astronomical pictures cannot be compared with the original sources directly; therefore, we need to resort to some other kind of information. For nearby galaxies, peculiar morphological structures like spiral arms or elliptical shapes may be evident, but this is not the case for very distant galaxies. The only fundamental fact we can rely upon is that stars are point sources, while galaxies are extended sources so that their images look, respectively, compact and diffuse (Figure 2, upper left and upper right). In the presence of noise, it is hard to discriminate between the two cases, so a decision has to be taken regarding the maximum noise tolerable for achieving reliable classification.

Therefore, we have chosen to define not only the training sets of stars and of galaxies, but also of the "background". The selection is performed mainly on the basis of: (i) the appearance of the images on the video display, (ii) the characteristics of the image histograms, (iii) the effects of grey-scale modifications, (iv) zooming, and (v) thresholding.

The preselection module displays the image in a standard format, while all the other operations are selected interactively by the user, and can be iterated, starting from any level and asking for different operations. The interactive option is selected for deciding whether an object has to be included in a training set or not. The visual preselection, applied to the 450 input images, resulted in a working catalogue of 399 images to be analyzed (51 spurious images have been excluded), of which 150 have been selected for the training sets (30 "background", 50 "galaxy", and 70 "star" images).

3.2. Preprocessing

First, a plus-sign median filter (5 pixels long) is applied to eliminate locally the spikes of the signal. Alternatively, we can apply two-dimensional digital filters [9]. Then a logarithmic conversion from transparencies to densities is applied to the filtered images, which are reduced to 45×45 pixels to avoid border effects caused by the filter. Lastly, the density levels are compressed into an $Ng = 16$ grey-level scale. This compression is used not only to reduce the computational time (the number of operations increases with Ng, as will be seen later), but also to obtain significant statistics.

4. ANALYSIS

To analyze the data and to derive the classification parameters, we apply the method of modified co-occurrence matrix introduced by Malagnini and Sicuranza [4] for the discrimination of astronomical objects. This modified co-occurrence matrix differs in two main respects from grey-tone spatial-dependence matrices, used to derive textural information [10], and successfully applied to different kinds of imagery [11]. First, as the images under study present a roughly circular symmetry, we ignore orientations in the computation of this matrix.

Second, as suggested by preliminary tests, the relative frequencies defining the entries of the matrix are suitably weighted, according to the distance between two neighbouring pixels. As weighting functions, rotated versions of one-dimensional B-spline functions are assumed.

Let us indicate with $B_h(r, d)$ such two-dimensional spline functions, where h indicates the order of the function, and r is the Euclidean distance between two neighbouring pixels. The distance r ranges from zero to a maximum value D, given by:

$$D = (h+1) \times d, \tag{1}$$

where d is a parameter which determines the actual interval D encompassing all the pixels to be considered as neighbours. The values ($h = 1$, $d = 2$) have been found to provide a good trade-off between performance and computational complexity. Therefore, each entry p(i, j) of the modified co-occurrence matrix, P, is the weighted number of times that two neighbouring pixels have, one, the grey level i, and the other, the grey level j (i, j = 1, Ng). From such a matrix, a set of textural features can be extracted, of the kinds listed in [11].

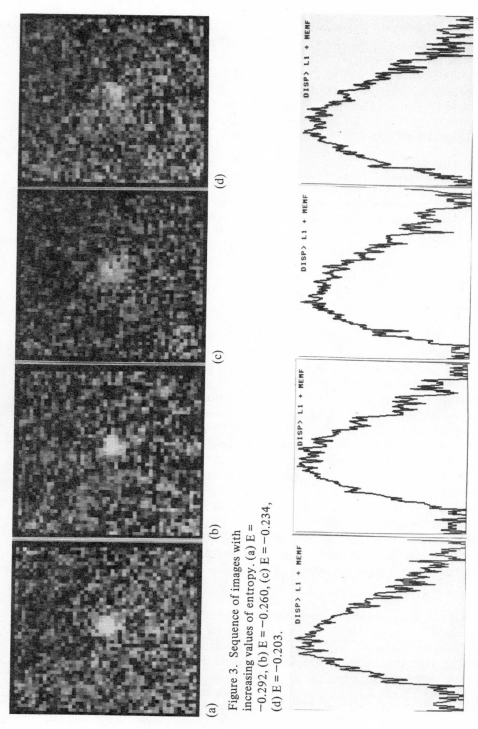

Figure 3. Sequence of images with
increasing values of entropy. (a) E =
−0.292, (b) E = −0.260, (c) E = −0.234,
(d) E = −0.203.

Figure 4. Histograms of the images of
Figure 3.

We have chosen the following strategy:

(1) First, to evaluate a parameter for automatic discrimination of the objects for which a reliable classification can be achieved.

(2) Second, to use the training sets to "learn" the form of the parameters best suited for characterizing the different categories.

These two moments are illustrated in Subsections 4.1 and 4.2.

4.1. The entropy feature

For noisy images, it may be desirable to perform an automatic selection in order to avoid passing to the classifier images that are unclassifiable because of a too low signal-to-noise ratio, and to avoid contaminating the training sets. A valuable feature for doing this is the entropy parameter, defined as follows:

$$ E = - \sum_{i}^{Ng} \sum_{j}^{Ng} p(i, j) \log p(i, j) \qquad (2) $$

(the actual values of E are normalized to Ng and to the image dimensions).

As an example, the sequence (a)–(d) of Figure 3 shows that increasing noise in the images corresponds to increasing values in entropy (E = −0.292, −0.260, −0.234, −0.203, respectively). The same conclusion can be drawn by comparing the image histograms given in Figures 4(a)–(d), which correspond to the images of Figures 3 (a)–(d), respectively. Therefore, the entropy feature is assumed as a measurement of the influence of the background on the image. Next, a threshold for E is defined, on the basis of the statistical behaviour for the training set labelled "background" by visual preselection.

The entropy distribution for this training set is displayed on the interactive graphic monitor, together with the mean value and with the values taken at an assigned confidence level (95 or 99%). The lower value taken at the chosen confidence level is assumed as the threshold value. Then, the images of the training sets of stars and of galaxies are tested against this threshold, and those having a lower value for E are eliminated from the training sets.

The entropy distribution for the training sets "star", "galaxy", and "background" are plotted together in Figure 5. It appears that stars are clearly separated from background, while galaxies verge on the background, as a consequence of the fact that diffuse images tend to have large tails merging into the surrounding background.

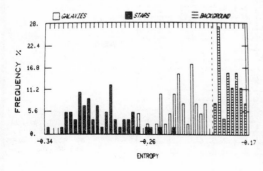

Figure 5. Entropy distributions for the training sets of stars, galaxies, and background (relative frequencies in ordinate).

4.2. Classification parameters

Once the training sets of stars and galaxies have been redefined by means of the test on entropy, we apply a feature selection procedure to select the classification parameters.

First, average modified co-occurrence matrices are computed for the two training sets, $\bar{P}_{stars}$, $\bar{P}_{galaxies}$, and then the difference matrix $\bar{C} = \bar{P}_{galaxies} - \bar{P}_{stars}$ is displayed. To facilitate the feature selection, the difference matrix is given in pictorial form (Figure 6): the ordinate represents grey levels from 1 to 16 (bottom to top), the abscissa represents the absolute difference in grey levels, from 0 to 15 (left to right). Blank cells are seen

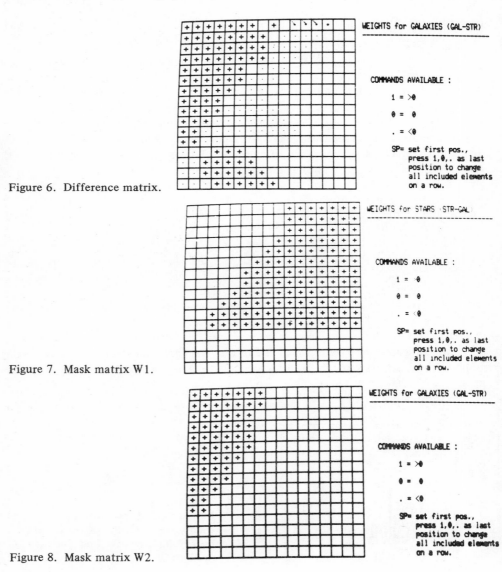

Figure 6. Difference matrix.

Figure 7. Mask matrix W1.

Figure 8. Mask matrix W2.

when the entry $c(i, |i-j|)$ of the difference matrix C is zero or, in absolute value, lower than a prefixed percentage of the corresponding entries in $\bar{P}_{stars}$ and $\bar{P}_{galaxies}$. Plus signs represent positive entries, and points represent negative entries, whenever the blank condition is not satisfied.

Roughly speaking, we can say that galaxy features dominate at small grey-level differences (left part of Figure 6), and star features dominate at high grey-level differences (right part of Figure 6). These results are related to the diffuse aspect of galaxies, and to the compactness of stars. This behaviour is checked on different random samples derived from the training sets, in order to test the stability of the results. This test permits us to conclude that: (i) the lower part of Figure 6 (grey levels in the range 1–5) fluctuates with the quality of the sampled images, evaluated by their entropy values; and (ii) the plus sign and point parts are almost stable.

Therefore, we choose a mask matrix, W1, to enhance the features corresponding to galaxies, assigning the value 1 to the entries in the positions of the most representative pixels (plus sign in Figure 6) for this class of objects, and 0 elsewhere (Figure 7). The same kind of operation is made for a mask matrix, W2, enhancing stellar features (Figure 8).

The masks W1 and W2 are then applied to the modified co-occurrence matrices, as weighting matrices, and two classification parameters are computed in the form:

$$A_k = \sum_i^{Ng} \sum_j^{Ng} \frac{p(i, j)}{f(i)} \, w_k(i, j) \qquad (k=1, 2) \tag{3}$$

(the entries $p(i, j)$ are normalized to the absolute frequencies $f(i)$ of each grey level i).

The form of the parameters and their number obviously depend on the typical features of the objects to be classified. For simplicity, and to reduce the computational cost, we tend to limit the number of parameters, so long as discrimination can be achieved. The selection of features in the form given in (3) produces almost unimodal distributions for the training sets, as shown in Figures 9 and 10; therefore, the final classification of all the images of the working catalogue is accomplished on the basis of these parameters, A_k, as described in the next section.

5. CLASSIFICATION

From the analysis of the training sets, it emerges that the three features, viz. the entropy, a smoothness feature represented by A1, and a contrast feature measured by A2, are sufficient to characterize noisy images, stars, and galaxies. Therefore, we apply the standard procedure summarized here:
(1) median filtering of the raw data,
(2) logarithmic conversion from transparencies to densities,
(3) grey-scale normalization,
(4) computation of the modified co-occurrence matrix, and
(5) evaluation of E, A1, and A2.

Then, two alternatives follow:
(1) $E \geq$ threshold: the image is defined as "unclassifiable".
(2) $E <$ E threshold: the image passes to the classification step.

A linear classifier, computed from the means and standard deviations of the training set, is chosen as a decision boundary in the plane of the parameters (A1, A2). A linear classifier, as is well known, is not, in general, the optimum one, and more complex ones will be introduced in future. The distribution of A1 and A2 for the images passing the test on

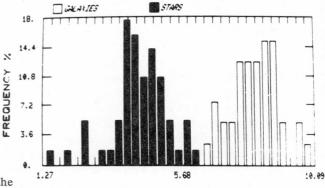

Figure 9. Histogram of A1 for the training sets (stars and galaxies).

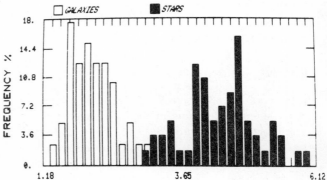

Figure 10. Histogram of A2.

Figure 11. Distribution of A1 and A2 for all the images passing the test on entropy. Ellipses are equiprobability contours derived from the training sets.

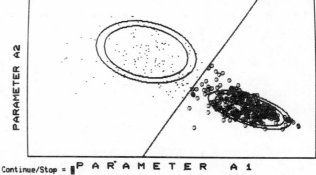

entropy is shown in Figure 11, where the decision boundary is also plotted. Stars are represented by points, and galaxies by circled points (to the left and to the right of Figure 11, respectively). Figure 11 also shows the equiprobability contours computed at the 95 and 99% confidence levels of the distributions of the training sets. The hypothesis of two-dimensional normal distribution is assumed. Under this hypothesis, a confidence level can be associated with the category assigned to each image of the working catalogue.

6. CONCLUSIONS

The present status and the possible applications of a project designed to perform a set of operations, varying from plate digitization to a standard classification procedure, have been illustrated. The overall structure has been tested, starting from digitized data, in particular for the phases of feature extraction, exploiting all the pictorial and interactive facilities of the system. The method of analysis through textural property description has proved efficient for discriminating faint astronomical objects. Different methods of feature selection, algorithms for nonlinear decision boundaries in multiclass problems, and, possibly, semiautomatic procedures for the definition of the training sets, require further development.

ACKNOWLEDGEMENTS

The authors wish to thank G. L. Sicuranza of the Istituto di Elettrotecnica ed Elettronica dell'Università di Trieste, for his kind collaboration in the development of this methodology. We also thank D. Nanni, G. Pittella, D. Trevese, and A. Vignato, of the Osservatorio Astronomico di Roma, for providing the digital images used for the present test, and for useful discussions. We also wish to acknowledge the support given to this project by G. Sedmak. This work has been partly supported by a CNR-GNA contract.

REFERENCES

1. N. M. Pratt, The COSMOS Measuring machine, *Vistas in Astronomy* 21 (1977) 1–42.
2. J. F. Jarvis and J. A. Tyson, FOCAS – Faint Object Classification and Analysis System, *Proceedings of the SPIE* 172 (1979) 422–428.
3. W. L. Sebok, A Faint Galaxy Counting System, *Proceedings of the SPIE* 264 (1980) 213–221.
4. M. L. Malagnini and G. L. Sicuranza, A statistical approach to the image discrimination problem, in: G. Sedmak, M. Capaccioli, R. J. Allen (eds.), *Image Processing in Astronomy*, (Trieste, Italy, 1979) 319–328.
5. M. L. Malagnini and G. L. Sicuranza, Second-order statistics for image classification, in: P. Crane and K. Kjärs (eds.), Proceedings of the *ESO Workshop on Two Dimensional Photometry* (Noordwijkerhout, Holland, November 21–23, 1979) 383–390.
6. P. Battistini, F. Bonoli, A. Braccesi, F. Fusi-Pecci, M. L. Malagnini, and B. Marano, Search for (globular) clusters in M31. I: Candidates in a 70' square field centered on M31, *Astronomy and Astrophysics Supplement Series* 42 (1980) 357–374.
7. M. Pucillo, Il polo Astronet di Trieste, Analisi degli sviluppi previsti, *Memorie della Società Astronomica Italiana* 53 (1982) 101–108.

8. D. Nanni, G. Pittella, D. Trevese, and A. Vignato, Photometry of galaxies in an extremely distant cluster, *Proceedings of the SPIE* 264 (1980) 180–185.
9. M. L. Malagnini and G. L. Sicuranza, Two-dimensional digital filtering techniques for astronomical plate processing, *Astronomy and Astrophysics* 70 (1978) 579–582.
10. R. M. Haralick, Statistical and structural approach to texture, *Proceedings of the IEEE* 67 (1979) 786–804.
11. R. M. Haralick, K. Shanmugan, and Its'Hak Dinstein, Textural features for image classification, *IEEE Transactions on Systems Man, and Cybernetics* 6 (1973) 610–621.

M. L. Malagnini M. Pucillo, and P. Santin
Osservatorio Astronomico di Trieste
Via G. B. Tiepolo 11-I
34131 Trieste
Italy

34 Discriminating between stars and galaxies in VDGC images

P Di Chio, S Di Zenzo and A Vignato

1. INTRODUCTION

The classification of a set of objects involves additional problems with respect to the classification of individual objects. We may tentatively divide them into three groups. In the first, one has problems such as that of the *a priori* probabilities and that of the biassedness of the collection with respect to the parent population. In a logical sequence, the second group comprises those problems which stem from the existence of constraints on the possible simultaneous classifications of the individual objects in the collection. In the third we have the problem of bringing into account the stochastic dependence which, in general, exists among the objects.

Problems of the first group are encountered in applications such as the estimation of cell populations in a cytology specimen or the recognition of galaxies in astronomical plates. These are examples of statistical analysis of biased samples.

We shall consider the problem of discriminating between stars and galaxies in deep astronomical plates. There are features which allow us to perform the discrimination: we shall review some of them and propose a new one.

For objects of magnitude up to 22 discrimination can be done with a relatively small probability of error and there is no need of the machinery of discriminant analysis (the magnitude is -2.5 times the decimal logarithm of the total luminosity of the object, hence the greater the value of magnitude, the fainter the object). Beyond 22, uncertainty affects discrimination. More data on the astronomical aspects of the problem are given in the following section.

We conclude this introductory section with some general remarks on the peculiar aspects of this problem as a pattern-recognition problem. It is a very special case of pattern recognition of collections. Problems of the second and third group in the above classification are absent, and we have a good case study for problems of the first group. We are given a set U of objects, namely, the objects in a given plate, and a set C of classes (or labels, if one prefers). C contains only two elements, say s and g (for stars and galaxies). To each object i there is associated a feature vector $x(i)$ (in our treatment we have just one feature). The two class probability density functions $f(x|s)$ and $f(x|g)$ are partially unknown. The objects in a plate can be ranked according to the total magnitude: as this increases, whatever the feature x may be, the estimation of the densities $f(x|s)$ and $f(x|g)$ becomes more and more unreliable. Besides, there is evidence that the degree of separation becomes worse.

So what we have is a problem which is trivial for low magnitudes (large objects) and gradually becomes untractable at high magnitudes (small objects). In a first approximation the set U of the objects in a plate could be considered as taken at random from a parent population, and the objects could be classified on the sole basis of their observed features. We shall consider U as a biased sample, which is far more realistic. Therefore, to recon-struct the nature of the bias shall be part of the job of the recognizer.

2. ASTRONOMICAL DATA

Discrimination between stars and galaxies is the final step in the data-reduction process of deep astronomical plates containing several very distant clusters of galaxies (VDCG). The plates we work with come from the 4 m Mayall telescope of the Kitt Peak National Observatory, and were taken in the North Galactic Pole direction, in a region surrounding the Kapteyn selected area 57; where enough stars with known values of the photoelectric magnitudes are present, thus providing an absolute magnitude reference system. The region near the North Galactic Pole was also chosen to avoid too heavy stellar contamin-ation, our interest being mainly in the distribution and variation of galaxies.

The discrimination algorithm described here has been applied to the Mayall Prime Focus (MPF) 1053 plate, with the following characteristic data:

Scale: 19''/mm.

Exposure time: 0.75 hours.

Seeing (HWHM): 0.8''.

Limiting magnitude: 25.0.

Band: J (centred at 4900 Å).

The plate was scanned, in transparency mode, by a PDS 1010A microdensitometer, with an aperture of 20×20 μm. A sampling rate (same as the distance between successive scan lines) of 15 μm has been adopted, corresponding to about one-third of the seeing. A digi-tal image of size 8000×8000, covering the unvignetted area of the plate, was obtained.

Before undertaking discrimination the data-reduction procedure went through the following steps:

(1) Background evaluation from the histograms of the transparency values over adjacent squares.

(2) Object detection via an iterative version of the algorithm in ref. [1] designed to avoid object merging.

(3) Plate calibration and photometry. The functional dependence of intensity on trans-parency has been deduced by the stellar profile method [2] and classical spot sensi-tometry. Photometry was done computing the magnitude of each object inside a suitable circular diaphragm. A complete description of the various steps of this procedure is given in ref. [3].

3. ASTROPHYSICAL AIMS

Our purpose is to obtain results on the distribution and properties of extragalactic objects. More specifically:

(1) To investigate the galaxy cluster structure in order to determine the presence and the characteristics of the secondary maximum in the density distribution of galaxies. A number of models have been proposed to explain this feature (see ref. [4] and the references quoted there); at present, however, it seems of importance to establish if the secondary maximum is a real feature or is due to observational effects.

(2) To test the cosmological models through differential counts of both galaxies and QSOs: the slope of the counts versus the magnitude is model dependent.

(3) To understand the colour variations of galaxies in clusters. After the discovery, by Butcher and Oemler [5], that a number of blue galaxies are present in the cores of two centrally concentrated distant galaxy clusters, in contrast with the properties of corresponding near clusters, there was confirmation of this result for many remote galaxy clusters [6], while other authors [7] found that this feature is, in one case, an effect induced by foreground galaxies. The presence in our field of some clusters, together with the availability of some plates in different colour bands, allows us to check the consistency of this characteristic.

(4) To compare our colour–colour results with the data given by Bruzual galaxy synthetic models [8], to obtain both galaxy types and redshifts.

4. DISCRIMINATING FEATURES

The features used to discriminate stars from galaxies are all based on the different nature of these objects: in fact, while stars are point-like objects, galaxies are diffuse, thus the image of a star on a plate is just the point-spread function of the imaging system, while that of a galaxy is the convolution of the actual object with the point-spread function.

Various features have been used as discriminators. Jarvis and Tyson [9] evaluate for each object the moments of the intensity

$$M_{ij} = \Sigma_A (x-\underline{x})^i (y-\underline{y})^j I(x, y), \tag{1}$$

where A is a suitable area around the centroid $P(\underline{x}, \underline{y})$ of the objects, and $I(x, y)$ is the intensity. From these moments is possible to compute a number of invariant quantities, e.g., the second moment

$$C_2 = (M_{20} + M_{02})/M_{00}, \tag{2}$$

which deal with the classification.

Kron [10] observes that star images have a maximum central compactness, and uses as a discriminator the quantity

$$r_{-2} = \frac{\int_0^{2\pi} d\theta \int_1^\infty \rho^{-2} I(\rho, \theta) \rho \, d\rho}{\int_0^{2\pi} d\theta \int_1^\infty I(\rho, \theta) \rho \, d\rho}^{-\frac{1}{2}}, \tag{3}$$

where $I(\rho, \theta)$ is the brightness of the object as a function of polar coordinates ρ, θ with origin in the centre of gravity of the object; this quantity weights the central light strongly. This discriminator avoids the overdependence of C_2 from noise in the outer parts of the objects, losing on the other hand the information contained there. Valdes [11] compares the object brightness profile with star templates.

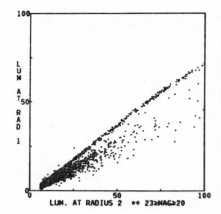

Figure 1. L_1-L_2 plot for MPF1053 objects. The magnitude can be obtained by $m=25-2.5 \log L$.

We found that the difference

$$Dm = m_1 - m_2 \tag{4}$$

between the magnitudes at diaphragms of radii r_1 (1″) and r_2 (2″) is a reliable discriminator between star-like and extended objects. This choice avoids certain problems found with other discriminators. Indeed, while still being strictly related to the profile, it seems to be computationally more efficient than the complete profile matching suggested by Valdes. Besides, with proper choice of r_2, it is possible to reduce the effect of the background noise fluctuations while not underweighting the periphery of the object. In the following we will call this feature 'radial magnitude difference' (RMD).

In order to avoid the logarithmic scale compression, we shall replace the subtraction of logarithms in eq. (3) by the ratio

$$\lambda = L_1/L_2 \tag{5}$$

between the corresponding luminosities. The luminosity for a star-like object is explicitly given by

$$L = I_0 \int_0^\infty f(r) r \, dr, \tag{6}$$

where I_0 is the central intensity and $f(r)$ is the point-spread function (circularly symmetric, hence dependent on r only)

$$f(r) = c_1 \exp(-r^2/\sigma_1^2) + c_2 \exp(-r^2/(\sigma_1^2+\sigma_2^2)). \tag{7}$$

The quantities c_1, c_2, and σ_2 of the plate are determined by the calibration method based on stellar profiles. Once these quantities are known, the expected value of the ratio $\lambda=L_1/L_2$ for point-like objects in the plate under investigation can be calculated. For plate MPF1053 we found a value of 0.76. For diffuse objects a smaller value is expected.

Figure 1 gives the plot of L_1 against L_2. There is a marked concentration of objects around a straight line with a slope equal to the stellar theoretical value; it is straightforward to identify these objects as stars. Well separated from that concentration (at least up to a certain critical value of luminosity) a more dispersed category of objects is present, with smaller values of λ; it seems natural to assume that these objects are galaxies. Beyond this limiting luminosity, the two categories of objects start to merge, thus making it impossible to draw a neat separation line between them.

There is evidence that the star-like objects are distributed around the theoretical value, with a variance which depends on noise and centroid errors while galaxies have a more disperse distribution due to their intrinsically different structures. Thus the total distribution of the objects can be written as

$$f(x) = f(x|s)f(s)+f(x|g)f(g), \tag{8}$$

where $f(s)$ and $f(g)$ are the fractions of stars and galaxies present in the plate, $f(x|s)$ is the fraction of those stars in the plate having a value of λ belonging to the interval x, and $f(x|g)$ is the corresponding fraction for galaxies.

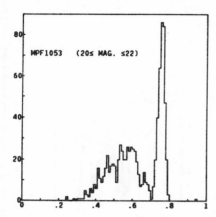

Figure 2. Histogram of λ values for objects up to magnitude 22. The situation of 'deterministic' separation is well evident.

Up to the critical luminosity we have very good separation between the two class probability density functions $f(x|s)$ and $f(x|g)$. This means that the supports of the two functions do not intersect, i.e., $f(x|s)$ vanishes almost everywhere, $f(x|g)$ is nonzero, and conversely. In this case the dicotomic inferred variable (which takes the values s or g) functionally depends on λ (at least in a measure theoretic sense). We may say that, up to this critical value of luminosity, separation is "deterministic". The histogram in Figure 2 describes this kind of situation.

5. DESCRIPTION OF THE METHOD

We consider a series of intervals of the total magnitude. All the intervals have the same width, Δm, and each interval overlaps the next by the same amount, $\Delta m - \delta m$:

$$s_k = (m_k, m_k + \Delta m), \tag{9}$$

where

$$m_k = m_0 + k\,\delta m, \qquad \delta m << \Delta m. \tag{10}$$

The set of objects having total magnitude belonging to the interval s_k will be referred to as the kth "stratum" of our sample, and we shall denote it by the symbol s_k.

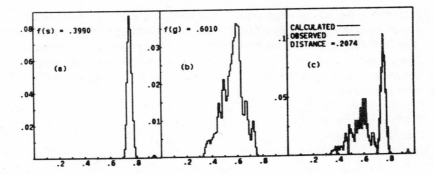

Figure 3. Theoretical and actual distributions of λ for the magnitude interval 21.5–22.5. (a) Stars only, (b) galaxies and (c) all objects.

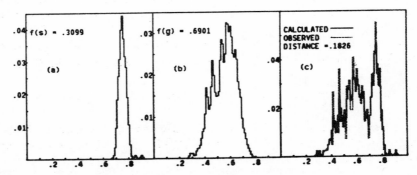

Figure 4. The same as Figure 3 in the magnitude interval 22.0–23.0.

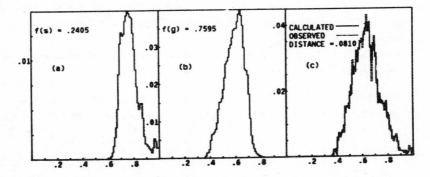

Figure 5. Distribution of λ in the case of total confusion (magnitude interval 23.0–24.0).

Basically our method is as follows. We start from a stratum where we have almost deterministic separation; we estimate all the relevant fractions pertaining to this starting stratum and use these data to estimate the class probability density functions for the next. The proportions of stars and galaxies found in the starting stratum are used as *a priori* estimates for the proportions in the next one. The process is iterated for the subsequent strata.

Quite obviously, however small we take δm, the expected errors of the estimates grow. There are, however, methods to estimate these errors, so we are able to understand when the process becomes meaningless.

In what follows, $f^{(k)}(s, x)$ is the fraction of the objects in s_k which are stars and exhibit the ratio λ in the interval x. Analogously for $f^{(k)}(g, x)$. $f^{(k)}(s)$ and $f^{(k)}(g)$ are the fractions of stars and galaxies in the kth stratum. We have

$$f^{(k)}(s, x) = f^{(k)}(x|s)f^{(k)}(s),$$
$$f^{(k)}(g, x) = f^{(k)}(x|g)f^{(k)}(g). \tag{11}$$

In the rest of this section we provide the mathematical details of our procedure. It is sufficient to specify how the $(k+1)$th iteration is carried on under the hypothesis that the kth has been completed.

We assume that all the relevant fractions pertaining to the stratum s_k have been estimated. We estimate the fraction $f^{(k+1)}(s, x)$ for stratum $(k+1)$th as follows:

$$f^{(k+1)}(s, x) = \frac{1}{N^{(k+1)}} \sum_i^{N^{(k+1)}} \delta(x-x_i)p^{(k)}(s, x_i), \tag{12}$$

where δ is the Kronecker symbol and $N^{(k+1)}$ is the number of objects in the $(k+1)$th stratum. The joint probability $p^{(k)}(s, x)$ in eq. (12) is estimated by the Parzen method with Gaussian kernel from $f^{(k)}(s, x)$, namely,

$$p^{(k)}(s, x') = \sum_x f^{(k)}(s, x) g(x-x'). \tag{13}$$

Here $g(x-x')$ is a Gaussian function with mean x'. According to Parzen [12], the variance must be very small as the cardinality of the sample (the objects in the kth stratum) is large. Notice that the proportion of stars varies smoothly from one stratum to the next as a consequence of the fact that there is a very large overlap between adjacent strata.

In order to assess the reliability of the iterative scheme as it proceeds to fainter magnitudes, at the end of each iterate the ℓ_1 distance between the calculated and the observed total distribution of the ratio λ is computed:

$$d^{(k)} = \sum_x |f^{(k)}(x) - h^{(k)}(x)|, \tag{14}$$

where $h^{(k)}(x)$ is the actual fraction of the objects in stratum s_k whose observed value of the ratio λ belongs to x, while

$$f^{(k)}(x) = f^{(k)}(s, x) + f^{(k)}(g, x). \tag{15}$$

Figures 3(c), 4(c), and 5(c) show the total distributions for three typical situations: the first, in the magnitude interval 21.5–22.5, with a good bimodal behaviour; the second, with considerable merging between the two classes of objects, in the interval 22.0–23.0,

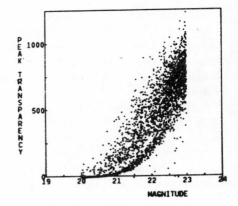

Figure 6. Plot of the peak transparency
against magnitude. Star–galaxy separation
up to magnitude 22 is well evident.

and the third (between 23.0 and 24.0), with a total confusion of the two classes of objects. The total theoretical distributions deduced by the RMD method have very small ℓ_1 distances with respect to the actual histograms: this fact is a good internal check of the separate galaxy and star-like distributions that we obtain, still in the case of the total merging (Figures 3(a), (b), 4(a), (b), 5(a), (b)).

Some comments on the results and perspectives are needed:

(1) The percentage of the stars decreases at faint magnitudes.

(2) Between these "stars" there are probably a growing number of QSOs; separation between a star and a QSO will be done in the near future by means of colours and variability.

(3) The number counts for galaxies are contained in the histograms of Figures 3(b), 4(b), and 5(b). From colours, type, and redshifts obtained for each galaxy, it will be possible to apply the corresponding K-correction; the so-obtained number counts will be directly comparable with the numbers given by cosmological models. With the same technique it is statistically possible to confirm or reject the foreground nature of blue galaxies in distant, compact clusters.

6. FINAL REMARKS

The astrophysical interest in establishing a classification for objects up to magnitude 24.0 is connected with the fact that a normal galaxy (i.e., a galaxy with a mean absolute magnitude) is seen at this magnitude as it was 10 billion years younger than a nearby galaxy; this time is nearly half of the age of the universe. The RMD method also seems, in this respect, very promising for further investigation involving new features. Figure 6 is the plot of the peak transparency against the total magnitude. The sequence of the stars is easily recognizable up to magnitude 22. Also, if correlated with radial magnitudes, this feature may contribute to discrimination because it is free from some sources of errors (e.g., centroid determination) present in the magnitude computation.

REFERENCES

1. G. Pittella, A. Vignato, FINDER — A program for detecting objects in a digitized image, *Memorie della Società Astronomica Italiana* 50 (1979) 537–541.
2. G. Agnelli, D. Nanni, G. Pittella, D. Trevese, A. Vignato, A new computerized method for plate calibration: an application to photometry of galaxies, *Astronomy and Astrophysics* 77 (1979) 45–52.
3. P. Di Chio, D. Nanni, G. Pittella, D. Trevese, A. Vignato, Multicolor photometry and classification of galaxies: clustering in color diagrams, European Colloquium "Clusters of galaxies" (Meudon, October 13–15 1982; in print).
4. D. Trevese, A. Vignato, The velocity distribution and the secondary maximum in galaxy clusters, European Colloquium "Clusters of galaxies" (Meudon, October 13–15 1982; in print).
5. H. Butcher, A. Oemler jr., The evolution of galaxies in clusters. 1. ISIT photometry of C10024+1654 and 3C295, *Astrophysical Journal* 219 (1978) 18–30.
6. H. Butcher, A. Oemler jr., P. Wells, Photometry of remote galaxy clusters, in G. O. Abell, P. J. E. Peebles (eds.), *International Astronomical Union, Symposium n. 92, Los Angeles, August 28–31 1979* (D. Reidel Publishing Company, Dordrecht, Holland).
7. R. D. Mathieu, H. Spinrad, Luminosity function and colors of 3C295 cluster of galaxies, *Astrophysical Journal* 251 (1981) 485–496.
8. G. Bruzual, Spectral evolution of galaxies, Ph.D. Thesis, University of California, Berkeley (1981).
9. J. F. Jarvis, J. A. Tyson, FOCAS — Faint Object Classification and Analysis System, *Proceedings Society Photo-Optical Instrumentation Engineers* 172 (1979) 422–428.
10. R. G. Kron, Photometry of a complete sample of faint galaxies, *Astrophysical Journal Supplement Series* 43 (1980) 305–325.
11. E. Valdes, The resolution classifier, (1982), to appear.
12. E. Parzen, On estimation of a probability density function and mode, *Annals of Mathematical Statistics* 33 (1962) 1065–1076.

P. D. Chio and S. Di Zenzo
IBM Scientific Centre
Via Giorgione 129
00147 Roma
Italy

A. Vignato
Osservatorio Astronomico
Viale del Parco Mellini 84
00136 Roma
Italy

35 Writer recognition by special characters

K Steinke

1. INTRODUCTION

Regarding text-insensitive methods [1–3], writer-specific information about handwriting is extracted from complete text. The features are averaged globally over all available letters. In addition to this summarized feature extraction, one suspects that there might be a gain in information if the characteristics of individual components of the handwriting, e.g., single characters, are investigated separately. In doing this, it does not make sense to use pixels of the quantized image of a character directly as components of a feature vector. Its dimension would be too high. With the chosen resolution, a character lies within a range of about 80×80 pixels. Preprocessing is necessary to represent the character in a more compact fashion, if possible without any loss of information.

In terms of the writing process, a character consists of one or more continuous line traces. The image over an area results from the width of the writing implement. It is convenient to choose, as representation of the character, the sequence of coordinates of the writing process. However, obtaining the sequence for an individual letter comprises the problems of localization, isolation, and reconstruction of the writing line.

Automatic localization of characters would presuppose recognition by the computer. Recognition of a character is equivalent to the capability of reading it. The problem of automatic reading of cursive writing, however, is unsolved at present and is not the goal of this paper. For isolation from background noise, e.g., preprinted sections and stamps on cheques, and for segmentation from other characters, *a priori* knowledge about the character is necessary. In our approach, it is provided by a human being.

From the mathematical point of view, the writing line has no extension in width. To come very close to this ideal, the writing could be peeled in layers until the width consists of only one pixel in the discrete grid. But the skeleton of a character has considerable disadvantages; thus closed loops are thinned out into a line, and a line crossing in principle is resolved into two branching points. An interactive method is specified below which solves the three above-mentioned problems and which extracts a writing line out of the image.

2. INTERACTIVE METHOD

The method, which can be seen as an intelligent line tracing, uses a circular mask with about the same diameter as the thickness of the handwriting. The starting point is deter-

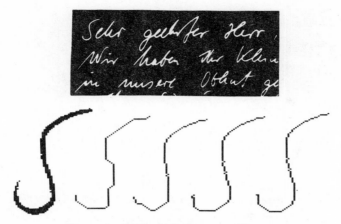

Figure 1. Original character and line
tracing with increasing degree of freedom.

mined by interactive means. From this marked point, the mask is shifted by one pixel so that as many black pixels as possible are contained. A prototype consisting of several support points indicates the approximate direction. Thus, the mask can be bound on its way like a slave to the prototype or can follow more freely the actual character (Figure 1). The degree of freedom is determined by an angle by which the mask can deviate from its preferred direction. Next to the starting and ending points of the writing line special points are marked serving as a help for segmentation, when there are connections with other characters.

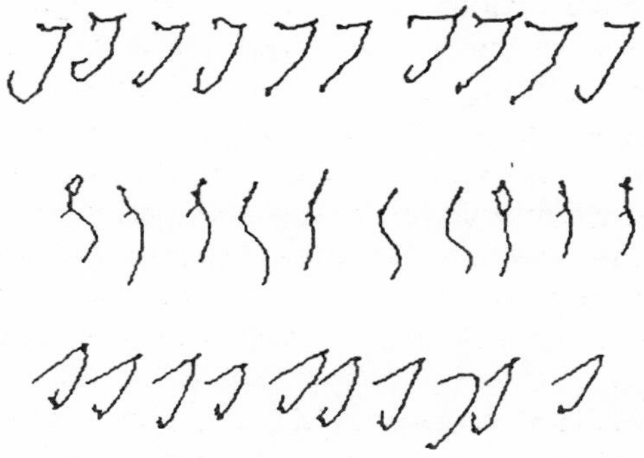

Figure 2. Extracted letters from three
writers.

Computation of the preferred direction at a point is always performed by using the two next support points. In this way, the resulting line trace is relatively independent of the choice of interactively marked points. An angle of +45° has proven favourable as the maximum deviation from the precalculated direction. A remarkable advantage of this procedure is the insensitivity with respect to background noise (e.g., with cheques), since human knowledge and context analysis also enter the process.

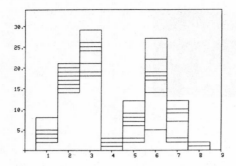 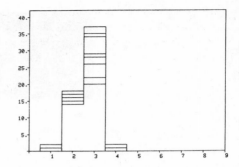

Figure 3. Frequency distribution of
Freeman code (two writers with 10
characters each).

3. STATISTICAL FEATURES

The produced sequence of coordinates is concatenated so that it can be described by the
Freeman chain code. A frequency distribution of eight basic directions forms a feature
vector with eight components. By normalization it becomes size invariant.

An alternative to a description in a discrete grid is to map the image into a continuous
space. Each pixel can be thought of as representing a single discrete point. A set of points
can be approximated by a continuous curve. If the desired curve is a polynomial, the
approximation can be performed easily by the method of least mean squares. A character
consisting of pixels with the coordinates x_i, y_i, $i=1, ..., n$, is approximated by a parameter
curve:

$$x(t) = \sum_{k=0}^{m} a_k t^k, \qquad y(t) = \sum_{k=0}^{m} b_k t^k.$$

Choosing polynomials of the fourth degree, a character is represented by 10 coefficients.
a_0 and b_0 only determine the position. From the coefficients $a_1, ..., a_4, b_1, ..., b_4$, the
character can be reconstructed (Figure 4).

Approximations using the integral square error as a norm have the disadvantage that

Figure 4. Representation by parameter
curves.

they may miss such details of the data as a spike of short duration. But the strong smoothing also involves advantages. Because it is a continuous curve, gradient and curvature can be computed for each point on the curve. The K curvature in the point (x, y) on the curve is defined by

$$K = \frac{\begin{vmatrix} x' & y' \\ x'' & y'' \end{vmatrix}}{(x'^2 + y'^2)^{3/2}} .$$

For all $(x(i), y(i))$, $i=1, ..., n$, the curvature and gradient is calculated and a frequency distribution is evaluated. For the frequency distribution of gradients 24 possible values are admitted. The frequency distribution of curvatures is established in dependence on six possible angles of the gradient. When describing parameters of the histograms centre, variance, skewness, and integral are chosen as features.

4. STRUCTURAL FEATURES

The coefficients of the polynomial fit lack any perceptible interpretation; nevertheless, they describe the structure of a character. They do not appear particularly well suited for classification. Minimal changes of the character can entail major changes in the coefficients. In addition to this fact, the importance of class discrimination is different for all coefficients.

Another representation of line reconstruction is a sequence of directional elements, which can be obtained from the fitted parameter curve. The curve is cut into s pieces of equal length and the direction is computed. The value of each directional element becomes a component of an s-dimensional feature vector. Characters of different writers are now comparable, since the features are size invariant. This structural representation of a character is the result of a nonlinear transformation.

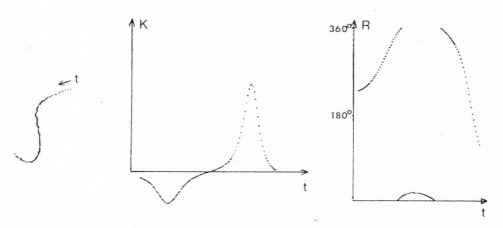

Figure 5. Character, sequence of curvature elements, and sequence of directional elements.

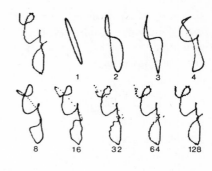

Figure 6. Lissajous fit with different numbers of coefficients.

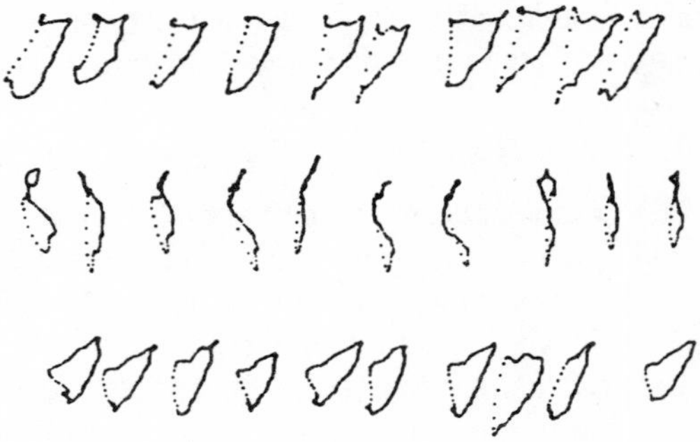

Figure 7. Representation by trigonometric polynomials.

To achieve further invariance, the elements of equidistant segmentation of the continuous curve are described by their curvatures. The curvature is independent of rotation. So the segmentation of curves in a fixed number of curvature elements provides a feature vector which has the quality of size and rotation invariance. This procedure makes different characters easily comparable (Figure 5).

Another analytical formulation is the development of the coordinate sequences in trigonometric polynomials (Figures 6 and 7):

$$x(t) = \frac{1}{2} a_1 + \sum_{l=1}^{m} (a_{2l} \sin(lt) + a_{2l+1} \cos(lt)),$$

$$y(t) = \frac{1}{2} b_1 + \sum_{l=1}^{m} (b_{2l} \sin(lt) + b_{2l+1} \cos(lt)).$$

The trigonometric representation has the advantage that the length is normalized to 2π. Without regard to size, characters can be compared by the power spectrum (Figure 8). It has the quality of shift invariance for each of the directions x and y. It turns out that phase information is very important for writer discrimination. The addition of the phase angle to the feature vector produces better results.

The representation of letters by Lissajous figures means operations such as translation, rotation, and magnification have very simple effects. Translation of a letter only changes the components a_1 and b_1. Rotation by the angle W implies a multiplication of the coefficients by the factor e^{iW}, and magnification of the letter by the factor R entails a

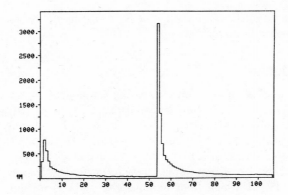

Figure 8. Power spectrum of a character (x and y direction).

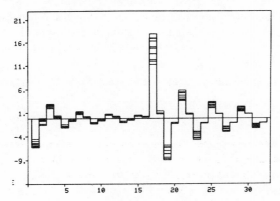

Figure 9. Coefficients of trigonometric fit.

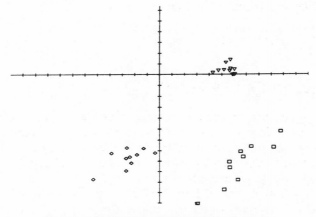

Figure 10. Representation of the sine and cosine components of the first harmonic.

multiplication of the coefficients by the factor R. Thus, the normalization of direction and size can be performed in simple fashion.

5. DIRECT COMPARISON

In order to obtain precise information concerning the quality of the above-described analytical representations, it is desirable to compare the extracted writing lines directly with one another. Because of the different chain lengths, the sequence of coordinates cannot be used as a feature without preprocessing. Different characters form polygonal curves which are geometrically distorted relative to one another. Burr [4] introduces a distance measure in regard to handprinted character matching. He uses a dynamic programming technique [5, 6] which finds minimal cost paths in weighted networks. The pairings (i, j) of the elements of both strings are regarded as nodes of a network. The sum of pairwise differences between string elements under certain constraints is minimized.

For our problem, the polygonal curve of x and y directions are examined separately. The independently chosen distortion function in both dimensions can be justified by the relative independence of local bending and stretching in both directions. Some practical restrictions of distortion can be assumed. An additional constraint ensures that the path

Table 1. **Recognition rates with character specific features**

Feature	Components	Recognition rate (%)	Invariances
Statistical			
Frequency distribution of Freeman code (normalized)	8	77	Size
Frequency distribution of Freeman code (unnormalized)	8	83	–
Frequency distribution of direction elements	24	88	Size
Frequency distribution of curvature and direction	24	91	Size
Structural			
Polynomial coefficients	9	67	–
Sequence of direction elements	64	79	Size
Sequence of curvature elements	64	66	Size, rotation
Sequence of direction elements (continued)	64	84	Size
Power spectrum of trigonometric development	16	77	Size, shift
Coefficients of trigonometric polynomials	32	98	Size
Direct			
Sequence of coordinates	50–120	86	–

connecting points P_1 and P_2 never has a slope greater than 2 and smaller than 1/2. The distance measure is the sum of the two corresponding paths in the network (Table 1).

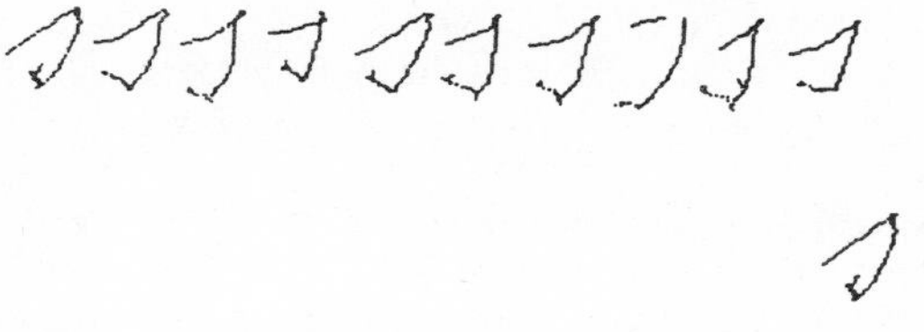

Figure 11. Nonlinear distortion of characters with a reference of the same writer (writer 1).

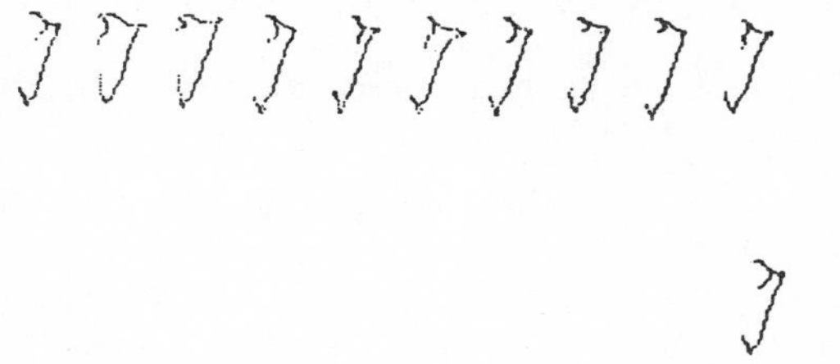

Figure 12. Nonlinear distortion of characters with a reference of the same writer (writer 2).

Figure 13. Nonlinear distortion of characters with a reference from another writer (writer 1).

The peaks of the character have a strong effect on distortion, so characters of the same writer are well matched to one another (Figures 11 and 12). Taking a character of another writer as reference, the distortion effect is so strong that the character can no longer be identified (Figure 13).

6. RESULTS

By the described method 100 characters of 10 writers were extracted. A classification experiment was done with a modified k-nearest-neighbour classifier. The experiment with the Freeman code frequency distribution (Table 1) shows a higher recognition rate for unnormalized features than for the normalized ones. The size of a character, therefore, seems to be a writer-specific feature.

The recognition rate of the polynomial coefficients is lower than those of derived features. Although coefficients contain the whole information, class separability is improved by a nonlinear transformation. Comparing the results of directional and curvature elements, a clear disadvantage can be seen with regard to rotation-invariant features (curvatures). An explanation for this fact seems to be the lack of directional information. This information, which is not contained in the rotation-invariant features, is an essential characteristic of a writer. The recognition rate for the directional elements can be increased when, during passing beyond the limit of 360°, the angle is not continued at 0° but, without jumping, continued steadily even beyond the 360° mark. Comparison with the recognition results of the direct method reveals that the trigonometric representation is an adequate description of the handwriting process.

REFERENCES

1. W. Kuckuck, B. Rieger, K. Steinke, Automatic writer recognition, in: *Proc. Carnahan Conference on Crime Countermeasures*, Lexington (1979) 57–63.
2. K. Steinke, Automatische Schreibererkennung mit textunabhaengigen Merkmalen, in: J. P. Foith (ed), *Angewandte Szenenanalyse, DAGM Symposium*, Springer Berlin (1979) 180–190.
3. K. Steinke, Scene analysis of handwriting images, *Microscopica Acta*, 4 (1980) 274–279.
4. D. J. Burr, A technique for comparing curves, in: *Proc. IEEE, Conference on Pattern Recognition and Image Processing*, Chicago (1979) 271–277.
5. R. Bellman, Dynamic Programming, Princeton University Press, Princeton, N.J. (1957) Chapter 11.
6. F. Itakura, Minimum prediction residual principle applied to speech recognition, in: *Proc. IEEE, Symposium on Speech Recognition* (1974) 72–76.

K. Steinke
Fachhochschule Hannover
Ricklinger Stadtweg 120
3000 Hannover 91
Federal Republic of Germany

36 Recognition and location of mechanical parts using the Hough technique

A Arbuschi, V Cantoni and G Musso

1. INTRODUCTION

Computer science is becoming the meeting point of many disciplines of the world's science, especially when we consider the field of artificial intelligence. Beyond the computational objectives characteristic of early computer generations, the capability, achieved in the last 10 years, of using the power of computers in performing high-level logical procedures, often very similar to human behaviour, have added to computer science several new areas of research in psychology, physiology, and neurology, together with the study of processing capable of introducing, in modern automation, "human-like" performances. Though computer modelling is a powerful tool of investigation, by simulating human intelligence in reasoning, learning, and interacting with the physical world, we are still far from the capability of understanding and implementing systems with as broad a capacity as the human being has. But many of these studies, though very far from their final goal, generate many new ideas which are finding direct application in the field of automation.

In particular, it seems that industrial automation will become one of the most important fields of application of human-like performances of modern electronic systems. Of course, it does not matter if the system obtains such performances using procedures similar to those of the human body or not. In the field of robot's vision many efforts are dedicated to investigating powerful recognition and location procedures that, even if very different from one's eyes, give to a machine the capability of self-adaptation to the environment in doing some operational work.

In this paper we describe a generalization of the Hough transform for recognizing and locating mechanical parts independently of their orientation. Beyond description of the analytical methodology, some space is dedicated to the problem of implementing such a recognition process with a speed of operation matched to the requirement of real-time execution, as in the case of robotic vision. In the last part of the paper some experimental results obtained by a first implementation of the process on a standard mini-computer are given.

2. A BRIEF DESCRIPTION OF THE HOUGH METHOD

In industrial applications of vision systems, the role of shape description and recognition is one of the most important and many generalized methods, e.g., scale rotation invariant, for shape description are continuously investigated by several researchers. Among them, a method based on the Hough transform [1] presents, in addition to a very useful methodological characteristic, the powerful capability of being implemented in the multiprocessing environment, in order to achieve the speed of operation generally required in industrial applications.

The basic concept at the origin of the Hough method consists of defining a mapping from the object space into a parameter space in which points belonging to the shape we are looking for give rise to a sharp distribution of mapped points around a coordinate representing the location of a selected reference point of the shape. As we shall describe, the above mapping process may be defined in a scale-rotation invariant way, and, moreover, in addition to shape recognition we can obtain location and orientation and scaling detection.

In order to achieve the capability to recognize and locate nonanalytical shapes, defined by pictorial models represented by a specified set of points, the Hough transform method [2] was generalized [3]. Let us consider a model C' and denote by C a set of points which can be obtained by translation, rotation, and uniformly scaling the model C'; we assume that C has the same shape as C', that is, C belongs to the family of possible representations of the same shape. The element C is described, with the Hough method, by a point in the four-dimensional parameter space P:

$$P = (x', y', \theta, s),$$

where x', y', θ, and s are the parameters of the quoted evolutions of the model C'; x' and y' measure the translation, θ measures the rotation and s is the scaling factor. The corresponding transformation is known as the generalized Hough transform and, in many practical recognition problems, allows a serious reduction in computational complexity.

Let us consider the case of a continuous shape C, given in polar coordinate $\rho\,(\phi)$, and assume the tangent direction of the model $\beta\,(\phi)$ is known. If we assume that no scaling is present ($s=1$), and that we are not interested in rotation detection (but rotation is, of course, permitted), the parameter space P is reduced to the bidimensional image space [4]. In this case, the points of the locus $\Gamma(e)$ in the parameter space P, corresponding to the shape C', can be obtained from the values that satisfy the equations for all values of ϕ:

$$x_{\Gamma(e)} = X_e - \rho(\phi) \cos[\phi + \psi_e - \beta(\phi)],$$

$$y_{\Gamma(e)} = Y_e - \rho(\phi) \sin[\phi + \psi_e - \beta(\phi)],$$

where X_e and Y_e are the picture edge element coordinates; ϕ is the independent variable; $\rho(\phi)$ is the polar description of the model; $\beta(\phi)$ is the tangent direction of the model in ϕ; and ψ_e is the contour direction relative to the edge element e.

For recognition purposes the Hough method is used by evaluating the distribution of points in parameter space obtained by putting, in the equations described, the observed values X_e, Y_e, and ψ_e in the image.

The model is completely described, given $\rho(\phi)$ and $\beta(\phi)$, and refers to a suitable choice of a polar system origin. In a digital computer, this information can be supplied by a table, normally named the R-table, which gives, for a suitably finite set of values of the independent variable ϕ, the corresponding values for ρ and $(\phi - \beta)$. In many cases, convenient selection of the model origin of a polar coordinate system, used to define the model, reduces the number of mapping operations and, consequently, a saving in computation time is obtained. This is particularly achieved when some geometrical symmetry is present in the shape.

3. IMPLEMENTATION OF THE HOUGH TECHNIQUE

In order to investigate performances of the recognition and location process based on the Hough technique, we have implemented, on a standard minicomputer, a system capable of recognizing and locating mechanical parts, such as those which can be found when designing some sort of intelligent manipulator in an industrial environment. In our experiment, the recognition process uses the following processing steps:

(1) An edge detector operates on the image data and determines whether a pixel belongs to an edge.
(2) For each edge pixel, the orientation ψ of the edge segment is estimated.
(3) For each edge pixel, the points mapped in the parameter space through the R-table, describing the shape we are looking for, are computed and corresponding counters are incremented.
(4) A decision rule is applied to the final counter values of the points of the parameter space P, to determine whether the shape sought is present and, if so, what are its particular location and orientation.

Implementation of the previous subprocesses requires some efforts in searching and defining methodologies to extract from the image the necessary set of data (i.e., edge pixel position and orientation), the procedure for obtaining the associated R-table of a given object, defining its shape model description, and, lastly, the decision rule for the parameter space.

The edge points on the scene under observation have been detected using the isotropic differential operator represented by the masks of Figure 1. In order to save computation

$$H1 = \begin{bmatrix} 1 & 0 & -1 \\ \sqrt{2} & 0 & -\sqrt{2} \\ 1 & 0 & -1 \end{bmatrix}, \qquad H2 = \begin{bmatrix} -1 & -\sqrt{2} & -1 \\ 0 & 0 & 0 \\ 1 & \sqrt{2} & 1 \end{bmatrix}$$

Figure 1. Isotropic differential operator. The two components G1 and G2 of the gradient can be computed convolving the image by the two operators H1 and H2, respectively.

time and to stress the reliability of the recognition process, only pixels above a prefixed threshold have been considered (for computational simplicity a nonlinear magnitude of the gradient is used: $G(x, y) = |G1(x, y)| + |G2(x, y)|$). For these edge points the edge orientation angle with respect to the horizontal axis is given by:

$$\psi(x, y) = \arctan[G2(x, y)/G1(x, y)]$$

(of course the transcendent function has been implemented by use of the suitable look-up table). This procedure allows us to test the recognition procedure on the same image with a variable number of detected edge pixels.

The model description has been obtained by automatic and interactive procedures. Of course, the higher the number M of points used to describe the model and the number E of edge points detected, the higher the number of votes ExM obtained in the parameter space (M of which are expected in the positions in which the object we are looking for is located), consequently the higher the computation cost required by the mapping operation: it depends, linearly, on the number of votes. Using the automatic procedure we obtain high values of M but it is not easy to reduce it to a cost-effective solution and maintain efficiency in recognition and location.

The interactive solution requires the following off-line procedure:

(1) Acquire a good image of the object of which we want to define a model (R-table).
(2) Define, using intuitive criteria, the reference point for the model (taking care that shapes composed by the same set of segments in different order have the same mapping rule!).
(3) Select a suitable set of points on the contour of the image of the object (the sequence depends on the relative luminance of object and background and is different for internal or external border). The interpolation between two consecutive selected points is linear; so a good definition of the model requires a point density inversely dependent on the local curvature of the boundary.

We observed that a very precise definition of the model is required when the shape to be recognized was similar to shapes of different objects present in the scene. A rough model is often sufficient when the context of the scene is friendly. Let us note that using a rough-model description means a considerable saving of time in running the recognition process. Of course, in doing these simplifications, care must be taken of the noise level on the image, because a compressed R-table gives rise to less evident distribution peaks in the parameter space, and the higher error consequent to the presence of noise in processing the edge pixels can obscure these peaks.

Finally, the accuracy required in inspecting the parameter space to investigate the presence of peaks and their precise position, depends on the trade-off solutions selected for all the problems quoted above. In our experiment, local analysis of the distribution near the maxima of the counts array was sufficient to detect similar shapes in a not particularly good environment.

4. EXPERIMENTAL RESULTS AND CONCLUSIONS

In this section we give some results obtained so far in implementing a system on which to investigate methodology. Only successive experiments, performed in real-time on multiprocessing environment, may give us statistical results about operating performances. In particular, attention was paid to measuring discrimination capabilities among different shapes and their sensitivity to changing the reference point selection, to the possibility of describing roughly both model and object, and to various alternative methods of detecting parts orientation.

The following figures illustrate some of the most significative results obtained; in their captions time-processing results, referring to a high-level language program running on a standard minicomputer, are given. We are confident that, using multiprocessing architecture not beyond the economical limit required by industrial application, it will be possible to achieve real-time operation. Further investigation will cover these applicative aspects.

Figures 2 and 3 show the processes to recognize and locate instances of objects with a model built interactively by the operator. In (a) the original image is presented and a cross shows the position of the reference point detected; (b) contains edges detected by

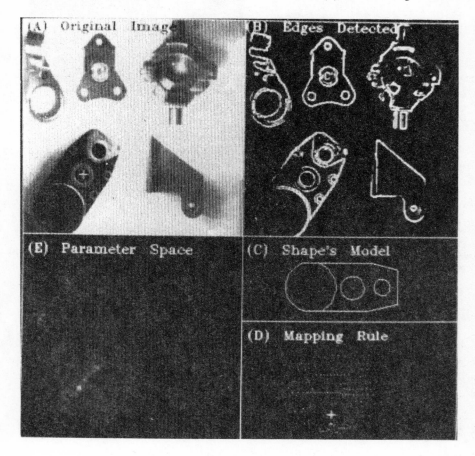

Figure 2. Recognition and location
process of a mechanical part.
Characteristics of the computation:
M=100; E=5453; computation time (on
a serial computer HP1000F) 231 s; the
position marked in (a) had 60 votes;
second maximum had 45 votes.

the isotropic operator quoted above; (c) shows the shape's model and the reference point selected; (d) represents the mapping rule for a horizontal edge located in the cross position; and finally, (e) shows the parameter space. Peaks corresponding to recognized shapes are quite evident in all cases. As can be easily seen in the figures, the recognition process has proved to be independent enough of the completeness of the contour of the shape sought, as long as the false edges present in the scene do not belong to the model (extraneous objects, edges due to light reflecting surfaces, shadows, etc.).

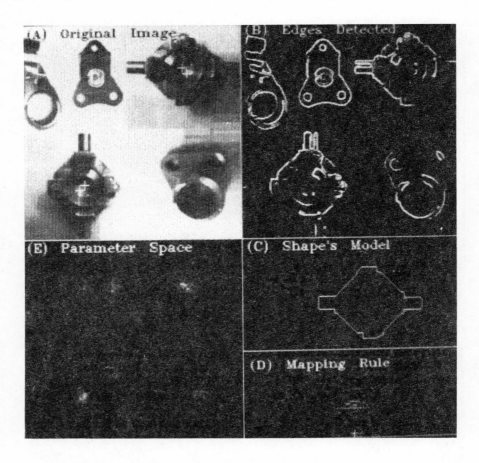

Figure 3. Recognition and location process of a mechanical part. Characteristics of the computation: M=99; E=4138; computation time (HP1000F) 175 s; the four positions marked in (a) had 57–59 votes, next maximum position (isolated) had 50 votes.

In a separate paper a process based on an hierarchical generalized Hough transform, to recognize, locate, and detect the orientation of an object, again using the parameter space of the image, but with reference to two different reference points (or to two segments of asymmetric shapes), is presented.

A significant improvement of the time performances of the method is expected when the location of the reference point in supervised searched areas is constrained. Of course, this context-dependent solution is not always possible. At present investigations concerning methodology have been carried out on the PRIMPS [5] system of Pavia University (built around a serial computer HP1000F); further studies will be dedicated to real-time problems, and the experimental technique will be implemented on a multiprocessing environment such as that of EMMA® of Elettronica S. Giorgio [6].

REFERENCES

1. A. Jannino, S. D. Shapiro (1978). A survey of the Hough transform and its extensions for curve detection. Proc. IEEE Comp. Soc. Conf. on Pattern Recognition and Image Processing, Chicago, p. 32.
2. P. V. C. Hough (1962). Methods and Means for recognizing complex patterns. U.S. Patent 306954.
3. K. R. Sloan, D. H. Ballard (1980). Experience with the generalized Hough transform. Proc. V Int. Conf. on Pattern Recognition, Miami, p. 174.
4. V. Cantoni, G. Musso (1981). Shape recognition Using the Hough transform. International workshop on Cybernetic System, Salerno.
5. V. Cantoni, E. Galasso, G. Previde (1982). Presentation of PRIMPS. RI 82.04 Istituto di Informatica e Sistemistica. Pavia University.
6. L. Stringa EMMA (1979) Multiprocessor on a field. Doc. DRC 011A.

A. Arbuschi and V. Cantoni
Istituto di Informatica e Sistemistica
Pavia University
Pavia
Italy

G. Musso
Elettronica S. Giorgio
Genova Sestri
Italy

Index